Distillation Control

Distillation Control

For Productivity and Energy Conservation

F. G. SHINSKEY
The Foxboro Company

McGRAW-HILL BOOK COMPANY
New York St. Louis San Francisco Auckland Bogotá
Düsseldorf Johannesburg London Madrid
Mexico Montreal New Delhi Panama
Paris São Paulo Singapore
Sydney Tokyo Toronto

Library of Congress Cataloging in Publication Data

Shinskey, F Greg.
Distillation control.

1. Distillation. I. Title.
TP156.D5S5 660.2′8425 76-19027
ISBN 0-07-056893-6

1234567890 KPKP 786543210987

*The editors for this book were Jeremy Robinson and Betty Gatewood,
the designer was Naomi Auerbach, and the production supervisor
was Frank P. Bellantoni. It was set in Primer
by Monotype Composition Company, Inc.*

It was printed and bound by The Kingsport Press.

Contents

Preface

Distillation may be considered a mature technology. At its present state, substantial improvements in tray efficiencies and operating range are not expected. Computer models for determining column sizing, tray selection, feed location, etc. give exact and rapid solutions, even for multicomponent systems. Yet two aspects of distillation technology are relatively unexploited: thermodynamics and control.

When the cost of hydrocarbon fuels began to rise sharply in 1973, distillation became the focal point of many plant studies. The reasons were twofold:

1. Distillation is the principal means for separating and refining hydrocarbon fuels. These products are now taking on higher and more changeable values.

2. Distillation is a heavy consumer of prime energy in the form of steam and hydrocarbon fuels. (Some columns consume well over a million dollars per year of fuel.)

The objectives of these investigations are obvious: to produce as much of these products as possible with existing plant equipment, while minimizing fuel consumed per pound of product.

Distillation columns are generally wasteful of energy. Every Btu introduced at the reboiler is rejected by the condenser, causing thermal pollution in the manner of a power plant. But vapor-compression systems only beginning to be introduced can provide the same amount of boiling and condensing with a fraction of the energy input and rejection. Feed preheaters common to many columns are only half as efficient as reboilers. And condensers are the greatest troublemakers of all. In most present-day installations, they are the stumbling blocks to stable control. Yet their proper use can bring about substantial reductions in energy requirements, increased column capacity, and reduced reboiler fouling, all at the same time.

Some improvements require addition or relocation of equipment. Yet substantial economies can be realized simply by enforcing closer control. The author has never had the luxury of designing or modifying separation or heat-transfer equipment in a distillation plant. My task has always been to improve regulation, stabilize operation, and enhance profitability *without shutting down the unit.* Consequently, the recommendations contained in the pages that follow will be found applicable to almost any operating unit.

This book is therefore dedicated to controlling distillation processes more profitably. It is not an introductory text on distillation. The reader is assumed to have had undergraduate training in the subject. Enough books have already been written describing vapor-liquid equilibria and mass transfer. What have been lacking are guidelines for operating these plants with a view toward economic return.

Traditionally, students have been drilled in design procedures. Using equilibrium data, they are taught how to estimate the number of mass-transfer units required to make a specified separation. Sizing procedures and selection of operating conditions are based on achieving a desirable rate of payout of capital costs.

But a column is designed only once, whereas it is operated for years. During the course of its life, demands change many times, without trays ever being added or removed. The many texts written on designing columns are then of little help to those who wish to alter the operating conditions of existing units profitably. In describing how to *operate* a column (and groups of columns), this book will demonstrate methods to increase production and reduce fuel consumption through better *control.*

By the same token, this is not an introductory text on control. In the distillation column and its assorted auxiliary equipment, all the common process variables are measured and controlled. The difficulties encountered in improving the performance of a column can rarely be overcome by attention to each individual control loop, but rather to their relationship to one another. Distillation is very much a multivariable control problem. Before attacking a problem of this complexity, the reader should have a familiarity with single-loop control. An earlier text by the author, "Process-Control Systems," also published by McGraw-Hill (1967), is recommended to provide this background.

Control over any process is most successfully achieved by those who understand that process. That this axiom is especially true of the distillation process may only demonstrate that it is not an easy process to understand. Or more specifically, it is not an easy process to represent in those rigorous mathematical terms requisite for the design of intelligent control systems. For these reasons, the mathematical modeling of both conventional and specialized distillation processes will be presented before their control is considered. That the process comes before the control system is a historical fact—James Watt had no need to invent a governor until he had invented a steam engine.

The control systems described here are based firmly on the mathematical representations of the process. Nonetheless, the characteristics and limitations of the actual equipment making the separation are not overlooked. In fact, limitations or constraints play a major role in the performance of separation units and to a great extent dictate the arrangement of their controls.

Once stable regulation over mass and energy balances is achieved, tight control of product qualities is possible. As deviations in product quality are reduced, a column may be operated with closer tolerances, which in itself can enhance product recovery and reduce heat requirements.

The next step is optimization. Whether optimization is possible for a given separation depends on the relative values of the products, and the costs of heating and cooling. If optimization is not being applied, it is probably because so few columns are being tightly controlled, as this is a definite prerequisite. Finally, throughput may be maximized after all the above goals have been achieved.

That the tenor of this book is economic is no accident. The primary incentive for applying control to distillation is ultimately profit. Cost reduction will be demonstrated in several ways, from letting the column pressure float on the condenser, to applying optimum allocation policies for utilities and feedstocks. In fact, the introductory chapter sets the tone of the entire text by setting forth the objectives to be sought and the constraints along the way. Only by thoroughly assessing our goals and accepting out limitations can we realistically expect to achieve success.

Let me express my appreciation to Carroll Ryskamp, whose wealth of experience contributed substantially to the text, and to Norma Wiklund for her exceptionally accurate typing.

Greg Shinskey

PART ONE

Introduction

CHAPTER ONE

Objectives and Constraints

Contests of any kind have goals to be reached and rules to be followed. The first steps in learning to play a game are to understand the objectives and learn the rules. So it is with distillation. Consequently before delving into the science of separating volatile mixtures or studying the equipment used, the objectives need to be set forth.

Making specification products is not enough—they must be made profitably. So there is more than a single objective. In fact there will be generally two more objectives than there are specifications on the products:

1. To minimize manufacturing costs
2. To set production at its most desirable value

By the same token, constraints are all about. While some are stationary, others are as changeable as the weather. In fact most of the changes are due to the weather. Atmospheric conditions seem to exert more effect on this process than any other. So it is important not only to recognize the constraint thus imposed but even to take advantage of it whenever possible.

PRODUCTS OF GUARANTEED QUALITY

Guaranteed products must meet or exceed specifications placed on them by either the buyer or the government. Actually, the only time these specifications must be met is when the product "changes hands" between seller and buyer. Therefore, the seller may blend the product to meet specifications rather than control the separations unit to meet them. But as will be seen, blending is one of those irreversible reactions that are best avoided if at all possible.

It is important to note the units placed on the specifications. Most chemical products are sold on a weight basis, i.e., by the pound or ton. Their specifications should therefore be in weight percent. Most petroleum products are sold by volume, i.e., gallons or barrels (42 gal). Their specifications are normally given in liquid-volume percent. Gaseous products such as natural gas may be metered in standard cubic feet. Their specifications are accordingly listed in gas-volume percent, which is the same as mole percent. Since most chemical analyzers report results in units of mole percent, conversion is often required for accounting purposes, as described in Appendix C.

Failing to Meet Specifications There is absolutely no relationship between the quality of a guaranteed product and its selling price as long as that quality meets or exceeds specifications. In other words, there are no rewards for making a better product (within a grade). But there are many rewards for *not* making a better product than necessary, because operating costs increase with purity. The greatest profit will then be realized at the lowest purity that will command a given selling price. Figure 1.1 sums up this relationship.

When a product fails to meet specifications, a penalty is incurred. It may take any of several forms, all of which are costly. First, the product may be sold as a lower-grade stock commanding a lower price. From Fig. 1.1, the slight reduction in purity that causes the product to slip into the lower grade can wipe out profits and even incur a debit. This alternative must be ruled out for small variations in quality because of the large loss in revenue with little reduction in operating cost. Consequently, this outlet ought only apply to start-up of a new unit, when an extended period of time may be required to reach expected efficiencies.

For plants having no market for a lower-grade product, an alternative route of disposal is needed. The usual solution is to rerun the product by recycling it back with the feed. The cost of rerun is proportional to its percentage of pro-

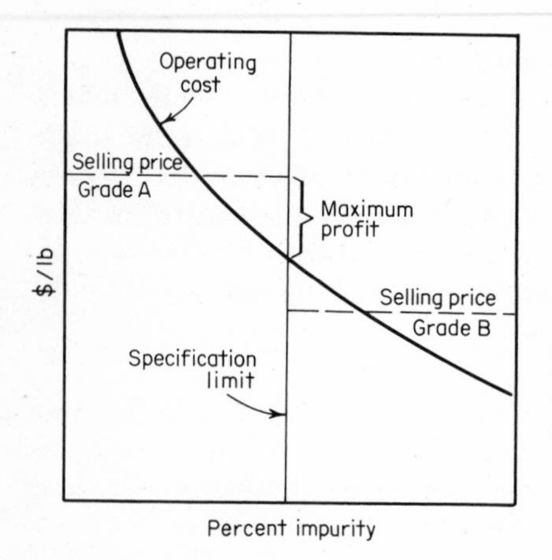

figure 1.1 *Cost of separation varies smoothly with purity, while the selling price of guaranteed products changes stepwise.*

duction. If 20 percent of the product must be rerun to meet specifications, the operating cost is 20 percent higher and the plant capacity 20 percent lower than they could be.

Blending to Meet Specifications Imagine a helium-filled balloon floating in the air—it owes its buoyancy to the low molecular weight of its gaseous contents. But the balloon leaks and its contents escape. How can the helium be recovered and forced back into the balloon? Only with great difficulty! The blending is easy—a downhill slide. The separation is quite the opposite.

From a thermodynamic point of view, blending is an irreversible reaction. Two streams of similar temperature and pressure but with different compositions can blend without producing useful work. The operation is adiabatic and isothermal, but the entropy increases. Described as a measurement of disorder, entropy is always increased by blending, whether helium with air, or 99 percent propylene with 98 percent propylene.

The entire purpose of distillation is to separate a feed mixture into two or more relatively pure products. That this is not easily done is borne out by the large amounts of energy and equipment necessary. And yet the final products contain no more energy than the feed—they are simply purer. Thus their entropy is reduced at the cost of increases in entropy elsewhere (by burning fuel and warming the environment).

Every Btu used to separate the components is rejected to the environment (if the products are in the same state as the feed). Hence distillation serves only to reduce the entropy of the system as blending increases it. As a consequence, blending must be viewed as decreasing the value of a product *just as much as distillation can enhance it.* And the cost paid in distilling is all lost in blending.

Blending can be accomplished in a number of ways. The products from two parallel columns may be blended if one product is above and the other below specifications. But the preceding argument indicates that this is a costly way to meet specifications. *Parallel processes should therefore be operated so that specifications are met individually by all product streams.*

The contents of tanks may also be blended. The cost goes up more in this case because of the space used and the maintenance required of the tanks, along with the loss in revenue of inventoried product.

Variations about an Average Another possibility—one that is in fact most common—is blending through a single tank. Variations in quality are smoothed by passing the product through a tank containing several hours of production. The tank essentially averages the quality over a time constant τ equal to the volume stored divided by the production rate. Fluctuations in quality $dx_i(t)$ entering the tank will be attenuated to fluctuations $dx_o(t)$ leaving, by the factor

$$\frac{dx_o(t)}{dx_i(t)} = \frac{1}{\sqrt{1 + (2\pi\tau/\tau_o)^2}} \tag{1.1}$$

where τ_o is the fundamental or longest period of fluctuations in x_i.

To achieve an attenuation factor of 0.1, the time constant of the tank must be typically 1.6 times the fundamental period of the fluctuations. If diurnal variations are experienced, a 40-h time constant would be required. All this assumes perfect mixing in the tank, too. If mixing is less than perfect, the attenuation factor will be reduced proportionately.

Nonetheless, this method of blending bears the same stigma as the others. Significantly more energy and tower capacity are needed to overpurify a product than are saved by underpurifying it to the same degree. Therefore the cost of maintaining an average purity which meets specifications increases with the amplitude of transient deviations from the average.

This principle may be demonstrated in the limit. Suppose a propylene product were to contain no more than 0.5 percent propane. If the propane content occasionally increased to as much as 1 percent or more (a plausible condition), the product would have to contain much less than 0.5 percent at other times to maintain the average. Since a propane content of 0 percent is impossible to achieve, the cost of reducing it to 0.1 or 0.2 percent may be expected to be all out of proportion to its value as blending stock.

Exceeding Specifications The penalties for failing to meet specifications are so great that most columns are operated so as to exceed specifications all the time. Others are allowed to fall below specifications for less than a certain interval related to the time constant of the receiving tanks.

How far beyond specifications the product quality will be depends on the variations in that quality. If the propane content of the product in the previous example were to vary ±0.2 percent during the course of normal operation, its control point would have to be set at 0.3 percent to meet specifications at all times. Depending on the period of the variations and the time constant of product receivers, some attenuation may be available. This will allow the average (control point) to be set that much closer to the limit.

It must be recognized, however, that variations in product quality incur a twofold debit:

1. The *average* purity must exceed specifications by the amplitude of variation, thereby increasing operating costs.

2. *Variations* about an average purity further increase the operating cost above that required to generate the *average* purity, in relation to the amplitude of these variations.

As an example, consider the propane-propylene separation with a specification of 99.5 percent propylene. If variations of ±0.2 percent are experienced, the control point will have to be set for 99.7 percent purity. Typically 10 percent more energy is required to produce 99.7 percent purity versus 99.5 percent. Further deviations to 99.9 percent require as much as 36 percent more energy.

If the variation between the levels of 99.5 and 99.9 percent took place in a rectangular wave, the average energy consumption would be 18 percent above that required to meet specifications. Since variations are more likely to be sinusoidal than rectangular, proportionately more time is spent near the mean than the extremes. Therefore the average energy consumption will fall somewhere between 10 and 18 percent higher than necessary. The payout most readily realized from improved control therefore stems from operating closer to specifications and with less variance.

Third Components Most products from distillation units have two specifications imposed on them, e.g., a higher and a lower boiler, or two lower boilers, etc. For example, methanol has both a water and an ethanol limit. However, in a simple distillation column, only one component impurity can be controlled.

Consider the example of a depropanizer separating a multiple-component feedstock into a propane overhead product and a residue consisting of butanes and heavier components as shown in Fig. 1.2. There may be an ethane specification (y_2) on the propane product. But if there is, it cannot be met by manipulation of the depropanizer since all the ethane in the feed (Fz_2) must leave with the propane product. If the ethane content is to be limited, control must be applied further upstream.

Specifications on guaranteed products can be written in a number of ways. A purity limit of 95 percent propane (y_3) may be established, for example, without regard to the relative amounts of other components present. With ethane and isobutane as the major impurities, control could be placed over the sum of their concentrations in the product ($y_2 + y_4$).

This arrangement would be satisfactory as long as ethane were the smaller impurity. But if it should approach the limit, isobutane content (y_4) may have to be reduced to the point where the tower could not make the separation economically.

Another possible combination would specify 95 percent propane purity with ethane content not to exceed 2 percent. Control in this case would still be based on the sum of the two impurities ($y_2 + y_4$), but ethane content would have to be controlled elsewhere.

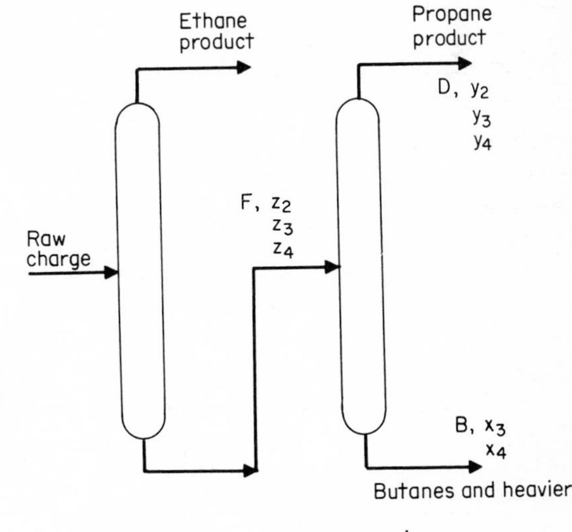

figure 1.2 *Control over ethane content of the propane product must be enforced at the deethanizer.*

A third possibility would require 95 percent propane with isobutane not to exceed 3 percent. In this case two controllers must be provided—one to hold $y_2 + y_4$ within 5 percent and the other to hold y_4 within 3 percent. Only one controller could function at a time, the choice depending on the ethane level. While y_2 is below 2 percent, y_4 would have to be held at or below 3 percent. Should y_2 rise above 2 percent, y_4 would have to be dropped to meet the 95 percent specification.

The additional degree of freedom in allowing the ethane content to vary between 2 and 5 percent offers the possibility of some optimization. If ethane has a lower worth than isobutane, increasing its level in the propane product will essentially replace an equal quantity of higher-valued isobutane. If this were the only criterion, the full 5 percent could be made up of ethane. But then the energy required to reduce the isobutane concentration to zero would be infinite. An optimum level of ethane does exist, however, that will minimize the cost of separating the 95 percent propane product from its two major contaminants.

PRODUCTS OF VARIABLE QUALITY

The firm specifications adhered to when controlling the quality of guaranteed products vanish when we turn to intermediate products and by-products. When products are not sold to outside parties, ordinarily no such rigid requirements exist. By the same token, the values of these products are more difficult to assign since they do not command a fixed selling price. Instead, values tend to change smoothly with composition rather than abruptly as with guaranteed products. Consequently, there will not be the concern over "meeting specifications" but rather of minimizing debits attributable to composition variations. Several illustrations will be used to enumerate products of this type.

Controlling Third Components In the overhead product of the depropanizer of Fig. 1.2, the ethane content is allowed to float within certain limits. In addition, the ethane content is not controllable at the depropanizer because all the ethane in the feed has to leave with the product. Actually, then, control over ethane content must be exercised on the feed, which is the bottom product of the deethanizer.

Observe, however, that the desired ethane concentration in the deethanizer bottom product is a function of the concentration of propane at that point. In other words, increasing propane content in the depropanizer feed will require increasing ethane content also if a constant propane product composition is to be maintained. This can be demonstrated mathematically with a material balance of both components. Let F, D, and B represent feed, distillate, and bottoms flow rates in and out of the depropanizer, in the usual units of liquid volume:

$$F = D + B \tag{1.2}$$

Using z, y, and x as the volume fraction of components in those streams, with subscripts 2 and 3 representing ethane and propane, respectively, we have

$$Fz_2 = Dy_2 + Bx_2 \tag{1.3}$$

$$Fz_3 = Dy_3 + Bx_3 \tag{1.4}$$

Being interested only in the distillate product, let us substitute $F - D$ for B in Eqs. (1.3) and (1.4). Then

$$F(z_2 - x_2) = D(y_2 - x_2) \tag{1.5}$$

$$F(z_3 - x_3) = D(y_3 - x_3) \tag{1.6}$$

Dividing (1.5) by (1.6) and solving for z_2 yields

$$z_2 = x_2 + \frac{y_2 - x_2}{y_3 - x_3}(z_3 - x_3) \tag{1.7}$$

In the case of the depropanizer, x_2 (the ethane content in the bottom product) is small enough to be neglected, leaving

$$z_2 = y_2 \frac{z_3 - x_3}{y_3 - x_3} \tag{1.8}$$

A further simplification—which may not be justified—would eliminate x_3, which is small compared with z_3 and y_3, yielding

$$\frac{z_2}{z_3} \approx \frac{y_2}{y_3} \tag{1.9}$$

If Eq. (1.9) is valid, it can be exceptionally useful, for it states that control over the ethane-propane ratio in the feed will yield the same ratio in the product. Therefore, it is worthwhile examining how accurate this simplification is.

example 1.1

Let the following specifications apply: $y_3 = 95$ percent, $x_3 = 1.0$ percent. Solve Eq. (1.9) for values of $y_2 = 3$ percent and 4 percent, with $z_3 = 20$ percent and 30 percent. Then, using the estimated z_2, find corresponding values of y_2, using (1.8) as a test for the accuracy of the approximation.

Desired y_2, %	z_3, %	z_2, %	Resulting y_2, %
3.0	20	0.632	3.13
3.0	30	0.947	3.07
4.0	20	0.842	4.17
4.0	30	0.126	4.09

Observe that y_2 resulting from the approximation of z_2 is high in every case due to the omission of x_3 from the estimate. This deviation can be distributed more equitably by solving Eq. (1.8) to obtain a correction factor to apply to the estimate.

First, factor y_3 and z_3 from Eq. (1.8) to fit the form of (1.9):

$$\frac{z_2}{z_3} = \frac{y_2}{y_3}\frac{1 - x_3/z_3}{1 - x_3/y_3} \tag{1.10}$$

Then the correction factor needed to match the approximation to the exact equation at one point is

$$C = \frac{1 - x_3/z_3}{1 - x_3/y_3} \tag{1.11}$$

The correction factor is then applied to improve the accuracy of the estimate:

$$\frac{z_2}{z_3} \approx C\frac{y_2}{y_3} \tag{1.12}$$

example 1.2

Repeat Example 1.1, using Eq. (1.12) instead of (1.9). Apply a correction factor based on a midrange value of z_3.

$$C = \frac{1 - 0.01/0.25}{1 - 0.01/0.95} = 0.97$$

Desired y_2, %	z_3, %	z_2, %	Resulting y_2, %
3.0	20	0.613	3.03
3.0	30	0.919	2.98
4.0	20	0.817	4.04
4.0	30	1.226	3.97

Observe that the correction factor has reduced the deviation in the final product quality to 1 percent of value. Since the deviation will increase with x_3 and variations in z_3, however, the approximation should be tested for accuracy in each application.

Obviously Eq. (1.8) can be used to arrive at an accurate set point for z_2 under all conditions. But being able to work with ratios of component concentrations simplifies the mathematical models of columns and the control systems applied to them. These ratios will appear again when material balances and separation factors are discussed.

Another point to be made under this discussion is that in most cases the third component does not have to be controlled rigidly. Typical specifications on the propane product of 95 percent propane would allow the balance to be any combination of ethane and isobutane. An *optimum* ethane content may exist, but the penalty for failing to control at that point would be smooth rather than abrupt.

Consequently variations in ethane content due to poor control of the deethanizer or inaccuracy in the estimate of its control point will not necessarily cause the propane to deviate from specification. If the depropanizer is controlled on the basis of ethane *plus* isobutane in the product, uncontrolled variations in the former can be compensated by controlled variations in the latter.

Intermediate Products and By-products Many by-products have no rigid specifications and yet have values assigned to them. In fact the value assigned to a by-product is likely to be a function of composition, as contrasted to guaranteed products where price within a grade is fixed.

If a by-product's principal use is as a fuel, its value should be based on Btu content. This would apply whether the by-product is sold or is used in the plant that produces it. If sold, its Btu content would have to be measured and multiplied by flow rate to arrive at a Btu billing. Otherwise the buyer would have no assurance of receiving his value for money paid, unless he insisted on a guaranteed Btu content. Then the fuel would fall into the classification of a guaranteed product, and control would have to be applied to minimize Btu "giveaway."

A value assigned to the Btu content of a fuel does not necessarily remove any dependence upon composition. Consideration still must be given to the values of components as products relative to their values as fuel. Ethylene, for example, has a high value as a monomer, but it is not a good fuel since it tends to form soot. So if ethylene is lost to the fuel system because of limited recovery in distillation, it may not receive as much fuel credit as its Btu content would indicate.

A similar relationship applies when separating butanes for alkylation to gasoline. Isobutane alkylates readily but *n*-butane does not. Consequently there is a penalty assigned to *n*-butane in the alkylation feed. However, the *n*-butane product from the separation is blended with gasoline to raise its vapor pressure for easy engine starting. If isobutane is used instead, less can be accommodated because of its higher vapor pressure. When the price of gasoline exceeds that of butanes (which is usually the case), *n*-butane has the greater value as a vapor-pressure additive because more of it can be used.

There is little reason to set any hard specifications on this butane separation. Instead, its operating cost should be minimized. That is, the products should be controlled at composition set points which minimize the sum of the penalties for *n*-butane in the alkylation feed, isobutane in the blending stock, and the costs of heating and cooling. Oddly enough, many towers ought to be controlled this way, but few are. Perhaps operators are more comfortable when they are given hard specifications to meet. Or perhaps the lack of optimization of these intermediate-product towers is due to the inability of engineers to define an optimum condition or the inability of management to demand it. Ample incentive exists—as little as 1 percent shift in composition between components differing in worth by only $1/bbl in a stream flowing at 10,000 bbl/d raises profit by $100/d. There are many streams meeting these requirements in a modern refinery.

Negative-valued Products Products with negative value are wastes. In essence, the manufacturer pays to have them removed as opposed to selling them for a profit. In times past, wastes were simply drained to the nearest waterway or burned at the pit or flare. That is, they were treated as having a value of zero.

Now, however, refineries and chemical plants are being forced to treat their effluents, and indiscriminate flaring is no longer tolerated. Treating effluents to the point where they meet environmental standards takes money. Therefore the treatment process, in upgrading an effluent to zero value, must have started with a negative-valued feed.

A methanol stripping column is a good example of this type of process. It is fed an aqueous stream containing perhaps 5 percent or less methanol. This stream cannot be discharged from the plant without treatment because its BOD (biochemical oxygen demand) is too high. So it must either be biologically decomposed or recovered.

As a waste product it has a negative value—essentially the cost of treatment required to bring its value to zero. But most of the methanol can be recovered by stripping with steam. The two products from the stripping column are typically a distillate of 90 percent or more methanol and a residue containing less than 0.1 percent. An economic analysis may be applied to the stripping operation. First, some specification must be set on the methanol product. Then the debit for methanol in the residue can be balanced against the cost of heating and cooling to arrive at an optimum residue concentration.

It is possible that a hard specification may be placed on the residue by a local regulatory agency. In that case, there is no opportunity for optimization, since the distillate also has to meet some purity limit in order to be useful in the plant.

Recycled Feedstocks Quite often, distillation is used to separate a product of a reaction from its unconverted feedstock. An example is the conversion of ethylbenzene to styrene by cracking. To maintain a reasonable yield of styrene, it is necessary to limit the conversion in the reactor to moderate levels. As a result, the reactor effluent may contain more unconverted ethylbenzene than styrene.

The two materials are separated in a vacuum column, with styrene leaving the bottom. This product must meet high purity specifications. An important factor in this separation is the tendency of styrene to polymerize in the column. This is the reason for vacuum operation: to minimize the temperature to which the product is exposed.

No standards exist for the ethylbenzene purity. If the styrene contained in it were to be recycled through the reactor without loss, the only penalty assignable to its presence would be a proportionate decrease in production rate. But a certain percentage of the recycled styrene is lost by overcracking. Its by-products include coke, whose value is negative in that it fouls the reactor. So substantial incentive exists for reducing the amount of styrene in the ethylbenzene.

But the cost of such reduction is also substantial. Because styrene quality must be controlled, improving ethylbenzene purity can be achieved only by increasing the energy input per pound of styrene produced. But as the energy input is increased, bottom temperature rises, promoting polymerization. The consequences are increased tars in the product, higher styrene losses, and more rapid fouling of the reboiler. These penalties must be combined with those arising from styrene in the distillate to arrive at the optimum distillate composition.

PRODUCT RECOVERY

The principal economic benefit of improved control is usually enhanced recovery of the more valuable product. There are exceptions to this rule, since the optimum operating conditions depend as well on the cost of heating and cooling. However, most products are more valuable than the energy used to separate them. This is not to say that energy costs are to be ignored; in fact, they are rising and more significance is being attached to them. But at this writing, rising energy costs have forced product values upward at a similar pace, so that their relative worth may be scarcely changing at all.

Consequently, product recovery will be examined before the contribution of energy costs. Although this discussion applies principally to towers yielding a guaranteed product, the concepts will be useful when considering others as well.

Recovery Defined For guaranteed products, recovery is defined as the amount of salable product generated per unit of that component in the feed. Mathematically, recovery of component i is defined as

$$R_i \equiv \frac{P}{F z_i} \tag{1.13}$$

where P is the product flow, F the feed rate, and z_i the fraction of the feed constituted by component i, in consistent units. Thus we speak of the "propylene recovery" of a propane-propylene column or "isobutane recovery" of a butane splitter.

Observe that Eq. (1.13) includes no term relating to the quality of the product. For guaranteed products, recovery applies specifically to the *product* and not to the components of the product. If, for example, the *only* specification on a propane product were 95 percent purity, the remaining 5 percent could be *anything* else. Then ethane, butanes, carbon dioxide, etc., could all command the propane selling price as long as their total did not exceed 5 percent.

Note also that it is possible for recovery to exceed 100 percent, depending on the specification on the guaranteed product. In fact, the upper limit on recovery is worth delineating at this point. Maximum recovery will be achieved when all the feed leaves as product, that is, when $P = F$. Then, from (1.13),

$$R_{\max} = \frac{1}{z} \tag{1.14}$$

Although this upper limit represents an absurd condition, it allows recovery to be evaluated over the entire range of possible product purities. The variation in recovery as a function of purity will be most important in assessing the benefits of improved control.

The recovery concept loses some significance when applied to variable-quality products whose value may change with quality. For them, optimizing programs will include flow rates, compositions, etc. The recovery factor actually is just a substitution for the group of terms on the right-hand side of Eq. (1.13). It will be most useful in evaluating the economic relationships involved in refining a guaranteed product and will later appear in product-quality control systems.

Recovery vs Purity To improve quality, quantity must somehow decrease. Or to look at it another way, reduction in quality will allow a commensurate increase in recovery. The improvement in recovery will be achieved by reducing purity closer to specifications.

Figure 1.3 describes a typical relationship between recovery and purity for a column separating a 50-50 mixture of two components. The maximum recovery will be 200 percent at 50 percent purity. The horizontal axis is scaled as the logarithm of the impurity in the product. Note that the distance between 99.9 and 99 percent equals that between 90 and 99—each represents a decade change in impurity. This scale was chosen both to cover a broad range of purities and to produce contours that are reasonably straight.

The contours appearing in Fig. 1.3 represent constant ratios of energy input to feed rate. The parameter is expressed as units of vapor flow V generated in the reboiler per unit of column feed F—hence V/F is a dimensionless number. The relationships upon which Fig. 1.3 was constructed are examined in Chap. 2. For the moment, accept these contours as being characteristic of a particular column.

Any purity up to about 98 percent can be achieved at $V/F = 2$. But most columns are designed to operate in the vicinity of 80 to 100 percent recovery. In this range, recovery falls off about 3 percent for each percent increase in purity. The deterioration is worse as purity is further increased.

Let us use one of these curves to evaluate the cost of poor control. Let the specification on the product be 90 percent, with a nominal V/F of 2. Poor control allows variations of ± 2 percent, so that the control point must be set at 92 percent. Recovery is 95.3 percent at 90 percent purity, falling off to 88.4 percent at 92 percent purity and 78.7 percent at 94 percent purity. A constant 92 percent purity results in a 6.9 percent loss in recovery. Rectangular-wave cycling between 90 and 94 percent purity will reduce the recovery to an average 87 percent. Consequently the loss in recovery for random or sinusoidal variations between 90 and 94 percent purity will be somewhere between 6.9 and 8.3 percent.

Cost of Unrecovered Product Whatever of the guaranteed product is not recovered leaves the column with the other stream. Its value is then reduced to that of the other stream or worse. In the propane-propylene separation, unrecovered propylene leaves with propane and may be sold as propane.

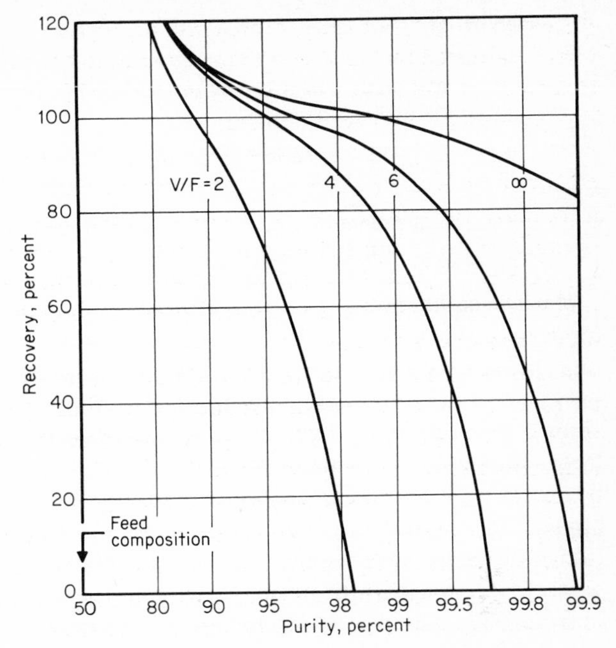

figure 1.3 *Product recovery can vary sharply with purity, attaching a severe penalty to overpurification.*

Occasionally, however, an additional penalty may be assigned, in that the unrecovered product may actually lower the value of the other stream. If the n-butane from the butane splitter is being used for gasoline blending, unrecovered isobutane finding its way into the blending stock will incur a double penalty. It suffers a loss beyond the value difference between the two product streams.

The cost of unrecovered product can be stated mathematically as

$$\$_P = \Delta v\,(P_0 - P) \tag{1.15}$$

where P_0 is the product flow possible if specifications are just met and not exceeded and Δv is the difference in value between the guaranteed and the other product. The cost may be placed on a per-unit-product basis:

$$\frac{\$_P}{P} = \Delta v\,\frac{P_0 - P}{P} \tag{1.16}$$

Substituting Eq. (1.13) into (1.16) places the cost in terms of recovery:

$$\frac{\$_P}{P} = \Delta v\,\frac{R_0 - R}{R} \tag{1.17}$$

where R_0 is the recovery attainable by just meeting specifications.

Equation (1.17) is useful in assessing the losses sustained by reduced recovery. The following example illustrates the method.

example 1.3

Using the curve $V/F = 2$ of Fig. 1.3, estimate the cost of operating at 92 percent purity instead of the specified 90 percent when the value difference between the two products is \$0.50/bbl.

$$\frac{\$_P}{P} = \$0.50\,\frac{95.3 - 88.4}{88.4} = \$0.039/\text{bbl}$$

For a nominal production rate of 10,000 bbl/d, estimate the daily loss.

$$\$_P = 0.039 \times 10{,}000 = \$390/\text{d}$$

Maximizing Recovery Figure 1.3 includes curves for higher ratios of energy input to feed rate (V/F). Where recovery is already high, there is little advantage in moving to a higher V/F ratio. However, low recoveries can be improved markedly by increasing the V/F ratio. Whether a given column will show a profit increase by such an adjustment is a subject which is explored later. The point to be made at the moment is that recovery can be maximized by operating a column at the highest attainable V/F ratio. This can be accomplished by maximizing V or minimizing F. But since feed rate F is determined by production requirements, maximizing V will maximize V/F for any given F. A point worth noting is that V/F being maximum implies that it is also variable. Hence when recovery is maximum it is also variable. This factor must be considered in the design of the composition controls because of the now variable relationship between flow rates and compositions.

Maximum recovery has historically been the operating mode for most columns. In general, energy V has been less expensive than feedstocks F, so there has been no reason not to maximize V/F. The principal exceptions have been sensitive products like styrene, where reducing energy input can reduce degradation and fouling.

Many engineers feel that in controlling the quality of their lower-valued product they will be maximizing recovery of their higher-valued product. Actually, however, controlling the quality of both products *fixes* the recovery. To maximize recovery means to minimize losses of the higher-valued product—in other words, to minimize the concentration of that component in the lower-valued product. It is achieved simply by controlling the quality of the more valuable product as closely as possible to specifications while holding energy input at its upper limit. No attention need be paid to the quality of the lower-valued product: it will float with feed rate, etc., while always being as high as conditions permit.

COST OF HEATING AND COOLING

Rising energy costs and short supplies have spurred conservation programs from homeowner to government and industry. A plantwide program [1] initiated by Dow Chemical Co. of Midland, Michigan, found opportunities for immediate savings in the following areas:

1. Excessive reflux in distillation columns
2. Excessive air to furnaces

3. High steam-oil ratios on reactors
4. Fouled heat exchangers
5. Poor control systems relating to the above

All but item 3 can apply to distillation columns. Dow's long-term program included using waste products as fuels, recovering useful heat from condensers, and reboiling one column with the overhead vapor from another. This last concept is called "cascading"; it can be applied more broadly by using compressors to boost energy levels.

The first step in an energy-conservation effort should be a survey of usage and costs. The following discussion will serve to place them on a quantitative basis.

Energy Sources The primary energy sources in a plant include electricity and the direct combustion of fuels. Although using electricity for most heating purposes is grossly inefficient, as a motive force for pumps and compressors it does contribute energy to the separation unit.

Secondary sources include process steam and also oil that has been heated by the combustion of fuel. A refrigeration unit which provides only cooling, or heating and cooling, would also qualify as a secondary source. In the past, these secondary sources have seemed—from the point of view of only one column—limitless in capacity. But this may have been simply because they serve several columns and could not be easily upset by an increase in demand by only one.

Tertiary supplies would have some energy already extracted from them. Examples are steam from turbine exhausts and waste-heat boilers, as well as hot effluent or coolant from reactors. Obviously, tertiary energy is much lower in cost than either primary or secondary supplies and therefore should be used when available. However, it does have two distinct disadvantages:

1. Its energy content tends to be low.
2. Both energy content and availability are variable.

The low energy content of tertiary supplies limits their usage to low-temperature operations. Water heated to 180°F by cooling a reactor effluent can supply heat only to towers separating close-boiling components (assuming a condenser temperature of about 110°F). Exhaust steam from a turbine could be used for preheating a naphtha feedstock, but it may not be hot enough for reboiling its heaviest components.

Variability of tertiary supplies creates other problems. As a reactor-effluent temperature varies with reaction rate, feed quality, poor control, etc., it will affect column boilup. Therefore, columns using these sources should have a limited amount of secondary energy available and a means of sensing and regulating the actual rate of heat being transferred.

Some heat-recovery schemes contain feedback loops within the process. Without tight control, they are capable of destabilizing a column or several columns. Here again, tight control can provide the needed stability while taking advantage of the energy savings. More is said on this subject in Chap. 6 and Ref. 2.

Energy can be conserved in many cases simply by proper management.

Irreversible processes should be avoided whenever possible. Throttling steam across a control valve is an example of an irreversible process. Although the steam has the same enthalpy as before, its lower pressure makes it less useful. Furthermore its increased superheat makes it a less efficient heat-transfer medium. Although desuperheating by injecting water may reduce pressure while maintaining saturation, it does not make up the available work that was lost.

The reversible processes of compression and expansion must take the place of many of the heat-transfer and throttling processes now being used. High-pressure steam expanded through a turbine can generate saturated steam for heating while extracting useful work. As energy-conversion systems like this become more complex, higher levels of control will be mandatory. Reference 3 cites the failure of industry to extract the maximum amount of work from the combustion of fuels.

Energy Sinks Rejecting energy to the atmosphere or to other streams within the plant is fraught with even more pitfalls. Energy rejection depends on the constancy of the sink. Primary sinks are environmental, i.e., atmospheric air and river water. Secondary sinks include cooling-tower water, refrigeration units, and waste-heat boilers, in that the cooling fluid is contained and dedicated to that function. The tertiary category includes such nondedicated sinks as other process streams.

Because all heat is ultimately rejected to the environment, its rate depends on the environment. Atmospheric cooling is especially precarious, since conditions change seasonally, diurnally, and abruptly with the weather. To a limited extent, all distillation columns reject heat to the environment through losses in column walls and piping. The reflux flow to a propane-propylene column of two 100-tray sections was seen to change 2.5 percent between day and night, with constant boilup and river-water cooling. This difference can only be attributed to increased heat losses to the surroundings during the night. Rain and high winds are certain to increase the losses far more, as are seasonal changes.

Although river and cooling-tower water may not change in temperature between day and night, their seasonal variations are still important. The most economical condensers—the air-fan units—are also the most variable. At this writing, much of the control effort in the distillation field has been applied to regulating the rate of cooling from these atmospheric condensers. As will be seen, overall column performance can be markedly improved if they are *not* controlled—a point which is developed to its fullest potential later in the book.

Great economies can be realized by using the energy released in condensation for reboiling. This principle has been used in evaporator technology for decades. Multiple-effect desalination units can distill 13 lb of water from the same amount of energy required to evaporate 1 lb with simple distillation [3]. Although these concepts are more difficult to apply to multicomponent systems, the future will see much development in this area [4].

Costs and Conversion Factors Even at currently rising rates, the cost of the energy required to boil a unit of a product is but a small fraction of the value of that product. Since this relationship will be used in later chapters on control and optimization, it is worthwhile quantifying now.

Fuel costs at this writing are quite variable, with natural gas cheaper than oil, particularly in those states producing it. By 1977, the cost of all fuels is expected to approach $2 to $2.50 per million Btu, with individual penalties assigned for impurities such as sulfur and ash.

Reference 5 gives the latent heat of vaporization of n-butane (typical of light hydrocarbons) as 148 Btu/lb at 100°F. With a specific gravity of 0.558 also at 100°F, the heat required to vaporize 1 gal is

$$148 \text{ Btu/lb} \times 0.558 \times 8.34 \text{ lb/gal} = 689 \text{ Btu/gal}$$

At 42 gal/bbl, the heat which will vaporize one barrel is

$$689 \times 42 = 28{,}930 \text{ Btu/bbl}$$

At a fuel cost of $1.60/million Btu, the cost of vaporizing a barrel is roughly 4.6 cents. Compare this to the value of the butane—$8/bbl or higher. The ratio of value to vaporization cost is in the neighborhood of 200:1.

If heat is supplied by steam saturated at 100 psia and condensing at 30 psia, each pound gives up about 968 Btu. Then 28,930/968 or 30 lb of steam is used to vaporize one barrel of butane. Polar compounds require considerably more energy for vaporization. Methanol, for example, consumes 473 Btu/lb in boiling at atmospheric pressure—over three times the latent heat of n-butane. Water, of course, has about six times the latent heat of n-butane.

One of the features of optimizing programs developed in later chapters is equations written in terms of heating and cooling costs and product values. These equations allow operators to update their programs easily as these figures change.

Cooling-water costs tend to be independent of temperature rise since they are primarily a function of pumping rate. The cost can be converted to equivalent Btu only by assuming or assigning a temperature rise. For a rise of 10°F, 1,000 gal of water will absorb

$$1{,}000 \text{ gal} \times 8.34 \text{ lb/gal} \times 10°\text{F} \times 1 \text{ Btu/lb-°F} = 83{,}400 \text{ Btu}$$

Using $0.05/1,000 gal, cooling cost for a 10°F rise comes to nearly $0.60/million Btu. At this rate, it would cost over 1.7 cents to condense the barrel of butane that was boiled for 4.6 cents.

Cooling and heating costs are probably not additive, since a reduction in heat input may not reduce cooling cost. The coolant flow rate (whether air or water) would tend to remain constant regardless of the heat transferred to it.

Minimizing Energy Consumption Maximizing recovery requires a maximum rate of energy input to the process. Minimizing energy consumption requires that recovery be reduced to some minimum acceptable value. There are several policies by which some "minimum" consumption can be achieved, but they involve an arbitrary assignment of product quality specifications or recovery. If the qualities of both products are specified at some minimum acceptable limits, then the energy required to make that separation is thereby established. Fixing a limit on recovery of a guaranteed product accomplishes essentially the same result. Whether either of these assignments can be justified is questionable.

The most meaningful method of determining energy usage must be an optimization program. Strictly speaking, minimum usage means zero usage, except in the rare case where two guaranteed products are made in a single tower. Thus, the goal of minimum energy consumption is not realistic unless arbitrary composition constraints such as those above are assigned.

For any given set of compositions, however, energy consumption may be reduced by operating the tower more efficiently. This requires moving in a different dimension, such as adding more trays to the tower, improving tray efficiency, or increasing the relative volatility of the components. Still other ways would be to trade feed heat input for reboiler heat input where it is more effective or to relocate the feed to a more optimum tray. Let us examine these items one at a time:

1. Adding trays cannot be accomplished while the tower is operating, so this method is discounted for the purpose of this discussion.

2. Improving tray efficiency is possible in cases where the vapor rate happens to be outside the most efficient range. Varying boilup outside these limits may create other difficulties such as weeping and entrainment, which are discussed later in the chapter.

3. Relative volatility can be increased by adding another component such as an extractant or by operating at minimum pressure. Although the former method is reserved for especially difficult separations, minimum-pressure operation is applicable to all. Energy consumption may be reduced as much as 5 percent at night and as much as 25 percent during winter operation by taking full advantage of atmospheric cooling to minimize tower pressure.

4. Heat introduced with the feed is only a fraction as effective as that applied to the reboiler. The reason is that the vapor it generates passes through only a fraction of the trays. The principal justifications for preheating the feed have to be usage of a lower temperature (and hence less costly) source of heat or equalization of vapor loading between top and bottom sections of the tower.

5. If the composition of the feed differs markedly from that of the tray it enters, separation efficiency will be lost through blending. Most columns have several alternate feeding locations, but they are rarely used.

Of all the foregoing points, improving relative volatility through minimum-pressure operation is the most interesting and shows the greatest potential for profit.

Although any of the means listed above could be used to maximize recovery as opposed to minimizing energy consumption, they seem more rightfully treated under this latter heading. Maximizing recovery is achieved through maximizing heat input in systems where heating and cooling costs are inconsequential. Furthermore, at maximum heat-input rates, savings achievable through minimum-pressure operation are limited.

Energy Consumption vs Purity For the column refining a single guaranteed product, it is possible to evaluate energy consumed per unit of product made. The function sought can be represented by the dimensionless ratio V/P. Fortunately, the information required to make such an evaluation is already available in the form of the recovery curves of Fig. 1.3. Recovery must first be substituted for product flow using Eq. (1.13):

$$\frac{V}{P} = \frac{V}{RFz} \tag{1.18}$$

Since Fig. 1.3 plotted recovery vs purity for fixed values of V/F, V/P is readily calculated. The results are given in Fig. 1.4.

The fact that the curves of Fig. 1.4 cross reveals an opportunity for optimization. Between 50 and about 97 percent purity, V/F of 2 results in the lowest energy consumption per unit product. Between 97 and 99.3 percent, V/F of 4 is best. Actually, however, an infinite number of V/F curves exist, forming an envelope of minimum V/P values as a function of purity. Consequently, for each value of purity a single V/F exists which will result in the minimum V/P

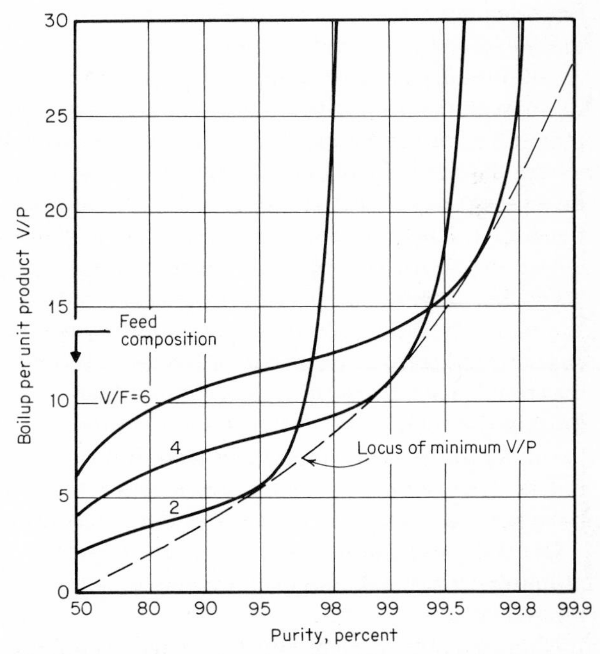

figure 1.4 *Recovery data from Fig. 1.3 were used to generate these curves of* V/P *versus purity.*

as traced by the envelope. An optimization program developed in Chap. 11 derives the optimum V/F as a function of product purity and other variables.

Having evaluated V/P, one only need multiply the cost of heating (and cooling) c, and the latent heat of vaporization H_V, to arrive at the energy cost $\$_Q$ per unit of product:

$$\frac{\$_Q}{P} = \frac{cVH_V}{P} \tag{1.19}$$

Equation (1.19) gives only part of the cost of making the separation, however. Reduced recovery incurs the twofold debit of increasing $\$_Q/P$ *and* $\$_P/P$. The sum of the two cost functions combines Eqs. (1.17) to (1.19):

$$\frac{\$_Q + \$_P}{P} = \frac{cVH_V/Fz + \Delta v(R_0 - R)}{R} \tag{1.20}$$

Only by completely evaluating Eq. (1.20) can the operating cost of a given tower be assessed. The profound effect of recovery then becomes apparent. Since V/P passes through a minimum as a function of V/F, the cost as calculated above will also. However, the weight of the $\$_P$ term will force the minimum toward a higher V/F ratio than indicated by the $\$_Q$ function alone.

LIMITATIONS ON ENERGY TRANSFER

The foregoing discussion related purity and recovery to energy consumed per unit feed or product, without regard to limitations. Actually the ratio of V/F or V/P is always subject to limits, not on F or P but on V. And when a limit is encountered, the two ratios must change if F or P is further increased. These limits will have a profound effect on the control-system arrangement.

Limitations on V appear at three different locations in a distillation system: where the vapor is generated, conveyed, and condensed. In this section limits on energy transfer are discussed, i.e., the maximum rates of boiling and condensing. The next section covers limits on conveying vapor through the column from reboiler to condenser and returning the condensed reflux.

Reboiler Constraints The following presentation is related strictly to heat transfer. As the author is not an expert in heat-transfer phenomena, the discussion will remain qualitative in tone. For simplicity in presentation and maximum usefulness, the relationships given are based on "The New Heat Transfer" [6]. Its author avoids heat-transfer coefficients altogether, proposing instead functional relationships between heat flow and thermal driving force. Many of the nonlinear properties associated with free convection and radiation are thereby accommodated with greater facility.

Heat transfer in the reboiler is almost always governed by free convection, although in special cases forced convection may be applied. Heat is transferred either from condensing steam or a hot liquid to a boiling bottom product. In this chapter, we are concerned only with the limit to heat flow, i.e., its maximum value, which is assumed to be transferred at the maximum flow or condensing rate of the hot fluid. At this upper limit, the hot side of the reboiler can be considered isothermal; i.e., the hot fluid enters and leaves at essentially the same temperature. This assumption greatly facilitates defining the operating limits, since the rate of heat transfer then is primarily determined by the difference in temperature between the heat source and the boiling liquid. In avoiding integration from inlet to outlet of the hot side of the reboiler, limits are easier to establish without sacrificing much in the way of accuracy. Although this is satisfactory for present purposes, more exact methods are used in the discussion on control over heat-transfer rates.

The thermal driving force, at the limit, is therefore the temperature difference between the hot fluid and boiling product. If the heating medium is steam, its maximum temperature is the saturation temperature at the available supply pressure. The temperature of the boiling product is a function of both composition and applied pressure. Knowing that no heat will flow when both fluids are at the same temperature allows one point on the boilup-vs-temperature or boilup-vs-pressure plot to be marked. The zero boilup point is indicated on Fig. 1.5 as the vapor pressure of the bottom product (in this case *n*-butane) at the saturation temperature of the steam or inlet temperature of the heating liquid (in this case 212°F).

Figure 1.5 is actually a vapor-pressure vs temperature plot for *n*-butane. As column pressure is reduced, the boiling point of the bottom product decreases, developing a temperature difference ΔT across the heat-transfer surface. If heat-transfer rate increases linearly with ΔT, the scale of ΔT in Fig. 1.5 could be replaced with an equivalent scale of molar or mass boilup rate, allowing for the variation of H_V with temperature.

Changing the heat-source temperature would move the boilup-vs-pressure curve left or right because heat-transfer rate is primarily influenced by ΔT rather than by absolute temperature. The slope of a boilup-vs-pressure curve would be a function of the area and condition of the heat-transfer surface. With a constant area, the slope should change only as the surface fouls, reducing the boilup attainable at a given pressure.

For steam-heated reboilers, the control valve may be placed on the steam

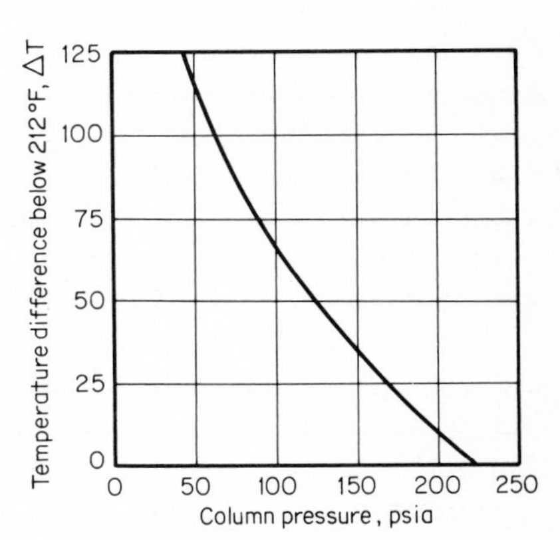

figure 1.5 *The maximum boilup of* n*-butane available using steam at atmospheric pressure is related to the above temperature difference.*

inlet or the condensate outlet. When manipulating steam into the reboiler, a trap must be provided for the condensate. Maximum heating will be indicated when the control valve is fully open. However, if the condensate trap cannot carry this full load, the steam chest will start to flood, causing the heat-transfer rate to fall because of the reduction in surface available for condensing. In this case, the transient heat-transfer rate is limited by the capacity of the control valve and steam piping, but the steady-state rate is limited by the trap and condensate piping. The limit imposed by condensate-removal capacity is fixed; it depends not on ΔT but on the difference between steam and condensate pressures. It is then a maximum boilup rate independent of column pressure.

With the control valve acting on the condensate, heat transfer is varied by flooding the surface area with condensate. Therefore maximum heat flow will be realized when the steam chest is drained, at which point steam will begin blowing through the valve. This condition will appear when the valve can carry more condensate than the reboiler can condense, and it is more likely to occur in fouled reboilers. If the capacity of the valve and piping is less than the reboiler can condense, some of the tube surface will always be covered. Then the condensate-flow limit described above applies. The operator must be alert to the possibilities of both steam blowing and condensate flooding in these systems.

Condenser Constraints Condensers have another dimension added to their heat-transfer limit: coolant-supply temperature is usually variable. Air cooling is most variable; refrigerant, evaporative (cooling tower), and river-water cooling are increasingly more constant and reliable. The cooling fluids themselves are usually not manipulated to control heat-transfer rate, except in the case of refrigerants. Thus the temperature rise of the coolant in passing across the heat-transfer surface is usually minimum.

As the heat load approaches zero, the condensing temperature will be the coolant-supply temperature. At this point, the pressure in the condenser will be the vapor pressure of the distillate product at that temperature. To increase the rate of condensation, the condensing temperature must increase. The limit on condensing rate is then determined by the temperature difference between the coolant and the dew point of the distillate product. For a given product composition, the limit on boilup rate that can be accommodated by the condenser is a function of column pressure. Figure 1.6 gives the ΔT across the condenser surface for isobutane condensed against a source of 63°F. Similar to Fig. 1.5, it is essentially a vapor-pressure plot of isobutane.

The greatest influence on the condenser limit is coolant-supply temperature. Ambient temperatures that vary from below freezing to as high as 100°F in temperate climates place severe demands on condensing systems. Wintertime operation requires protection from freezing and shutting down of some of the condensing capacity. But even day-to-night and rain-to-sun variations can impose significant disturbances on a column.

Air-cooled condensers which normally reject heat to ambient dry-bulb tem-

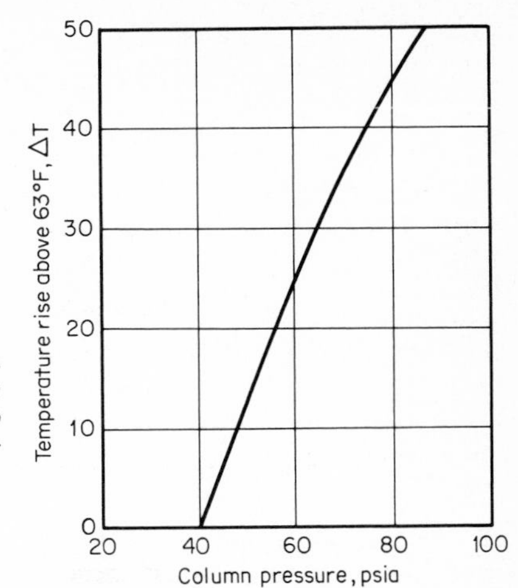

figure 1.6 *The maximum condensing rate for isobutane using a 63°F cooling source is related to the above temperature difference.*

perature may operate closer to wet-bulb temperature in the rain. Thus the onset of a sudden rain may cause a pronounced increase in condensing rate, forcing column pressure downward. A great deal of attention will be given to controlling these highly variable condensers.

The condenser-constraint curve is also shifted by changes in distillate composition. Very small amounts of noncondensable gases such as methane or nitrogen can severely limit column operation. Occasionally a gas is injected to raise the pressure for control purposes. However, this practice is discouraged since the gas contaminates the product, carries away product when released, and makes separation more difficult.

Cooling can be reduced below the maximum by flooding the tubes with condensed liquid or bypassing part of the vapor around the condenser. Bypassing lowers vapor velocity past the tube surface, thereby reducing the heat-transfer rate. For atmospheric condensers, fans may be reduced in speed or pitch or simply turned off to reduce cooling.

Floating-pressure Operation There is no law making column-pressure control mandatory. Some upper limit of pressure must be respected to avoid equipment damage, but no similar lower limit exists. It is entirely possible—in fact advantageous—to let the column pressure float on the condenser. By not throttling overhead vapor or condensate in any way, column pressure will ride the condenser constraint. Increasing boilup rate or coolant temperature will cause pressure to rise to a new equilibrium level, but it will always be minimum for the existing conditions.

Minimum-pressure operation has the advantage of maximizing the relative volatility of most pairs of components. This allows either an increase in recovery or reduction in energy required to make a given separation. Although the energy savings during wintertime operation may exceed 25 percent, even day-to-night variations are worthwhile exploiting in this way.

Other advantages include reduced reboiler fouling and increased reboiler capacity. The capacity improvement is obvious, but the reduction in fouling may require some evaluation. A rule of thumb is often quoted to the effect that reaction rates generally double with every 10°C increase in temperature. If reboiler fouling rate obeys this rule, then an 18°F (10°C) reduction in reboiler temperature should cut it in half. Reboilers normally operated at summertime temperatures year-round ought to require cleaning half as often with floating-pressure operation.

Resistance to floating-pressure operation is largely traditional. Temperature measurements have been used for quality indication and control for most columns down through the years. Naturally, tight pressure control is prerequisite to quality control using temperature measurement. Analyzers are increasingly replacing temperature measurements, removing this obstacle for many separations. But where analyzers are not available or justifiable, temperature measurements may be corrected for pressure, eliminating the effect of its variation. A differential-vapor-pressure sensor [7] is also available which provides inherent compensation by comparing column pressure to the vapor pressure generated by a reference fluid at column temperature.

Sudden changes in column pressure are to be avoided. A sudden reduction can cause flashing in the liquid on column trays and severely upset the internal material balance. A rapid rise in pressure could momentarily disrupt boilup, but it is more difficult to achieve. Weather changes impose the most severe upsets on cooling systems, and weather usually deteriorates much more rapidly than it improves. As a result, columns need only be protected against a sharp increase in cooling.

COLUMN CONSTRAINTS

The trays in the column have limitations which are more restrictive than those of the reboiler and condenser. Column performance may be measured as overall tray efficiency, which is the ratio of the number of theoretical equilibrium stages observed to the number of trays in the column. For a given tray configuration, efficiency will be maximized only over a specified range of liquid and vapor flow rates. The maximum efficiency and the range over which it extends vary with the type of tray. Perforated trays tend to exhibit efficiency that increases with superficial vapor velocity [8] up to a point and then decreases sharply. Bubble-cap and valve trays have a broader range of high efficiency, but they also suffer a loss at very low or high rates. Causes of these losses in efficiency are examined below in an effort to define the limits of efficient operation.

Weeping and Pulsation Liquid on a perforated plate is supported by the flow of vapor through the holes. As vapor flow is reduced, turbulence at each orifice decreases, which hinders vapor-liquid contact and lowers efficiency. Eventually a point is reached where some holes begin to weep liquid, which then substantially escapes vapor contact.

Valve trays are less sensitive in this regard since the area exposed to vapor flow is decreased by the valves as flow falls off. Bubble caps do not tend to weep since liquid is retained by the riser inside each cap. However, combinations of high liquid and low vapor rates can submerge some caps, causing them to dump liquid, while others carry all the vapor load [9]. Since this dumping essentially releases the liquid head, normal operation may again resume until the head returns to its previous level. Consequently tray operation becomes cyclic and is classified as unstable.

To avoid both weeping and instability, limits must be placed on the liquid and vapor rates. To be consistent with other constraints applied to the reboiler and condenser, a simple low limit will be applied to boilup to satisfy the vapor constraint. However, the limits described relate to vapor velocity. Consequently, lower-pressure operation, in which vapor density is reduced, can tolerate a lower molar or mass rate of boilup. Reference 9 indicates that minimum vapor rate varies with the square root of vapor density, as might be expected in this turbulent-flow regime.

Figure 1.7 shows a contour of molar or mass boilup as a function of column pressure, illustrating the form of a low-boilup constraint. It follows the relationship for isobutane:

$$V = V_M \sqrt{\frac{\rho_{VM}}{\rho_V}} \tag{1.21}$$

where V and ρ_V are the mass or molar boilup limit and vapor density at any operating pressure, with subscript M referring to maximum or design pressure.

Flooding The term "flooding" has been used to describe various conditions which cause a loss of tray efficiency at high vapor rates. In any given column, the actual mechanism could be entrainment, foaming, or downcomer flooding. The first two are most common and are both due to poor disengagement of liquid and vapor. The last is an extreme condition where vapor velocity and hence pressure drop are so high that liquid is prevented from flowing down the column.

Entrainment and Foaming Entrainment and foaming are both sensitive to the difference between the densities of vapor and liquid and to the surface tension

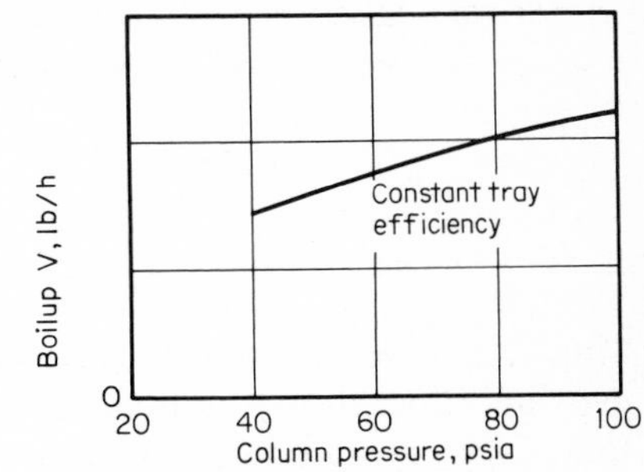

figure 1.7 *The low limit of boilup varies with vapor density and therefore with pressure.*

of the liquid. Consequently both are related to the critical point of the components since disengagement deteriorates as this point is approached.

Entrainment obviously increases with the velocity of the vapor, which tends to carry liquid droplets from one tray to the next. Tray efficiency is thereby decreased since higher-boiling materials are physically lifted up the column. Liquid loading is also increased, because all the entrained liquid must be returned through the downcomers. In the extreme, entrainment can be increased to the point where all the liquid cannot be returned, resulting in a flooded condition. This increased liquid loading raises the head encountered by the vapor in passing through the trays and is thus detectable by measuring column pressure drop. Many columns have differential-pressure controllers which manipulate boilup to maintain some maximum or optimum tray loading. The differential-pressure measurement is particularly valuable when there is no other way of determining the rate of heat input.

Failure of vapor to become disengaged from the liquid phase causes *foam* to accumulate on the trays and in the downcomers. Its effect is similar to entrainment in that the foaming liquid is physically lifted from one tray to the next, reducing overall efficiency. Because foaming lowers the density of the liquid phase, pressure drop does not change and so the problem is not detected except in off-specification product. As with entrainment, lower vapor and liquid loading will avert the loss in efficiency.

Predicting Flooding Limits Reference 10 presents a correlation by J. R. Fair for vapor velocity at flooding versus liquid–vapor flow and density ratios for columns with various tray spacings. A correction factor for surface tension below 20 dyn/cm is included. The same correlation is also applied to foaming limits and to predict fractional entrainment.

To demonstrate how flooding limits change with pressure, the cited correlation was evaluated for selected hydrocarbons at temperatures from 80 to 120°F. The data should be representative for columns yielding relatively pure distillate products and rejecting heat into air-cooled condensers. For each tray spacing, Fair's correlation yields

$$u_{VN}\sqrt{\frac{\rho_V}{\rho_L-\rho_V}}=f\left(\frac{L}{V}\sqrt{\frac{\rho_V}{\rho_L}}\right) \tag{1.22}$$

where u_{VN} = velocity of vapor based on net area, ft/s
L, V = mass-flow rates of liquid and vapor, lb/h
ρ_V, ρ_L = vapor and liquid densities, lb/ft^3

The velocity thus found may be converted to mass boilup V by multiplying by vapor density and net area A:

$$V = Au_{VN}\rho_V \tag{1.23}$$

Then the correction factor $(\sigma/20)^{0.2}$ is required for surface tension σ below 20 dyn/cm. Combining all these relationships yields the boilup limit as a function of vapor and liquid densities and surface tension:

$$V = A\left(\frac{\sigma}{20}\right)^{0.2} \sqrt{\rho_V(\rho_L - \rho_V)}\, f\left(\frac{L}{V}\sqrt{\frac{\rho_V}{\rho_L}}\right) \tag{1.24}$$

Table 1.1 presents values of V/A calculated using ρ_V and ρ_L for the cited hydrocarbons, with 18-in. tray spacing and assuming that $L/V = 1.0$. Densities and vapor pressures were taken from Ref. 11; surface tension was calculated using the parachor in Ref. 12.

TABLE 1.1 Flooding Calculations for Selected Light Hydrocarbons from 80 to 120°F

	Propylene			*Propane*			*Isobutane*		
	80°F	100°F	120°F	80°F	100°F	120°F	80°F	100°F	120°F
$p°$, psia*	174.7	227.6	291.2	143.6	188.7	243.4	53	73	96
ρ_V, lb/ft³	1.603	2.119	2.778	1.342	1.792	2.347	0.59	0.81	1.063
ρ_L, lb/ft³	31.52	30.30	28.90	30.59	29.50	28.31	34.2	33.2	32.3
σ, dyn/cm	6.95	5.47	4.04	6.68	5.38	5.15	9.11	7.84	6.79
V/A, lb/h-ft²	1.160	1.175	1.163	1.067	1.095	1.094	0.905	0.964	1.017

* $p°$ is the vapor pressure of the component at the indicated temperature.

Note that the allowable boilup for both propylene and propane overhead products passes through a maximum as a function of column temperature. Isobutane and higher-boiling products exhibit a boilup limit increasing with temperature whereas ethane and lower-boiling products have a decreasing limit, as confirmed by Ref. 8.

Three factors contribute to this variation. Increasing temperature (pressure) reduces vapor density, thereby reducing velocity for a given mass-flow rate. However, increasing temperature also decreases the difference between liquid and vapor densities and lowers surface tension. The latter effects are more pronounced for the lighter hydrocarbons, since at ambient temperatures they are closer to their critical pressure. These effects more than offset that of decreasing vapor density for propane and propylene above 100°F. Data from the last row of Table 1.1 are plotted against temperature in Fig. 1.8.

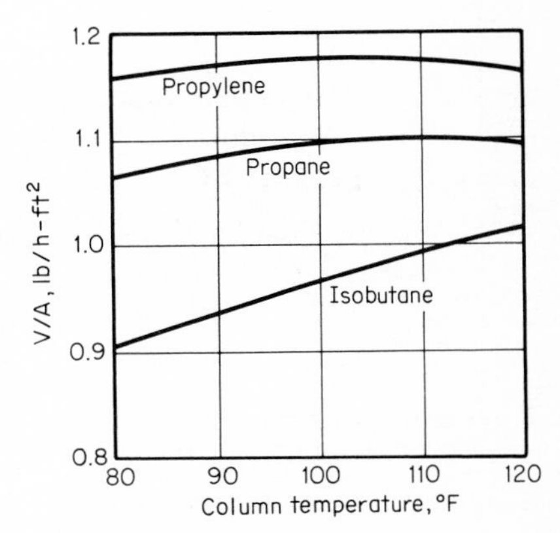

figure 1.8 *Calculated flooding limits vs column temperature for selected light hydrocarbons.*

The Operating Window Another constraint to be considered is the maximum allowable operating pressure. Vessels must be protected from overpressure by relief devices set safely below test limits. Operating pressures must not be allowed to approach these relief settings very closely or some leakage could result. Consequently each column will have some upper limit of pressure without regard to boilup rate. This limit will appear as a vertical line on the right side of the boilup-vs-pressure diagram of Fig. 1.9.

There may be additional limitations on pressure to ensure transfer of feed to the tower or products from it. These limits could vary somewhat with flow but not appreciably. Flow from the reflux pump is also limited, but by suction and discharge heads and vapor pressure. If column pressure floats on the condenser, all three of these variables should change together. Therefore there should be no net variation of reflux pumping limit with pressure, except in the case where a significant pressure drop exists between the column and the reflux accumulator.

Plotting all the significant constraints on the same figure forms an "operating window." Boilup must always be controlled within this window or along any constraint if normal operation is to be expected. Figure 1.9 illustrates a typical window for a column separating normal and isobutane.

No vertical scale is given on the figure because actual boilup limits depend on tray and column design, etc. But the relative position of the curves and their shapes are representative. Note that the positions of the condenser and reboiler constraints are variable, the former with coolant-temperature changes, the latter due primarily to fouling. Under changing conditions, maximum boilup could be limited by the condenser, by the reboiler, or by flooding. Whenever the reboiler is not limiting, maximum boilup for the

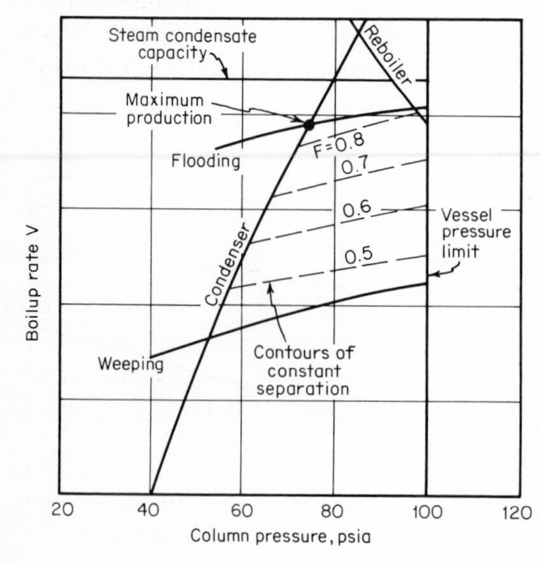

figure 1.9 *Typical operating window for a butane splitter, illustrating that production may be maximized at the condenser constraint.*

butane splitter will be coincident with maximum pressure. But this is not true for propane or propylene products whose flooding constraints pass through a maximum. For these and lighter products, maximum boilup will be achieved where the flooding curve meets either the condenser or reboiler constraint or at the maximum point of the flooding curve.

Maximum boilup is not necessarily consistent with maximum *production*, however. Separation is favored at lower pressure because of increased relative volatility, yet Ref. 13 indicates a slight drop in tray efficiency at lower pressure. Nonetheless, as pressure is reduced, less boilup is required to separate a given feedstock.

Within the operating window are contours of constant separation (i.e., constant product qualities), one for each feed rate. If these contours lie parallel to the flooding curve, maximum production and most favorable economy will be achieved where the flooding curve crosses the condenser constraint. Only if the slope of the flooding curve is more positive than the contours of constant separation is maximum production to be achieved at maximum pressure. This is likely to be the case only for distillate products heavier than isopentane.

By operating the column depicted in Fig. 1.9 at the condenser constraint, a significant reduction in boilup may be realized. Reduction in boilup should have no direct effect on tray efficiency because superficial vapor velocity actually increases somewhat as pressure is reduced along a separation contour. This is borne out by the divergence of separation contours and the weeping line as pressure is reduced. Recall that the weeping line is essentially a contour of constant superficial velocity and therefore virtually constant tray efficiency.

An additional benefit of minimum-pressure operation is that more heat may be extracted from the reboiler heating medium. If heating costs are assessed per pound of steam regardless of the heat extracted, fewer pounds will be required at lower column pressure because of reduced boilup and lower reboiler temperature.

Overall Operating Cost Now that the important constraints have been evaluated, the cost function may be completed. Return on investment for a column and its accessories improves with the rate of production. Production rate is, in turn, affected by feed rate and product recovery. If we wish to maximize production at all times, then boilup must be at a constraint, that is, V_{max}. The cost function of Eq. (1.19) and Fig. 1.4 evaluated the cost of boilup per unit product. The same function may apply to column return on investment simply by substituting V_{max} for V. Consequently, anything that can be done to reduce V/P will reduce V_{max}/P and increase the return by increasing P.

Labor costs may be assessed in the same manner. Operators are customarily placed in charge of certain groups of capital equipment. As the productivity of the equipment increases through better control, that of the operator does also. Note that all costs are profoundly affected by product recovery. This should provide sufficient incentive for each engineer to investigate the economic benefits obtainable through improved control of his columns.

REFERENCES

1. Kline, P. E.: Technical Task Force Approach to Energy Conservation, *Chem. Eng. Prog.*, February 1974.
2. Lupfer, D. E.: Distillation Column Control for Utility Economy, presented at 53rd Annual GPA Convention, Denver, Mar. 25–27, 1974.
3. Keenan, J. H., E. P. Gyftopoulos, and G. N. Hatsopoulos: The Fuel Shortage and Thermodynamics, *Proc. MIT Energy Conf.*, February 1973.
4. Tyreus, B. D., and W. L. Luyben: Two Towers Cheaper Than One?, *Hydrocarbon Process.*, June 1975.
5. "Technical Data Book—Petroleum Refining," 2d ed., pp. 7–32, American Petroleum Institute, Washington, D.C., 1970.
6. Adiutori, E. F.: "The New Heat Transfer," Ventuno Press, Cincinnati, 1974.
7. The Foxboro Company: Differential Vapor Pressure Cell Transmitter, Model 13VA, *Technical Information Sheet* 37-91a, Foxboro, Mass., April 1965.
8. Smuck, W. W.: Operating Characteristics of a Propylene Fractionating Unit, *Chem. Eng. Prog.*, June 1963.
9. Van Winkle, M.: "Distillation," pp. 527–532, McGraw-Hill Book Company, New York, 1967.
10. *Ibid.*, pp. 525–526.
11. "Matheson Gas Data Book," 4th ed., The Matheson Company, East Rutherford, N.J., 1966.
12. "Technical Data Book—Petroleum Refining," 2d ed., pp. 10–11, American Petroleum Institute, Washington, D.C., 1970.
13. Doig, I. D.: Variation of the Operating Pressure to Manipulate Distillation Processes, *Aust. Chem. Eng.*, July 1971.

PART TWO

Conventional Distillation Processes

CHAPTER TWO

Binary Separations

In actual practice, few binary mixtures exist. Virtually every distillation column is fed a multicomponent mixture of sorts. Yet this chapter is dedicated to binary separations in an effort to simplify the learning process. This by no means restricts the principles or mathematical models developed herein to binary systems. The fundamentals of this chapter will serve as the foundation for the next, wherein multicomponent feeds, multiproduct columns, and trains of columns are presented.

RELATIVE VOLATILITY

Relative volatility is the parameter that identifies the ease of separating components by distillation. While graphic methods of tray selection, etc., use equilibrium curves without resorting to numerical values of volatility, they cannot successfully be applied to difficult separations requiring many trays or to multicomponent separations. As a result, numerical procedures are used throughout this text. They are based on the relative-volatility parameter, even for nonideal and multicomponent mixtures.

Ideal Mixtures A review of the fundamental relationships existing in ideal mixtures will be helpful in laying the groundwork for later derivations. To begin, Dalton's law gives the total pressure in a gaseous mixture as the sum of the partial pressures of the components:

$$p = \sum_{1}^{n} p_i \tag{2.1}$$

In an ideal mixture, the partial pressure exerted by a component is proportional to its concentration in the vapor:

$$p_i = y_i p \tag{2.2}$$

Here y_i is the mole fraction of component i in the vapor.

Raoult's law states that in an ideal mixture the partial pressure exerted by a component at equilibrium is the product of its molar concentration x_i in the liquid times the vapor pressure p_i° of the pure component at that temperature:

$$p_i = x_i p_i^\circ \tag{2.3}$$

Combining the last two relationships allows us to express the concentrations of vapor and liquid relative to one another:

$$\frac{y_i}{x_i} = \frac{p_i^\circ}{p} = K_i \tag{2.4}$$

The ratio y_i/x_i or K_i is known as the "equilibrium vaporization ratio" or "K factor" for component i. For nonideal systems, experimentally determined values of K should be used instead of vapor-pressure data to evaluate Eq. (2.4).

Relative Volatility Defined Equation (2.4) may be solved for each component in a mixture to allow comparison of their contributions to the liquid and vapor at equilibrium. Dividing (2.4) in terms of component i by the same equation in terms of component j yields the "relative volatility" of component i with respect to j:

$$\alpha_{ij} \equiv \frac{y_i/x_i}{y_j/x_j} = \frac{p_i^\circ}{p_j^\circ} \tag{2.5}$$

Again, for nonideal systems, actual values of K will give more exact results:

$$\alpha_{ij} = \frac{K_i}{K_j} \tag{2.6}$$

References 1 and 2 are recommended sources for K factors for mixtures of light hydrocarbons and common volatile nonhydrocarbons. Since these factors vary with temperature, pressure, and composition, their tabulation requires extensive nomography. As a result, they are not tabulated here, although values taken from Refs. 1 and 2 are used to enumerate examples. Since this book is not intended as a column design guide, exact equilibrium data are not required. However, the role of α in relating compositions, and the degree of its variation with pressure and temperature, are crucial.

If component i in Eqs. (2.5) and (2.6) is more volatile than j, then K_i will exceed K_j and α_{ij} will exceed unity. When $\alpha = 1.0$, separation of the two components by simple distillation is impossible.

In a binary system, graphic column-design methods are based on an equilibrium curve of y versus x. For ideal binary mixtures, the points on the equilibrium curve may be calculated by substituting $1 - y_i$ for y_j and $1 - x_i$ for x_j. The subscripts may then be dropped:

$$\alpha = \frac{y/x}{(1-y)/(1-x)} = \frac{y(1-x)}{x(1-y)} \tag{2.7}$$

Then y may be found in terms of x:

$$y = \frac{\alpha x}{1 + x(\alpha - 1)} \tag{2.8}$$

or x may be evaluated in terms of y:

$$x = \frac{y}{\alpha - y(\alpha - 1)} \tag{2.9}$$

For purposes of illustration, Fig. 2.1 gives equilibrium curves for $\alpha = 1.2, 1.5,$ 2, and 4.

Unfortunately, α usually varies somewhat with composition, indicating nonideal behavior. When Eqs. (2.8) or (2.9) are used to make tray-to-tray calculations or to construct an equilibrium curve, α should be adjusted for compositions at each calculation.

To evaluate deviation from ideality, α was calculated for the butane-isobutane system from vapor pressures interpolated from Ref. 3 and K values taken from Ref. 2. The results appear in Table 2.1. Relative volatility calculated as K_i/K_n is only given to three significant figures since K values cannot be read more accurately from the charts given in Ref. 2.

Although α does not seem to vary significantly with composition, it is lower

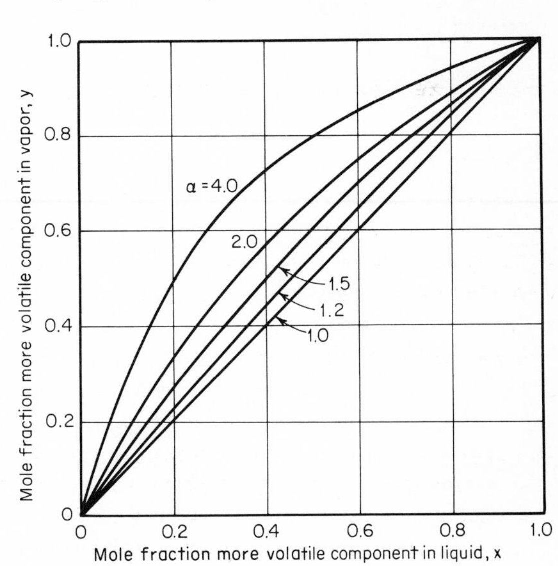

figure 2.1 *The equilibrium curve is symmetrical if α does not vary with composition.*

TABLE 2.1 Relative Volatility Calculated for the Butane-Isobutane System at 100°F

p_i°, psia*	p_n°, psia*	p_i°/p_n°	x_i	K_i	K_n	K_i/K_n
71.9	51.7	1.391	0	1.35	1.00	1.35
			1.0	1.00	0.74	1.35

* Subscripts i and n refer here to isobutane and normal butane.

than the ratio of the vapor pressures. Use of the ratio of vapor pressures should be restricted to close-boiling homologs below 30 psia where vapors do not appreciably deviate from ideal-gas behavior.

Variation of α with Temperature Vapor-pressure curves of similar components tend to converge as temperature is increased. This property causes α to diminish with increasing temperature. Separation is therefore most favorable at the minimum available temperature.

To evaluate this property, let us begin with vapor-pressure vs temperature relationships for two ideal components:

$$\log p_i^\circ = a_i - \frac{b_i}{T + c}$$

$$\log p_j^\circ = a_j - \frac{b_j}{T + c} \tag{2.10}$$

Being interested in the ratio of the vapor pressures, we can subtract the two logarithms:

$$\log \alpha_{ij} = \log \frac{p_i^\circ}{p_j^\circ} = \frac{b_j - b_i}{T + c} + a_i - a_j \tag{2.11}$$

Then Eq. (2.11) may be differentiated with respect to temperature:

$$\frac{d \log \alpha_{ij}}{dT} = \frac{b_i - b_j}{(T + c)^2} \tag{2.12}$$

Equation (2.12) can be linearized by dividing both sides by $d \log \alpha/d\alpha$:

$$\frac{d\alpha_{ij}}{dT} = \frac{\alpha_{ij}\,(b_i - b_j)}{(\log e)(T + c)^2} \tag{2.13}$$

The fractional change in α is simply

$$\frac{d\alpha_{ij}}{\alpha_{ij}\, dT} = \frac{b_i - b_j}{(\log e)(T + c)^2} \tag{2.14}$$

From Eqs. (2.12) to (2.14), if coefficient b for the more volatile component is the smaller of the two, α_{ij} will increase as temperature is reduced.

example 2.1

For the butane-isobutane system, estimate the fractional change in α with temperature in the vicinity of 100°F, given the following vapor-pressure equations fitted to data from Ref. 3:

$$\text{Isobutane: } \log p^\circ = 4.913 - \frac{1{,}473}{T + 382}$$

$$n\text{-Butane: } \log p^\circ = 5.033 - \frac{1{,}600}{T + 382}$$

(These equations are accurate within 0.15 psi over the range of 40 to 100 psia.)

$$\frac{d\alpha}{\alpha\, dT} = -\frac{127}{0.434(T + 382)^2}$$

At 100°F, $d\alpha/\alpha\, dT = -1.26 \times 10^{-3}\,°\text{F}^{-1}$ or 1.26 percent increase for every 10°F reduction in temperature.

Doig [4] reports variations in α for the butane-isobutane system as proportional to $p^{-0.055}$, as measured by the ratio of vapor pressures of the components.

Table 2.2 gives relative volatilities for the butane-isobutane system as a function of temperature. The calculations were made using K factors from Ref. 2 evaluated at the vapor pressure of the more volatile component. Smuck [5] reports α for the propane-propylene system to vary from 1.15 at 175 psia to 1.135 at 225 psia and 1.12 at 300 psia, as obtained from actual plant operating data. Although these variations seem small, later examples show their profound effect on operating costs.

Total-reflux Operation "Total reflux" describes that operating condition in which all vapor is condensed and returned to the column, feed and product rates being zero. At total reflux, all trays may reach equilibrium since there is no net upward or downward flow of products. If Eq. (2.5) is solved successively for each theoretical tray in a column, the relationship between the vapor composition y_n leaving the top tray (n) and the liquid composition x_1 leaving the bottom tray (1) is

$$\frac{y_{in}/x_{i1}}{y_{jn}/x_{j1}} = \bar{\alpha}_{ij}{}^n \tag{2.15}$$

where n is the number of theoretical trays and $\bar{\alpha}_{ij}$ is the average relative volatility across the column. In the case of a binary separation, (2.15) may be written in terms of the more volatile component:

$$\frac{y_n(1 - x_1)}{x_1(1 - y_n)} = \bar{\alpha}^n \tag{2.16}$$

TABLE 2.2 Relative Volatilities of Butane-Isobutane as a Function of Temperature

T, °F	p, psia	α
60	38.1	1.41
80	53.1	1.38
100	71.9	1.35
120	95.2	1.32

Van Winkle [6] arithmetically averages the relative volatilities at the top and bottom of a column:

$$\bar{\alpha}_A = \frac{\alpha_n + \alpha_1}{2} \tag{2.17}$$

whereas Treybal [7] recommends a geometric average:

$$\bar{\alpha}_G = \sqrt{\alpha_n \alpha_1} \tag{2.18}$$

example 2.2

Consider a six-tray column across which α varies linearly from 1.35 to 1.40. Calculate y_n for $x_1 = 0.2$ using Eq. (2.8), and from this determine the true $\bar{\alpha}$ using Eq. (2.16). Compare $\bar{\alpha}$ against $\bar{\alpha}_G$ and $\bar{\alpha}_A$ as estimated above.

$\alpha_1 = 1.35$	$x_1 = 0.2$	$y_1 = 0.25234$
$\alpha_2 = 1.36$	$x_2 = y_1$	$y_2 = 0.31460$
$\alpha_3 = 1.37$	$x_3 = y_2$	$y_3 = 0.38606$
$\alpha_4 = 1.38$	$x_4 = y_3$	$y_4 = 0.46460$
$\alpha_5 = 1.39$	$x_5 = y_4$	$y_5 = 0.54674$
$\alpha_6 = 1.40$	$x_6 = y_5$	$y_6 = 0.628074$

$$\bar{\alpha} = \left[\frac{0.628074\ (0.8)}{0.2\ (1 - 0.628074)}\right]^{1/6} = 1.37489$$

$$\bar{\alpha}_A = \frac{1.35 + 1.40}{2} = 1.3750$$

$$\bar{\alpha}_G = \sqrt{1.35(1.40)} = 1.37477$$

The true average falls almost exactly between the arithmetic and geometric averages. The best approximation would then seem to be the arithmetic average of the two:

$$\bar{\alpha} = \frac{\bar{\alpha}_A + \bar{\alpha}_G}{2} \tag{2.19}$$

Equation (2.15) is known as the Fenske-Underwood equation [8]. It is useful in estimating the number of theoretical trays required to effect a given separation at total reflux, i.e., the minimum number of trays within which that separation may be achieved:

$$n_{\min} = \frac{\log \dfrac{y_{in}/x_{i1}}{y_{jn}/x_{j1}}}{\log \bar{\alpha}_{ij}} \tag{2.20}$$

Total-reflux operation is incapable of making any product. However, when feed is introduced and products are withdrawn, separation between the products will decrease. For any production condition, then, n must exceed $n_{\min}$ if desirable values of y_n and x_1 are to be realized.

SEPARATION

At this point it is useful to define what is meant by "separation." The term will henceforth be used to describe the left-hand side of Eqs. (2.15) and (2.16). For a system of any number of components, then,

$$S_{ij} \equiv \frac{y_{in}/x_{i1}}{y_{jn}/x_{j1}} \tag{2.21}$$

And specifically for a binary system, in terms of the more volatile component,

$$S \equiv \frac{y_n(1 - x_1)}{x_1(1 - y_n)} \tag{2.22}$$

As mentioned above, S reaches a maximum value of $\bar{\alpha}^n$ at total reflux, decreasing as feed and product flow rates increase. Its minimum value is 1.0, when both products in a binary system have the same composition. In multicomponent systems, equal ratios of y/x for a pair of components indicate that no separation is being achieved.

Factors Affecting Separation Actually any finite value of reflux flow may satisfy the total-reflux situation. But since tray efficiency does vary with liquid and vapor flow rates, varying these rates can affect separation by altering tray efficiency. So one factor affecting separation is tray efficiency. The influence of overall tray efficiency E may be incorporated by multiplying by the number of trays in the column so that separation at total reflux is $\bar{\alpha}^{nE}$.

At total reflux, the liquid-vapor flow ratio L/V at any point in the column is unity. As feed is introduced and products are withdrawn, L/V above the feed tray decreases and L/V below the feed tray increases. An examination of the effect of a change in L/V on the separation of a single tray is worthwhile. Figure 2.2 shows the top tray of a column with a condenser and associated streams. Equilibrium on the top tray gives

$$\frac{y_n(1 - x_n)}{x_n(1 - y_n)} = \alpha \tag{2.23}$$

A material balance of the light component across the top tray follows:

$$V_n y_{n-1} + L y_n = V_n y_n + L x_n \tag{2.24}$$

If the L/V_n ratio is set at 1.0, Eq. (2.24) yields $x_n = y_{n-1}$. Then the separation achieved across the top tray, i.e., between y_{n-1} and y_n, is simply α.

But when $L/V_n < 1.0$, solution of these two equations yields values of $x_n < y_{n-1}$, causing a reduction in y_n and hence in separation.

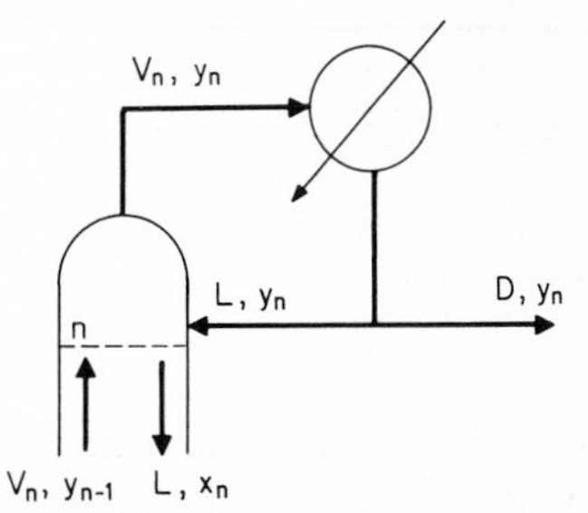

figure 2.2 *Withdrawal of distillate product reduces L/V in the top section of the column.*

example 2.3

Let $y_{n-1} = 0.9$ and $\alpha = 2.0$. Calculate the separation across the top tray when L/V_n is reduced to 0.9, using Eqs. (2.23) and (2.24).

Rearranging (2.23),

$$x_n = \frac{y_n}{\alpha - y_n(\alpha - 1)}$$

Substituting into (2.24),

$$V_n y_{n-1} + L y_n = V_n y_n + \frac{L y_n}{\alpha - y_n(\alpha - 1)}$$

Solving for y_n,

$$\left(1 - \frac{L}{V_n}\right) y_n^2 - \left(y_{n-1} - \frac{L}{V_n} + 2\right) y_n + 2y_{n-1} = 0$$

This quadratic is then evaluated for the stated conditions:

L/V_n	y_n	x_n
1.0	0.9474	0.9
0.9	0.9446	0.895

Separation across the top tray is then evaluated:

$$S = \frac{y_n(1 - y_{n-1})}{y_{n-1}(1 - y_n)} = 1.895$$

Reducing L/V_n from 1.0 to 0.9 has reduced separation from 2.0 to 1.895.

The procedure followed in the preceding analysis is the basis for the tray-to-tray calculations used in column design. Progressing beyond one tray or two components complicates the solution, and iterative methods become necessary.

In an effort to obtain a simplified method for determining separation for conditions other than total reflux, the influence of the boilup-feed ratio V/F will be examined. As opposed to reflux or L/V ratios, it connotes energy consumption per unit throughput, which is meaningful in light of discussions presented in Chap. 1. As will be seen, the V/F ratio has a dominant effect on separation.

However, data reported by Luyben [9] indicate that the V/F ratio required to effect a given separation varies somewhat with feed composition. It is maximum when feed composition z is roughly centered between that of the two products. Increasing or decreasing feed composition improves separation presumably by bringing it closer to one or the other product. Luyben also claims a small reduction in V/F brought about by selecting the optimum feed tray n_F. These parameters broaden our functional relationship to

$$S = f\left(\frac{V}{F}, \bar{\alpha}, nE, z, n_F\right) \tag{2.25}$$

The effect of the last two parameters is significant, but minor in comparison to the other terms. Typically a variation in z from 0.5 to 0.3 will reduce V/F by perhaps 15 percent in achieving the same separation if the feed is relocated to the optimum tray, and only 10 percent if it is not. Although the reduction is significant from an economic point of view, it is not normally controllable. In other words, feed composition does not normally change substantially nor can it be made to change to improve column operation. As a consequence, the relationship between separation and feed composition will only be described qualitatively.

Feed enthalpy also has an effect similar to feed composition and feed-tray location. Rather than consider this as a separate input, however, it will be lumped together with the boilup. Energy introduced with the feed in the form of vapor will simply be added to the reboiler vapor with a weighting factor relative to the number of trays through which its vapor passes:

$$V = V_r + Fq\,\frac{n - n_F}{n} \tag{2.26}$$

Here q is the fractional vaporization of the feed and V_r is the reboil rate. The total vapor rate V may then be used in the correlation given below.

Separation versus *V/F* A functional relationship between separation and V/F has been formulated to fit data generated by the tray-to-tray simulation given in Appendix A. In an attempt to include all possible values of $\bar{\alpha}$ and n, the functional group Y chosen to represent separation is

$$Y \equiv \frac{\log S}{nE \log \bar{\alpha}} \tag{2.27}$$

The denominator will be recognized as the logarithm of the maximum separation, that is, $\bar{\alpha}^{nE}$.

As it happens, Y has another significance. From Eq. (2.20) it can be seen that $\log S/\log \bar{\alpha}$ represents the minimum number of theoretical trays required to make a stipulated separation. Then

$$Y = \frac{n_{\min}}{n} \tag{2.28}$$

where the n's may be either theoretical or real trays.

It is intended that Y be given primarily as a function of V/F. For this purpose, Fig. 2.3 was prepared. In order that a single curve could be applied for all values of nE and $\bar{\alpha}$, the abscissa X had to be made a function of those two parameters:

$$X = \frac{V\bar{\alpha}^{1.68}}{F(nE)^{0.32}} \tag{2.29}$$

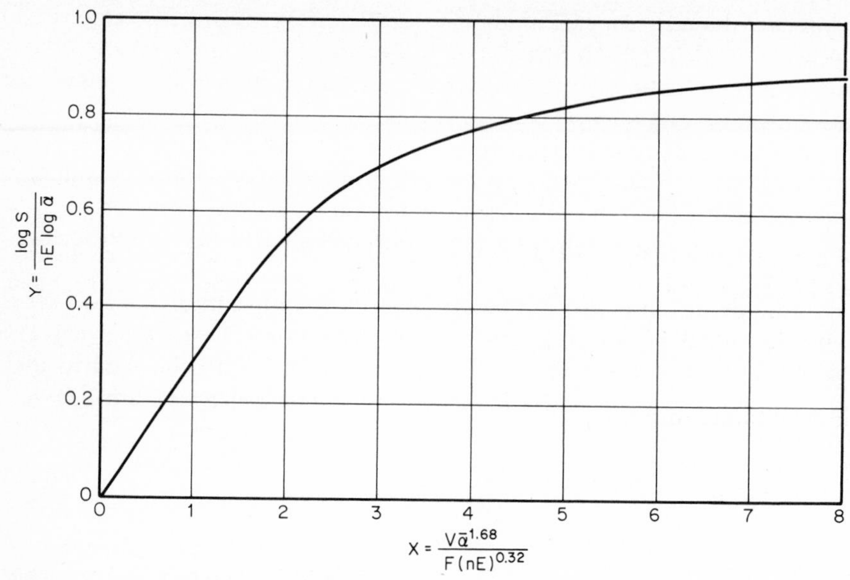

figure 2.3 *Separation as a function of* V/F, nE, *and* $\bar{\alpha}$.

The curve of Fig. 2.3 is somewhat complex, such that it cannot be fit precisely by a single equation. It is best sectioned into two algebraic relationships:

$$Y = \begin{cases} 0.286X & \text{for } X < 1.825 \\ \dfrac{1}{\sqrt{8.35/X^2 + 1.163}} & \text{for } X > 1.825 \end{cases} \tag{2.30}$$

At $X = 1.825$, the two solutions are identical. Group X can also be found in terms of Y:

$$X = \begin{cases} \dfrac{Y}{0.286} & \text{for } Y < 0.522 \\ \sqrt{\dfrac{8.35}{1/Y^2 - 1.163}} & \text{for } Y > 0.522 \end{cases} \tag{2.31}$$

Figure 2.3 was developed using a 50 percent feed composition and equal purities for both products. As mentioned above, a shift in feed composition or imbalance between product purities may cause a slight improvement in Y for a given X. The lowest separation seems to coincide with a centrally located feed tray. Table 2.3 illustrates the effects of departures from a balanced set of compositions. The first row gives a value of Y which falls on the curve.

The relationship shown in Fig. 2.3 is not intended to be accurate enough for column design purposes. Instead, it is a convenient method to evaluate the effects of V/F and $\bar{\alpha}$ on separation.

TABLE 2.3 Values of *Y* for a 20-tray Column at *V/F* of 3 and α of 1.5 (X = 2.274)

x_1	y_{20}	z	n_F	Y
0.082	0.920	0.5	11	0.600
0.062	0.930	0.3	9	0.650
0.181	0.975	0.5	6	0.638
0.145	0.950	0.6	11	0.582

The potential for reducing V/F with an increase in $\bar{\alpha}$, n, or E or a reduction in S can be evaluated for the linear portion of Fig. 2.3. Since in that region

$$\frac{\log S}{nE \log \bar{\alpha}} = 0.286 \frac{V\bar{\alpha}^{1.68}}{F(nE)^{0.32}}$$

V/F may be found directly:

$$\frac{V}{F} = \frac{\log S}{0.286(nE)^{0.68}\,\bar{\alpha}^{1.68}\log \bar{\alpha}} \tag{2.32}$$

The percentage change in V/F for different values of S, nE, or $\bar{\alpha}$ may then be estimated. Note that in the nonlinear portion of Fig. 2.3 the effect of these variables on V/F will be even more pronounced.

example 2.4

Estimate the potential reduction in V/F for the propane-propylene separation if the column pressure is reduced from 300 psia (122°F) to 175 psia (80°F). This increases α from 1.12 to 1.15.

α	$\frac{1}{\alpha^{1.68}\log\alpha}$
1.12	16.80
1.15	13.02

The percentage reduction in V/F is

$$100\,\frac{16.80 - 13.02}{16.80} = 22.5\%$$

Contours of Constant Separation In Fig. 1.9, contours of constant separation were plotted to illustrate how boilup could be reduced at lower pressures. Each contour represented a different feed rate such that V/F at a given pressure was the same for all contours. Using Fig. 2.3, the V/F ratio required to achieve a given separation will now be related to α and therefore to pressure.

To illustrate the construction of these contours, Table 2.4 gives values of V/F required to separate a 50-50 mixture of butanes into products each 97 percent pure in a column of 40 theoretical trays. Observe that nearly a 1 percent reduction in energy-feed ratio is attained for every 1°F reduction in condenser temperature. Since condenser temperature follows ambient temperature, about 1 percent energy savings are possible with every 1°F reduction in ambient if column pressure is allowed to float on the condenser.

TABLE 2.4 Boilup-Feed Ratios for Constant Separation of Butanes

T, °F	p, psia	α	Y	X	V/F
60	38.1	1.41	0.506	1.77	3.23
80	53.1	1.38	0.540	1.92	3.64
100	71.9	1.35	0.579	2.14	4.21
120	95.2	1.32	0.626	2.45	5.00

Boilup rate V is plotted in Fig. 2.4 in units relative to feed rate F, as a function of pressure, using the data in Table 2.4. The contours appear as straight lines within the accuracy of the relative-volatility data. Recall that the slope of these contours relative to that of the flooding curve in Fig. 1.9 determines the point of maximum production for the column.

For conditions of constant separation, overhead and bottom-product compositions are related to one another by Eq. (2.22). Either may be found in terms of the other by rearranging that equation:

$$x_1 = \frac{y_n}{S - y_n(S-1)} \qquad y_n = \frac{Sx_1}{1 + x_1(S-1)} \tag{2.33}$$

This is but a single relationship between two compositions—it cannot determine absolute compositions by itself. Another equation in x_1 and y_n is required; it is derived in the next section.

MATERIAL BALANCES

The most neglected aspect of distillation-column operation has been control over material balances. Although columns had been separating products to

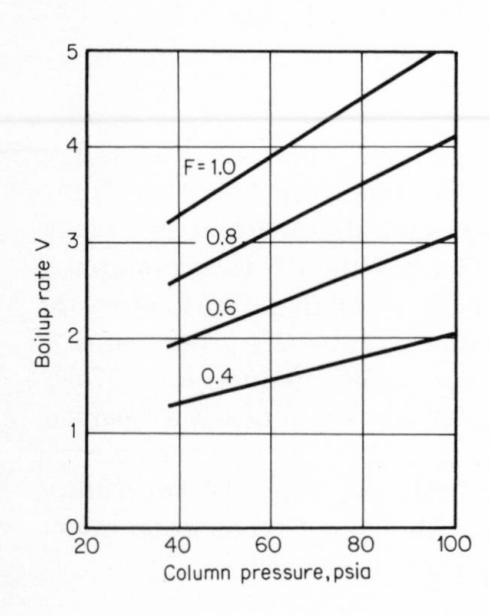

figure 2.4 *Contours of constant separation between butane and isobutane, plotted from the data in Table 2.4.*

some degree of satisfaction for decades, only after 1960 did any literature appear citing material-balance control. Since that time much success has been reported using this new technique applied specifically to towers that could not be controlled in any other way. Yet at this writing, most towers are still controlled traditionally, and some plant managers are still not aware that there is a better way.

Because this concept is essential not only to the control of columns but also to the understanding of how they function, it is introduced at this point in the book. Notwithstanding the favored positions given to reflux ratio and liquid-vapor ratio in the past, control over product quality cannot be achieved without manipulating the material balance, either directly or indirectly. Those systems which manipulate reflux ratio or boilup and reflux for product-quality control affect the material balance indirectly. But as will be seen, much tighter control can be achieved if direct manipulation is enforced.

Overall and Component Balances Figure 2.5 represents a column as a simple block without reboiler, condenser, accumulator, etc. Its purpose is to frame the column as it appears to other units in the plant. In this frame of reference we are concerned only with inputs and outputs.

To satisfy the overall material balance,

$$F = D + B \tag{2.34}$$

where feed (F), distillate (D), and bottom-product (B) flow rates are in mass or molar units. Ordinarily one of these variables—usually F—is uncontrolled or set independently to establish throughput. Another—either D or B—may be manipulated to control the quality of one or the other product. The last variable is dependent on the first two and must be manipulated to maintain liquid inventory in the system. Selection of the *best* stream to manipulate for control of product quality is covered under the heading "Material-balance Control" in Chap. 8.

Next a component balance may be written:

$$Fz = Dy_n + Bx_1 \tag{2.35}$$

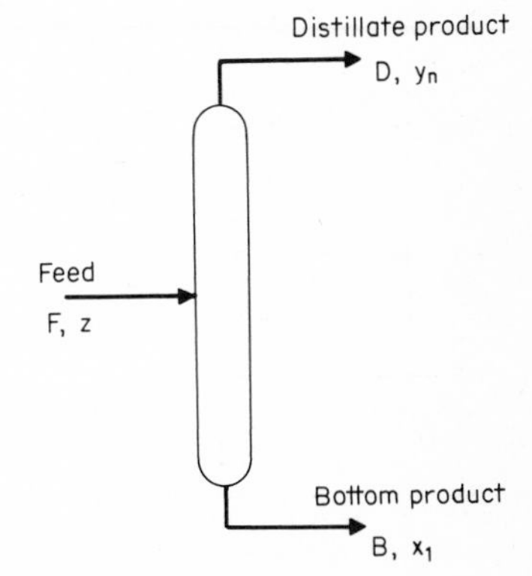

figure 2.5 *The split between distillate and bottom product is the primary factor which determines product quality.*

Here again, compositions are given in terms of the more volatile component, in units consistent with those chosen for F, D, and B.

These two equations may be combined by eliminating the flow rate selected to control liquid inventory. The solution then appears as the ratio of the two remaining flows:

$$\frac{D}{F} = \frac{z - x_1}{y_n - x_1} \tag{2.36}$$

$$\frac{B}{F} = \frac{y_n - z}{y_n - x_1} \tag{2.37}$$

$$\frac{D}{B} = \frac{z - x_1}{y_n - z} \tag{2.38}$$

These three equations are all dependent, of course; only one is applicable to a given column. But it provides the relationship needed in addition to Eq. (2.33) to define both product compositions.

Quality vs Quantity Whatever the separation, product compositions may be adjusted over a broad range simply by adjusting the product distribution (for example, D/F, B/F, or D/B). As an example, consider splitting a 50-50 feed mixture into products each 95 percent pure. Separation S for these products is $(95/5)^2$ or 361. Using Eq. (2.33), values of x_1 corresponding to selected values of y_n may be calculated at this constant separation. Then D/F may be evaluated for each set of product compositions. The results are plotted in Fig. 2.6. Using logarithmic scales for product compositions demonstrates how widely they are affected by the D/F ratio.

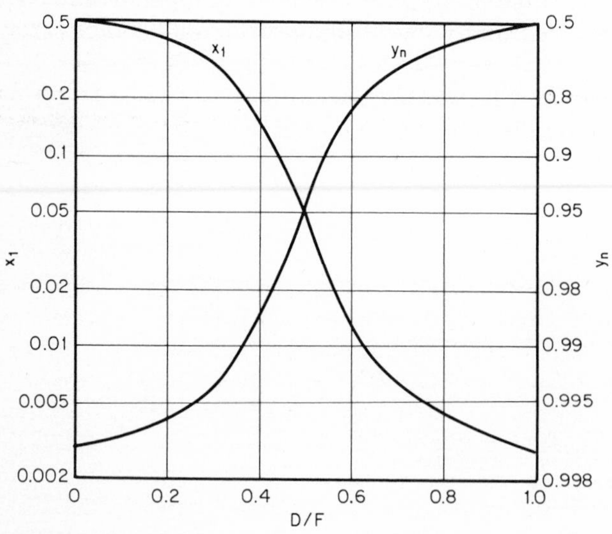

figure 2.6 *At constant separation, both compositions are varied over a wide range by manipulating the product split.*

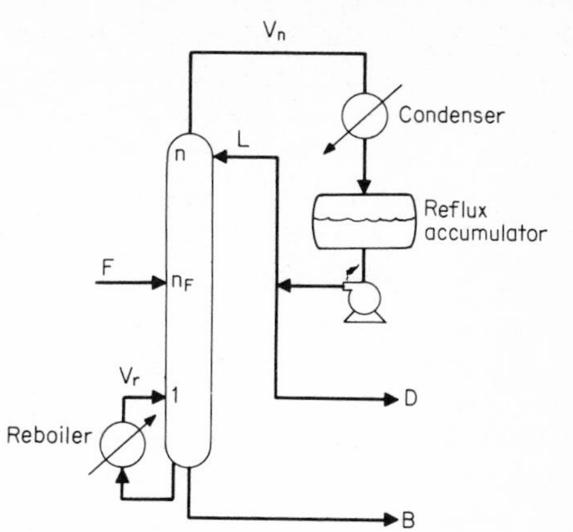

figure 2.7 *The difference between overhead vapor and distillate product is returned as reflux.*

Having defined the relationships between compositions and flow rates, it is now possible to estimate product recovery. Starting with Eq. (1.13), distillate-product recovery can be described by

$$R_D = \frac{D}{Fz} = \frac{z - x}{z(y - x)} \tag{2.39}$$

and bottom-product recovery by

$$R_B = \frac{B}{F(1 - z)} = \frac{y - z}{(1 - z)(y - x)} \tag{2.40}$$

The recovery curves shown in Fig. 1.3 were plotted for the same column as the composition curves of Fig. 2.6. (The column has 20 theoretical trays; the relative volatility of the components is 1.52.)

Internal Balances Unfortunately, a real column is not so simple as Fig. 2.5 depicts. If it were, quality control would be much easier. In fact, addition of a reboiler and condenser as shown in Fig. 2.7 adds to the variables requiring manipulation.

Reflux L is returned as that part of the condensed overhead vapors not withdrawn as distillate product:

$$L = V_n - D \tag{2.41}$$

Whether the distillate product is totally condensed as in Fig. 2.7 or removed as a vapor is inconsequential. In fact, the actual flow of distillate product at any point in time has no direct bearing on product quality since it has *already left the tower.* The influence which the D/F ratio has on product quality, then, is actually exerted through the difference between vapor and liquid flow rates in the top section of the tower.

In the steady state, Eq. (2.41) will be satisfied so that manipulation of D is the same as manipulation of $V_n - L$. But with a place to store product between the point where vapors are removed and reflux returned to the column, significant departures from the steady-state balance can exist.

Obviously, the equilibrium being approached at every tray is a function of the relative flow rates of liquid and vapor passing through it. In other words, compositions of streams leaving an individual tray are only affected by the flow rates and compositions of the streams approaching it. Yet in the overall analysis, the difference between these liquid and vapor rates may be forced to fit a desirable product split in order to control quality.

Hydraulic Responses Small variations in liquid and vapor rates have essentially the same magnitude of effect on product quality in the steady state. However, the speed with which compositions respond to their manipulation varies widely.

Each tray may be covered with perhaps an inch or two of liquid (excluding froth) with up to 2 ft of vapor space between trays. Since the ratio of liquid to vapor densities may be upward of 30:1 (see Table 1.1), each tray contains much more liquid than vapor. To increase the vapor flow leaving a tray, it is only necessary to increase the vapor flow approaching it. Because the pressure drop may be less than 2 in. of water per tray, the compression of the vapor brought about by increasing its flow (and hence pressure drop) is insignificant except in vacuum towers.

For a column separating close-boiling components, an increase in reboiler heat input will therefore cause an increase in vapor flow entering the condenser within a few seconds. The dynamic response of the reboiler (5-s dead time) is probably of the same order of magnitude as that of a 50-tray column. With a large difference between component boiling points, however, considerable sensible heat may be stored in the temperature gradient across the trays. This has the effect of both delaying and reducing the effect of boilup on overhead vapor flow.

Considerable delay is encountered in transferring liquid from tray to tray, however. To increase the liquid flow leaving a tray, the head on the tray must first be raised by increasing liquid rate entering. Thus higher flow rates cause more liquid to be stored on the trays. Then the change in storage required for each new liquid rate delays the transfer of the new rate to trays farther down the column.

Beaverstock and Harriott [10] measured the hydraulic lag of liquid flow down a 12-in.-diameter, 15-sieve-tray column at an average of 6.5 s per tray. This was over three times as great as results predicted using the Francis weir equation. The time required for a selected tray to respond to a liquid flow change introduced n trays above it was approximately $6.5n$ s. This verifies that the trays are essentially noninteracting capacities as described in Ref. 11.

From the foregoing discussion, it follows that the liquid-vapor ratio at any point in a column except the top tray may be changed much more rapidly by manipulating boilup than reflux.

Composition Time Constants Only *after* a new L/V ratio has been established will tray compositions *start* to change. In manipulating reflux rate to assess its effect on tray composition, Beaverstock and Harriott observed a single composition time constant of 9.65 min, superimposed on the 6.5 s per tray hydraulic time constants. Recognize that 9.65 min is so much larger than

6.5 s that more than a single tray must contribute to it. Actually a change in L/V affects *all* tray compositions. This induces a secondary wave front where composition variations travel from tray to tray long after L/V has stabilized. Whereas hydraulically the tray capacities exhibited little interaction, with respect to composition the interaction is extensive.

Increasing reboil rate, for example, will start to reduce the mole fraction of the more volatile component on all trays of the column. But as compositions of liquid and vapor leaving each tray start to fall, they tend to lower compositions on neighboring trays. Each composition thus affects each other composition such that substantial time may elapse before all trays in the column reach a new steady state. In the column tests cited above, the composition lag was essentially the same for all trays; only the hydraulic lag increased with number of trays away from the top. The composition lag then seems to be a function of the *total* holdup in the system.

The composition time constant will increase with the number of trays in the tower, since holdup is thereby increased. Van Kampen [12] reports a 100-min composition time constant in response to reflux manipulation for a 96-tray propane-propylene column. It also changes with reflux-distillate ratio: McNeill and Sacks [13] report a 20-h time constant for a ratio of 70:1, compared to van Kampen's 25:1 ratio. This last observation would make the composition time constant dependent on the total column holdup divided by throughput. A column with a large reflux ratio would then have a relatively large ratio of holdup to throughput.

The composition time constant does not seem to depend on the selection of the manipulated variable. McNeill and Sacks found that the response of distillate composition to a feed-rate change was virtually the same as to a distillate-rate change (reflux was returned from the accumulator under liquid-level control).

Lag in the Reflux Accumulator To illustrate the effect of liquid holdup in the reflux accumulator, reconsider the relationship between compositions at the top tray only. Figure 2.8 shows the top tray and reflux accumulator, assuming a fixed vapor rate V_n and composition y_{n-1} entering the top tray. When equilibrium exists on the tray,

$$\frac{y_n(t)[1 - x_n(t)]}{x_n(t)[1 - y_n(t)]} = \alpha \tag{2.42}$$

The functional (t) indicates that y_n and x_n will change with time.

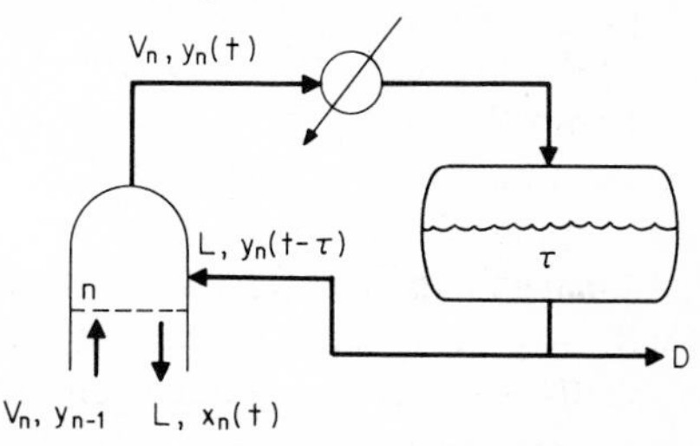

figure 2.8 *A reflux composition change is delayed in its effect on the column.*

A material balance of the light component on the top tray gives

$$V_n y_{n-1} + L y_n(t - \tau) = V_n y_n(t) + L x_n(t) \tag{2.43}$$

Functional $(t - \tau)$ for reflux composition indicates a delay or dead time τ existing between the composition of liquid entering and leaving the reflux drum. Although some mixing ordinarily takes place there, such that its dynamic response is not pure dead time, this example will more easily illustrate the effect of accumulator capacity. Equations (2.42) and (2.43) may be solved for the steady state by letting $y_n(t) = y_n(t - \tau)$ at time zero as in Example 2.3. Then a change in L/V may be imposed and $y_n(\tau)$ calculated for the first time interval τ, using $y_n(0)$ for reflux composition. After τ elapses, reflux composition will change to $y_n(\tau)$ and so on. Table 2.5 illustrates the time sequence of overhead vapor compositions following a change in L/V_n from 1.0 to 0.9. In this example $y_{n-1} = 0.9$ and $\alpha = 2.0$.

The percent response is very close to the familiar exponential step response of a first-order lag $(1 - e^{-t/\tau})$. Consequently the effect of the material stored in the reflux drum is that of an exponential time constant, even without mixing. The stepwise response indicative of dead time as plotted in Fig. 2.9 has been observed in certain poorly mixed systems, although with some rounding of the corners. With perfect mixing, the response curve following the initial step is smooth.

In the foregoing example, constant vapor composition y_{n-1} was assumed. But since the liquid composition x_n decreases following a decrease in L/V_n, y_{n-1} will also decrease, as will all the y's down the column. Eventually they will have their cumulative effect on y_n in a manner similar to the effect of the delay in the reflux drum. This explains the exponential series of steps observed in composition even following a step increase in boilup. Due to the capacity distributed through the trays, and partial mixing in the reflux accumulator, only the first step may be apparent, the remaining smaller steps being smoothed into a continuous curve.

Controllability The ability to control a variable such as composition depends on the ratio of its dominant time constant to the sum of all the other delays in the loop [14]. Control will be improved whenever the sum of these secondary

TABLE 2.5 Dynamic Response of Top Tray and Reflux Accumulator

L/V_n	t/τ	$1 - y_n(t)$, %	Response, %	$1 - e^{-t/\tau}$
1.0	0	5.263	0	0
0.9	1	5.444	65.6	0.632
	2	5.506	88.0	0.865
	3	5.527	95.7	0.950
	4	5.535	98.6	0.982
	5	5.537	99.3	0.993
	6	5.538	99.8	0.998
	∞	5.5385	100.0	1.000

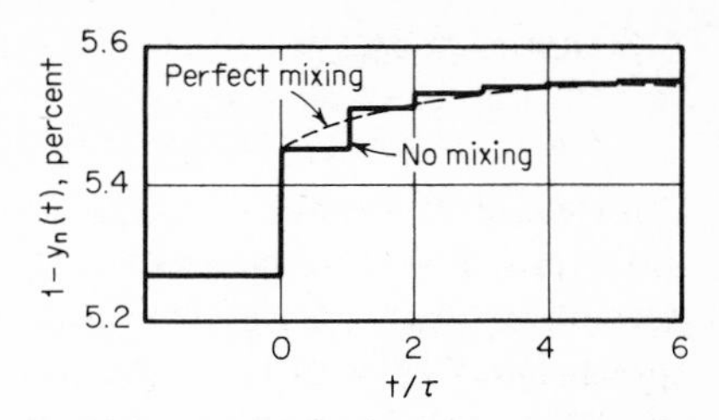

figure 2.9 *Overhead vapor composition traces an exponential response following a change in* L/V_n.

delays can be reduced. An operator manipulating reflux to control composition several trays below the top will encounter a substantial hydraulic delay—composition at the top tray is much easier to control. And composition at any tray except the top will ordinarily be easier to control with boilup than reflux.

Additional delay is introduced when the measuring device is moved to the distillate-product stream. Composition of the distillate will lag behind that of overhead vapor by the residence time τ of liquid in the accumulator (liquid volume divided by total liquid flow rate).

The dynamic response observed in tray or product compositions following a manipulation of one of the streams is also affected by the functioning of whatever control loops may be in operation. If, for example, boilup is adjusted with reflux under accumulator level control, a different response will appear than if reflux is held constant. An increasing boilup will cause reflux to increase, thereby reducing its ultimate effect on composition and promoting a more rapid reestablishment of equilibrium. Although the composition time constant of a column is fixed by holdup and throughput, careful control-loop selection can coordinate the various streams in such a way as to bring about superior regulation of composition. Much more is said on this subject in Part 4.

ENERGY BALANCES

To this point, vapor rate has been considered virtually an independently manipulable variable. In this context it has been useful in establishing separation and in forcing an internal mass balance within the column. The time has come now to frame it in terms of the various sources of heat entering the column.

Overall Energy Balance In the steady state, all the energy entering a column through heat transfer and with the feed must be balanced by an equal rate of removal through heat transfer and with products. This balance may be expressed mathematically as

$$(Q_{\text{in}} + FH_F) - (Q_{\text{out}} + DH_D + BH_B) = Q_s \qquad (2.44)$$

In this statement, Q_{in} and Q_{out} represent rates of heat transfer and the H's are the enthalpies of their associated variables. In the steady state, the imbalance term Q_s is zero; in the unsteady state, it represents the change of energy level of the contents of the column brought about by a difference between inflow and outflow. This imbalance principally converts liquid into vapor or vapor into liquid because of the equilibrium existing between the two phases. In the constant-volume system, it appears as a rising or falling pressure.

If more vapor is generated than condensed or removed, pressure will therefore rise at a rate proportional to Q_s and to the volume of the system. An effect of rising pressure is an increase in the boiling temperature on all the trays; therefore part of Q_s will reside in changing the sensible heat of the liquid on the trays. Because vapor pressure varies almost linearly with temperature over normal operating ranges, Q_s can be related to rate of pressure change with rate of temperature change implied:

$$Q_s = C \frac{dp}{dt} \tag{2.45}$$

Coefficient C then includes vapor volume and latent heat of vaporization along with the inventory of stored liquid, its heat capacity, and the variation of boiling point with pressure.

Any scheme which independently manipulates both heat input and removal rates will cause column pressure to vary uncontrollably. In short, pressure control requires manipulation of heat input or removal, either directly, i.e., by means of a controller, or indirectly through self-regulation. The latter method is employed in floating-pressure operation—an increase in heat-input rate causes pressure to rise until a sufficient temperature drop is created across the condenser to produce an equivalent heat-removal rate. The overall energy balance for the process is thereby closed through pressure control.

Liquid Inventory The overall column material balance requires a constant liquid inventory in the steady state. But this inventory is distributed among trays, reboiler, tower bottom, and accumulator. Liquid levels are ordinarily controlled both at the bottom and in the accumulator to hold inventory constant. But gross variations in boilup approaching flooding or weeping can significantly affect tray inventory. Columns with many trays may contain more liquid than the reflux drum. Ordinarily the tower bottom holds much less material than the reflux drum, and its liquid level is therefore more sensitive to upset and more difficult to control. It also reflects the turbulence present in the reboiler and therefore tends to be unsteady as well.

Referring to Fig. 2.7, it may be seen that accumulator level l_a varies with the flow of streams entering and leaving:

$$\rho_D A_a \frac{dl_a}{dt} = V_n - L - D \tag{2.46}$$

Here A_a is the cross-sectional area of the accumulator and ρ_D is distillate liquid density. The left-hand term then represents rate of change of accumulator mass contents. Accumulator level is most sensitive to the largest flow rate, that is, V_n, and less to the liquid streams in order of their magnitude.

Bottom level l_b is determined by

$$\rho_B A_b \frac{dl_b}{dt} = L_1 - V_r - B \tag{2.47}$$

where A_b = cross-sectional area of tower bottom
ρ_B = bottom-product density
L_1 = liquid rate leaving bottom tray
V_r = boilup

The variable having the greatest effect on bottom level, that is, L_1, is unavailable for manipulation so that either boilup or bottom flow must be used for level control. Boilup has the greater effect.

Whereas column pressure is capable of self-regulation, ordinarily the two levels are not. Gravity return of reflux has occasionally been used to regulate accumulator level, but only on small columns. Most accumulators are mounted only one level above grade such that reflux must be returned by pump, thereby removing any possibility of self-regulation. As a consequence, a column cannot be operated without level controls. This is why dynamic-response tests cannot be conducted purely "open loop." As will be shown, dynamic performance depends on which streams in Eqs. (2.46) and (2.47) are selected for level control.

Reflux Enthalpy Internal energy balances turn out to be just as important as internal material balances. To point out a problem area in this regard, few columns receive reflux at its bubble point. In most installations, most of the time cooling availability exceeds demand such that reflux enters the column below its bubble point. If the condenser is not flooded, this subcooling is usually slight; in fact further condensation may take place on the liquid surface within the accumulator so that some subcooling is lost. But if the condenser is partially flooded with condensate to reduce the surface area available for condensation, reflux temperature tends to approach coolant temperature.

When the subcooled reflux encounters vapor at the top tray, it is heated to its bubble point by condensing some of the vapor. Thus the vapor rate leaving the top tray is less than that entering, in relation to the sensible heat of the reflux:

$$V_n - V_{n-1} = L(T_V - T_L)\frac{C_p}{H_V} \tag{2.48}$$

where T_V = bubble point
T_L = temperature of reflux
C_p = heat capacity of reflux
H_V = latent heat of vaporization

Some C_p/H_V ratios for selected hydrocarbons are given in Table 2.6.

TABLE 2.6 C_p/H_V Ratios for Selected Hydrocarbons at 100°F

Hydrocarbon	H_V, Btu/lb	C_p, Btu/lb-°F	C_p/H_V, °F^{-1}
Propylene	132	0.714	0.00541
Propane	136	0.667	0.00490
Isobutane	135	0.633	0.00469
n-Butane	149	0.602	0.00404
Cyclohexane	195	0.500	0.00256
Benzene	180	0.436	0.00242

From Eq. (2.48) it follows that liquid flow from the top tray exceeds reflux by the stated difference between V_n and V_{n-1}. Its value may be calculated:

$$L_n = L\left[1 + (T_V - T_L)\frac{C_p}{H_V}\right] \tag{2.49}$$

Note that subcooling cannot develop when column pressure floats on the condenser.

Feed Enthalpy A similar situation exists at the feed tray. Feed enthalpy is customarily identified by the factor q, which represents its fractional vaporization. When $q = 1$, all the feed is vaporized; when $q = 0$, it is all liquid at the bubble point. Negative values indicate subcooling in the amount of vapor which the feed is capable of condensing in rising to its bubble point.

Accordingly, vapor flow below the feed tray V_{F-1} will be augmented by the vapor in the feed Fq in generating the vapor flow leaving the tray, V_F:

$$V_F = V_{F-1} + Fq \tag{2.50}$$

In accordance with the definition of q given above, Eq. (2.50) is valid for all values of q.

An energy balance on the feed tray shows that the liquid flow leaving that tray is also affected by Fq:

$$L_F = L_{F+1} + F(1 - q) \tag{2.51}$$

From these relationships, it becomes apparent that vapor and liquid flow rates at various points in a column are subject, to a limited extent at least, to external influences. While these disturbances are principally thermal, they can exert an influence over the internal material balance of a column.

INTERACTION BETWEEN MASS AND ENERGY

Compositions of vapor and liquid streams leaving the various trays in a column have been shown to be associated with their respective flow rates. Although vapor flow is fundamentally determined by the rate of energy input to the column, liquid flow is not. Consequently, compositions are affected by both *energy* and *mass* inputs. This is evident in the two sets of equations relating x_1 and y_n: the separation factor is energy-related whereas the product split is strictly material-related. Although it is desirable to keep these two relationships separate and distinct, by nature they interact within the process. It is this interaction which is to be explored.

Interaction and Control Although a separate chapter is later devoted to interaction between control loops, a qualitative introduction to the subject is in order at this point. Equations (2.44) to (2.47) suggest the existence of a problem area. Vapor inventory (pressure) and liquid inventories are all subject to several influences. Some of these forces are part of the energy balance and others belong to the material domain. A case in point is reflux-accumulator level (2.46). It responds to boilup, reflux, and distillate flow rates, allowing it to be controlled by any of the three.

The choice of which variable to manipulate for control is paramount. If, for example, distillate flow is selected for level control, it must be manipulated to make up for imposed differences between vapor and reflux rates. If, because of its relatively small flow range, it is occasionally incapable of filling the gap between the two larger streams, level control will be lost during these intervals.

Multiplication by Reflux Ratio Much of the responsibility for poor control in existing distillation processes can be associated with amplification by the reflux ratio (L/D). In the example just described regarding control of accumulator level, the reflux ratio played a dominant role. With a reflux ratio of 1, either distillate or reflux can control level with equal ease. However, variations in vapor rate would require twice as much percentage change in L or D to maintain level. As the ratio increases, the selection between L and D becomes more important. At $L/D = 10$, D must change drastically to control level. A 5 percent variation in L would command a 50 percent variation in D, and a 5 percent variation in V would cause it to change by 55 percent.

Compositions within the column have been shown to vary with the relative values of L and V. As L and V become large relative to their difference (i.e., as L/V approaches 1), control over composition becomes more difficult because slight variations in L and V exert proportionally greater influence over their difference.

Columns are designed to achieve some optimum balance between the number of trays and the reflux ratio. More difficult separations, i.e., where α is lower or where products must be purer, require more trays. The optimum balance then tends to shift toward higher reflux ratios. The result is that L/D tends to be increased with n. Thus difficult separations typically use more trays *and* higher reflux ratios. As a result, the more difficult separations cannot be controlled by independently manipulating L and V.

References 12 and 13 describe production columns where composition control could not be achieved by manipulating reflux because the reflux ratio was too high (>25). Relatively minor variations in L and V upset composition extensively; at the same time, accumulator level could not be controlled by the small distillate flow rate. When distillate flow was then manipulated to control composition with reflux controlling accumulator level, product quality was stabilized. In this configuration, variations in L and V upset accumulator level but are quickly corrected by its controller. Thus the controlled distillate rate is forced upon the column internal material balance as the V-L difference. Compositions cannot start to change until the internal material balance is adjusted. Fortunately, the response of the level-control loop is sufficiently rapid to restore the V-L balance following upsets before significant composition variations can develop. In this way the external material balance is forced on the column internals.

A secondary effect of interaction among the streams at the top of the column is the propagation of upsets to downstream processes. As mentioned above, the distillate flow may be unable to control accumulator level when the reflux ratio is high. Minor variations in L and V promote major variations in D. If D is the feed to a downstream column, that column must cope not only with its own energy-mass interactions but with gross variations in feed rate as well. However, if D is manipulated to control composition, it will not tend to vary any faster or wider than the composition it controls.

Enforcement of the material balance of a column is relatively easy to impose. Flow control over feed and distillate, with level control over bottom product, determines the product split. Each flow measurement is accurate to perhaps ±2 percent of full scale and is repeatable to better than ±0.5 percent. Although this may not be accurate enough to control compositions within fractions of a percent, it is much more accurate than regulation over the energy rates.

Reflux flow can be controlled to the same accuracy in terms of percentage of full scale as distillate. But if reflux flow is controlled and distillate is manipulated by a level controller, the percentage error in reflux will be imposed on distillate with an amplification of L/D.

Disturbances in the Energy Balance Equation (2.48) showed how reflux subcooling affects the vapor flow leaving the top tray. The loss in vapor flow to the condenser resulting from a reduction in reflux temperature is imposed on the distillate if it is under level control. Differentiating (2.48) with respect to reflux temperature yields dV_n, which produces an equivalent dD:

$$dD = -L \frac{C_p}{H_V} dT_L \tag{2.52}$$

The fractional change in distillate flow can be obtained by dividing both sides of (2.52) by D:

$$\frac{dD}{D} = -\frac{L}{D}\frac{C_p}{H_V} dT_L \tag{2.53}$$

Thus the material balance is sensitive to reflux temperature in proportion to the reflux ratio and C_p/H_V. For the propane-propylene splitter cited in Ref. 12, L/D is 25 with C_p/H_V (from Table 2.6) at 0.00541°F^{-1}. With distillate under level control, then, each 1°F change in reflux temperature could alter distillate flow by 13.5 percent.

A particularly severe interaction between energy and mass balances is encountered when column pressure is controlled by manipulating distillate flow. The rate of condensation—which regulates column pressure—is changed by varying the level of condensate in the condenser. An increase in cooling caused by the onset of rain, for example, will cause pressure to fall. The pressure controller then reduces distillate flow to raise the level in the condenser. This increases the inventory of light components in the column, raising reflux purity and therefore distillate purity. It also increases reflux subcooling, which further reduces vapor flow into the condenser, tending to augment the disturbance. Columns with high reflux ratios cannot be controlled in this way for three reasons:

1. Distillate flow is too small relative to the heat load to control pressure effectively.
2. Variations in reflux temperature impose too great an effect on the product composition.
3. Distillate-flow variations may be too severe for downstream units to accept.

Subcooling is always encountered with flooded condensers. Note that it varies not only with coolant temperature but with vapor loading as well. By contrast, subcooling is impossible under floating-pressure operation.

Boilup may also be subject to some variations. If constant-pressure steam is used for heating, control over its flow is essentially the same as control over heat-input rate. This assumes a constant latent heat with no subcooling of steam condensate. In situations where the reboiler is partially flooded with condensate, more heat is released per pound of steam with increased flooding (i.e., at lower heat-transfer rates). Even then, the steam-flow measurement gives a reproducible, if somewhat inaccurate, indication of heat flow.

Reboilers heated by combustion of a fuel are subject to more upsets. Fuel quality (either density or heat of combustion) may change if the fuel contains waste products from the plant. Hydrogen, carbon monoxide, and ethane and higher hydrocarbons all have different heating values and combustion-air requirements than natural gas or methane. Natural-draft heaters are also subject to variations in wind velocity and direction.

But the heating sources which exert the most pronounced upsets on boilup are liquids such as hot oil, reactor effluent streams, etc. The primary influence over heat-transfer rate is the temperature difference between the boiling product and the heat-transfer fluid. Relatively minor temperature changes can cause substantial variations in boilup. Therefore flow control over the heat source is usually insufficient to provide constant boilup. Naturally, variations in boilup are amplified by the V/D ratio in their effect on distillate flow if it is under level control.

But if distillate flow is controlled and reflux closes the level loop, variations in boilup will be returned to the column as equal variations in reflux. Or if the flow of the heating medium itself is manipulated to control level, then variations in boilup will be corrected directly by the level controller without affecting liquid rates at all.

A third source of variable heat input is the enthalpy of the feed. Equation (2.50) indicated that vapor flow across the feed tray varies directly with Fq.

Again, if distillate is under accumulator level control, variations in V are directly imposed on D such that

$$dD = dV = F\,dq \tag{2.54}$$

The fractional variation of D with respect to q can be obtained by dividing both sides by D. Then

$$\frac{dD}{D} = \frac{F}{D}\,dq = \frac{y - x}{z - x}\,dq \tag{2.55}$$

The amplification factor F/D is not ordinarily of a magnitude comparable to L/D, but feed enthalpy could be more variable than that of reflux.

Actually Eq. (2.54) only relates the effect of feed enthalpy on the material balance. Feed rate has a similar effect:

$$dD = q\,dF \tag{2.56}$$

Only if q happened to correspond to D/F would the material balance be unaffected by changes in feed rate. If the feed were completely vaporized, an increase in its flow would be entirely imposed on the distillate; if it were all liquid, it would all be imposed on the bottom product.

In all the foregoing discussion, the effects of energy-balance variations were reflected on distillate flow. A complementary set of equations can also be derived for bottom-product rate if so desired. The purpose of the development is to illustrate the pronounced influence that energy inputs can have over the product split, if allowed. Obviously both product flows will reflect these disturbances.

Many distillation columns use some of the heat of the bottom product to preheat the feed as shown in Fig. 2.10. While desirable from a heat-recovery standpoint, it forms a feedback loop through the column. An increase in bottom-product flow will transfer more heat to the feed, thereby tending to drive more vapor overhead. If no control is applied, bottom-product flow will eventually fall. The sense of the feedback is negative, and its response is delayed by the hydraulic lag between the feed tray and the bottom of the tower. Lupfer [15] indicates that it can cause an undamped oscillation in bottom-product flow of several minutes period if the bottom product is under level control. He found that temperature control of the feed by using a bypass valve around the exchanger was sufficiently responsive to dampen the cycling.

Partial Condensers If a vapor-phase overhead product is useful or desirable, partial condensation may be applied. As shown in Fig. 2.11, only the reflux is condensed. This practice is common in binary separation only where cooling is costly, as in cryogenic processing. It is often encountered in multicomponent distillation, however.

Since the distillate product is a vapor, it tends to be richer in the more volatile component than the reflux. Consequently, the condenser and accumulator act as an additional equilibrium stage, where

$$y_D = \frac{\alpha x_D}{1 + (\alpha - 1)x_D} \tag{2.57}$$

Product vapor y_D and liquid x_D compositions now differ from that of top-tray vapor y_n. Again, the equilibrium relationship is but a single equation with two variables. A material-balance equation is also needed to provide a solution. It is analogous to the column external balance:

$$\frac{D}{V_n} = \frac{y_n - x_D}{y_D - x_D} \tag{2.58}$$

With a partial condenser, then, even if y_n is constant, distillate-product quality will vary with flow. Naturally, the effect goes deeper: variations in D also affect reflux flow and composition and are therefore propagated into the tower.

Distillate vapor flow is often used to control column pressure. In a true binary system, this choice is unfortunate. Variations in coolant temperature affect the temperature difference across the condenser surface and therefore the rate of condensation. A pressure controller detecting variations in the rate of condensing will adjust distillate flow, changing the inventory of the lighter component in the system. Eventually a new equilibrium will be reached where both the rate of condensation and the condensate temperature have been altered to satisfy the energy balance. But the new condensate temperature at a controlled pressure requires a different product composition. Therefore product composition tends to change with coolant temperature and heat load, both of which affect condensing temperature.

With high-purity products, this effect is likely to be severe since a small change in temperature commands a relatively large change in purity. For best results, distillate flow should be manipulated for composition control while column pressure is allowed to float on the condenser.

Double Columns Double columns, used primarily to effect difficult separations or to obtain part-per-million purities, are resorted to when the required number of trays cannot be supported in a single structure. At this writing, 100 trays is about the limit found in a single column. Therefore separations

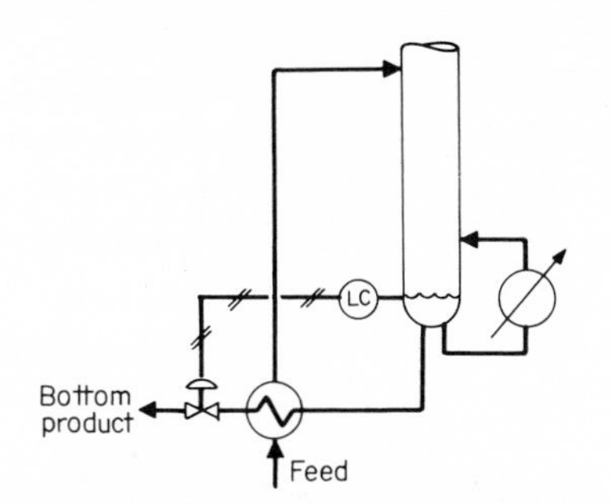

figure 2.10 *Using bottom product to preheat the feed forms a feedback loop through the column.*

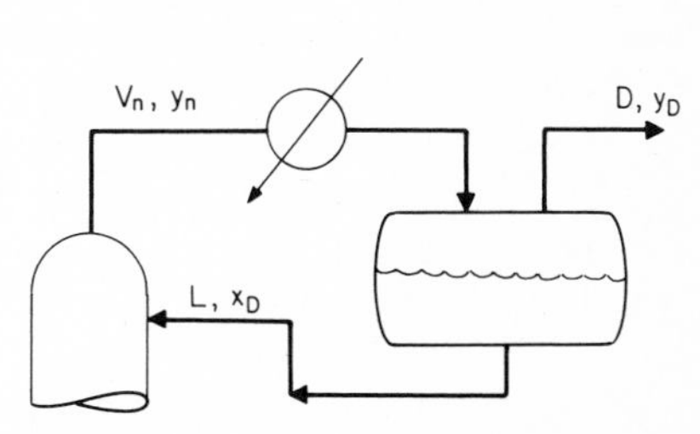

figure 2.11 *The partial condenser adds another equilibrium stage to the column.*

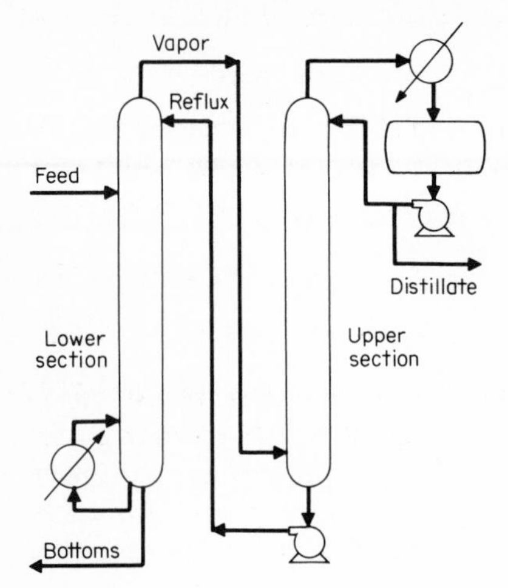

figure 2.12 *The double column is no more than a single column built in sections.*

like propane-propylene and ethylbenzene-xylene, which can require 200 trays, are carried out in double columns.

The double column, as described in Fig. 2.12, is simply a sectioned single tower. Vapor from the top of the lower section is piped to the bottom of the upper section, while liquid is pumped from the upper section to reflux the lower. Feed may be introduced into either section as dictated by feed and tray compositions. The liquid level in the bottom of the upper section must be controlled by manipulating the reflux to the lower section. There is no alternative—vapor rate has no direct effect on this level, and product is rarely withdrawn at this point.

There is a possibility of heat loss between the two sections through uninsulated liquid and vapor piping. Some vapor will condense and some subcooling of the reflux will occur, reducing upper-section vapor rate. When applying controls, this factor must be considered because the upper section will be more sensitive to weather changes than the lower. Vapor rate in either section may be controlled by selecting the appropriate measurement to manipulate reboiler heat input.

Theoretically, there is no limit to the number of sections that may be connected to form a column—as many as four have been used. The outstanding characteristic of these multiple-column units is their extremely slow response to control action. However, this has little to do with the sectioning of the columns: it is rather a property of the number of trays.

REFERENCES

1. Van Winkle, M.: "Distillation," McGraw-Hill Book Company, New York, 1967.
2. "Technical Data Book—Petroleum Refining," 2d ed., American Petroleum Institute, Washington, D.C., 1970.
3. "Matheson Gas Data Book," 4th ed., The Matheson Company, East Rutherford, N.J., 1966.

4. Doig, I. D.: Variation of the Operating Pressure to Manipulate Distillation Processes, *Aust. Chem. Eng.*, July 1971.
5. Smuck, W. W.: Operating Characteristics of a Propylene Fractionating Unit, *Chem. Eng. Prog.*, June 1963.
6. Van Winkle, M.: *op. cit.*, p. 261.
7. Treybal, R. E.: "Mass-Transfer Operations," p. 299, McGraw-Hill Book Company, New York, 1955.
8. Fenske, M. R.: Fractionation of Straight-run Pennsylvania Gasoline, *Ind. Eng. Chem.*, May 1932.
9. Luyben, W. L.: Feed-plate Manipulation in Distillation Column Feedforward Control, *Ind. Eng. Chem. Fundam.*, August 1968.
10. Beaverstock, M. C., and P. Harriott: Experimental Closed-loop Control of a Distillation Column, *Ind. Eng. Chem. Process Des. Dev.*, vol. 12, no. 4, 1973.
11. Shinskey, F. G.: "Process-Control System," pp. 38–44, McGraw-Hill Book Company, New York, 1967.
12. Van Kampen, J. A.: Automatic Control by Chromatographs of the Product Quality of a Distillation Column, presented at the Convention on Advances in Automatic Control, Nottingham, England, April 1965.
13. McNeill, G. A., and J. D. Sacks: High Performance Column Control, *Chem. Eng. Prog.*, March 1969.
14. Shinskey, F. G.: *op. cit.*, p. 35.
15. Lupfer, D. E.: Distillation Column Control for Utility Economy, presented at 53rd Annual GPA Convention, Denver, Mar. 25–27, 1974.

CHAPTER THREE

Multicomponent Separations

In the world of industrial processing, ideal conditions are rarely encountered and theory must be tempered with experience. To restrict the design of a column and its accessories (including controls) to a binary separation, even when the feed is guaranteed to be binary, would be foolish. Eventually the feedstock may change because of market conditions, availability, etc., so that additional components must be accommodated. But even if this were not so, trace quantities of contaminants must not be allowed to accumulate. Stories are told of small amounts of components boiling between the overhead and bottom products, accumulating in a column over a period of time until separation is impaired or dangerous concentrations are reached. Trays have been corroded by accumulation of trace quantities of acids, and columns have exploded when acetylenes reached detonable levels.

In short, every process—but particularly distillation processes—must be designed to remove potential accumulations of components boiling lower, higher, and between the principal products. In essence, no feed stream is purely binary. But even if one existed, start-up conditions must be faced where the column and its auxiliaries contain other components such as air, water, nitrogen, lubricating oils, solvents, etc., if only in trace amounts.

Because noncondensable gases seem to be the most common contaminants, their presence will be evaluated first. The subject is then expanded into multicomponent products with sequential separation in a train of columns.

Next, single columns yielding multiple products are considered, and the chapter concludes with a treatment of distilling crude oil composed of hundreds of components.

THIRD COMPONENTS OVERHEAD

The binary mixture of normal and isobutane was used in many examples throughout the first two chapters. Relative volatilities were given as a function of condenser temperature and were later plotted against vapor pressure of the pure isobutane. Quantities of n-butane in the overhead product tend to reduce the column pressure corresponding to a given temperature, but with little effect on relative volatility. A lighter component—propane—is commonly present in that system. Its presence raises column pressure and limits condenser capacity. In some cases these volatiles can be totally condensed; in others they are not. This section delineates the limits of total condensation and the benefits and penalties associated with partial condensation.

Solubility of Gases in Liquids Gases, even those generally considered noncondensable, can be condensed with less volatile vapors, depending on their characteristics and concentrations. Air is soluble in water, for example, to a small but visible extent. The solubility x_i of gases in liquids increases with their partial pressure p_i in the vapor according to Henry's law [1]:

$$x_i = \frac{p_i}{H_{ij}} \tag{3.1}$$

where H_{ij} is the Henry's law constant for component i in solvent j. Henry's law is considered applicable where p_i does not exceed 1 atm for gases above their critical temperature. According to Treybal [2], it is also applicable to vapors below their critical temperature, up to a partial pressure of about half their vapor pressure at the prevailing temperature.

Solubility data for light hydrocarbons in heavier hydrocarbons are customarily reported as K values:

$$x_i = \frac{y_i}{K_i} \tag{3.2}$$

Since, according to Eq. (2.2),

$$p_i = y_i p \tag{2.2}$$

where p is total pressure, it follows that

$$K_i = \frac{H_{ij}}{p} \tag{3.3}$$

Reference 1 lists values of H_{ij} as a function of temperature but not of total pressure. Therefore K values change with both temperature and pressure.

To complete the picture, the contribution of the liquid solvent to the system pressure must be included. Fortunately, Raoult's law may be applied to the solvent (j) since its concentration is high:

$$p_j = p_j^\circ x_j \tag{3.4}$$

where p_j and p_j° are the partial and vapor pressures of the solvent. Total pressure p is then the sum of p_i and p_j:

$$p = p_i + p_j = H_{ij}x_i + p_j^\circ x_j \tag{3.5}$$

In a binary system, $x_j = 1 - x_i$, reducing (3.5) to

$$p = H_{ij}x_i + p_j^\circ(1 - x_i) \tag{3.6}$$

Henry's law constants for some common systems are given in Refs. 1 and 3.

Where Henry's law constants are available, the calculation of column pressure is straightforward. But if K values are used instead, a trial-and-error procedure must be used since they change with pressure.

Combining Eqs. (3.6), (3.2), and (2.2) can yield Henry's law constant in terms of K_i and p_j°:

$$H_{ij} = \frac{K_i p_j^\circ(1 - x_i)}{1 - K_i x_i} \tag{3.7}$$

If both H_{ij} and K_i are known, p may be calculated by combining Eqs. (3.1), (3.2), and (2.2):

$$p = \frac{H_{ij}}{K_i} \tag{3.8}$$

Or column pressure may be calculated from K_i and p_j° by combining the last two equations:

$$p = \frac{p_j^\circ(1 - x_i)}{1 - K_i x_i} \tag{3.9}$$

Atmospheric Operation and Pressurizing with Gas The fractionation of liquids whose atmospheric boiling points are below condenser temperature can present certain problems. Since the vapors tend to draw a vacuum when condensed, some processors apply a "pad" of inert gas or air to keep pressure at or above atmospheric. Other columns are simply vented to the atmosphere through a condenser. These practices suffer the following disadvantages:

1. The distillate is contaminated with gas.
2. Its boiling point is depressed (affecting column temperatures).
3. Vented gas carries away product with it.
4. Relative volatility is lower than under vacuum.

To assess the effect of atmospheric or padded operation, Eq. (3.9) must be solved for x_i at the given total pressure:

$$x_i = \frac{p - p_j^\circ}{H_{ij} - p_j^\circ} \tag{3.10}$$

Then vapor composition may be found by using Eqs. (3.1) and (2.2):

$$y_i = \frac{p_i}{p} = \frac{H_{ij}x_i}{p} \tag{3.11}$$

In a binary system, y_j may be calculated as $1 - y_i$. It can perhaps be more accurately estimated by combining (3.10) and (3.11):

$$y_j = \frac{1/p - 1/H_{ij}}{1/p_j^\circ - 1/H_{ij}} \tag{3.12}$$

example 3.1

A benzene-toluene column is held at atmospheric pressure by padding with nitrogen. Estimate the concentration of nitrogen dissolved in the benzene product at 80°F and the benzene content of the gas phase. Assume that toluene concentration is essentially zero. From Ref. 3, H for nitrogen in benzene is 34,622 psia/mol frac. The vapor pressure of benzene at 80°F is 1.97 psia.

The mole fraction of nitrogen in the liquid is

$$x_i = \frac{14.7 - 1.97}{34{,}622 - 1.97} = 3.68 \times 10^{-4}$$

The mole fraction of benzene in the vapor is

$$y_j = \frac{1/14.7 - 1/34{,}622}{1/1.97 - 1/34{,}622} = 0.134$$

Although the amount of gas dissolved in the liquid tends to be small, it has a pronounced effect on boiling point, causing errors in relating temperature to composition at the top of the column. In Example 3.1, reflux returning at 80°F is far below the atmospheric boiling point of benzene, 178°F. Consequently, there will be a substantial temperature gradient across the first two or three trays.

Changes in heat load, ambient temperature and pressure, and coolant temperature all affect condensing temperature. And as condensing temperature changes, distillate vapor pressure will change, tending to raise or lower total pressure. If the column is vented to the atmosphere, increasing temperature will force the release of some vapor, which as the example indicates can contain a substantial amount of product. A reduction in temperature will draw in air or inert gas, which eventually must be released; then the temperature again rises. Every breath of this system carries away some product in proportion to the vapor volume and magnitude of the temperature variations. In essence, the column inhales nitrogen and exhales 13.4 percent benzene.

Within a tray or two of the top of the column, normal boiling points will be reached. The relative volatility from there downward is then a function of the equilibrium temperature at the applied pressure. For the benzene-toluene system, α at the atmospheric boiling point of benzene is about 2.6. But if the tower is operated under the vacuum achievable with an 80°F condenser temperature (1.97 psia), α increases to about 3.2. The reduction in V/F allowed by this increase in α is estimated to be 42 percent using Eq. (2.32). Vacuum operation does carry some penalties, however—the tower diameter must be greater to carry the increased volume of vapor. But with rising energy costs, the increased cost of the column may be justifiable.

If vacuum operation is unacceptable, pressure may be maintained at or above atmospheric by using a higher-temperature coolant such as steam condensate or recirculated hot water. The heat thus rejected by the column can be at a level high enough to be extracted, in contrast to the lower level of the energy in cooling water. Pressurizing with gas is seen, in this context, as reducing the energy level of the coolant, essentially by increasing the entropy (impurity) of the overhead mixture.

Total Condensation In a totally condensing system containing a single component lighter than the principal overhead component (light key), essentially all of that lighter component entering the feed will leave with the distillate. To state the relationship mathematically,

$$Fz_i = Dx_{Di} \tag{3.13}$$

where z_i and x_{Di} are the fractions of the most volatile component in the feed and distillate. No control can be applied in that column over the concentration of the most volatile component in the distillate. The ratio D/F is adjusted to control the concentration of a heavier component (heavy key) in the distillate. As a consequence, x_{Di} is an uncontrolled variable dependent on the composition of the feed.

Control of third components was discussed in Chap. 1. Equation (1.8) was derived to show how feed composition could be adjusted upstream to hold a desirable level of the third (most volatile) component in the distillate. At this point, however, we consider feed composition to be uncontrolled and assess its effect on the distillate product. Rearranging (1.8) and substituting symbols consistent with the foregoing development yields a steady-state relationship between concentrations of the lightest (i) and light-key (j) components:

$$x_{Di} = z_i \frac{x_{Dj} - x_{Bj}}{z_j - x_{Bj}} \tag{3.14}$$

Here subscript B is used to indicate bottom-product composition. The term to the right of z_i will be recognized as the F/D ratio.

In essence, the value of x_{Di} is forced on a column by the feed. It is then necessary to evaluate its effect on column pressure. For a multicomponent hydrocarbon system such as propane-isobutane-butane, a trial-and-error solution seems best. Approximate K values are first found from the condenser temperature and an assumed pressure. They are then multiplied by the liquid concentrations to generate equilibrium vapor concentrations. The latter should total 1.0 if the assumed pressure is correct. If the total is less than 1.0, select a lower pressure and try again; if it is above 1.0, select a higher pressure.

example 3.2

Given a liquid mixture containing 5 mole percent propane, 90 percent isobutane, and 5 percent n-butane, estimate its vapor pressure and equilibrium vapor composition at 100°F. As a starting point, assume that $p = 72$ psia, the vapor pressure of isobutane.

		$p = 72$ psia		$p = 76$ psia	
	x	K	y	K	y
Propane	0.05	2.25	0.113	2.13	0.107
Isobutane	0.90	1.00	0.900	0.95	0.855
n-Butane	0.05	0.74	0.037	0.70	0.035
			1.050		0.997

This example is applicable only to floating-pressure operation. If pressure is controlled, top-tray temperature will be affected by the third component. In this case, various values of *temperature* may be selected at the controlled *pressure* until the set of K values is found which makes all the y's add to 1.0.

In Example 3.2, the relative volatility of the butanes does not appear to change between 72 and 76 psia. This is because both sets of K values are taken at 100°F. However, the presence of propane elevating the pressure will raise tray temperatures throughout the column. The reboiler temperature, for example, will approach the boiling point of n-butane at 76 rather than 72 psia. This increase in temperature actually lowers α in a manner similar to that observed following Example 3.1.

As mentioned in Chap. 1, the presence of the lighter than light-key component shifts the condenser constraint toward higher pressure. As a consequence, it limits the reduction in V/F achievable with floating-pressure operation. This factor will be worthwhile considering later when we attempt to determine the optimum product compositions for a series of columns.

Because no vapor product is withdrawn, the material balance requires that the composition of the vapor leaving the top tray be the same as that of the distillate. This is true of any system employing total condensation.

Partial Condensation The partial condenser shown in Fig. 2.11 added another equilibrium stage to the column. As mentioned earlier, the composition of the overhead product is determined by the specifications for that product (as to light and heavy keys) and the relative concentration of lighter components in the feed. Example 3.2 indicates how the composition of the vapor is over twice as rich in propane ($K = 2.13$) as the liquid. It stands to reason, then, that if the overhead product is withdrawn as a vapor, the liquid reflux will contain roughly half the amount of propane. As a consequence, column pressure will be reduced for a given condenser temperature and α will be improved. Condenser duty will also be lower. This type of operation is frequently used to minimize the pressure on deethanizer and depropanizer columns that are cooled with water. The overhead product may then be condensed with refrigerant or compressed and condensed against cooling water.

Because reflux composition differs from that of the vapor withdrawn, it also differs from that of the vapor leaving the top tray. Equation (2.58) relating top-tray vapor composition and flow for a binary system may be restated for each component of a multicomponent system:

$$\frac{D}{V_n} = \frac{y_{ni} - x_{Di}}{y_{Di} - x_{Di}} \tag{3.15}$$

A trial-and-error solution must be applied to find reflux composition x_{Di} in equilibrium with the known vapor product. This is the reverse of that done in Example 3.2—now a set of K values must be found so that all the x_D's add to 1.0. Once the set of x_D has been found, the set of y_n may be determined from (3.15).

example 3.3

Assume that the mixture of Example 3.2 is being withdrawn as a vapor rather than a liquid product at 100°F. Estimate the composition and vapor pressure of the reflux. Then calculate top-tray vapor composition for D/V_n of 0.2.

		p = 72 psia		p = 73 psia	
	y_D	K	x_D	K	x_D
Propane	0.05	2.25	0.022	2.22	0.023
Isobutane	0.90	1.00	0.900	0.99	0.906
n-Butane	0.05	0.74	0.068	0.73	0.068
			0.990		0.997

$$y_{ni} = x_{Di} + \frac{D}{V_n}(y_{Di} - x_{Di})$$

Propane: $y_n = 0.023 + 0.2(0.05 - 0.023) = 0.028$
Isobutane: $y_n = 0.906 + 0.2(0.90 - 0.906) = 0.905$
n-Butane: $y_n = 0.068 + 0.2(0.05 - 0.068) = 0.064$
0.997

The benefit of the extra equilibrium stage provided by the partial condenser is apparent in the results above. With a total condenser, the ratio of butane concentrations in the vapor leaving the top tray is 0.90/0.05 or 18. For the case of the partial condenser, the ratio is 0.905/0.064 or 14.14. This ratio comprises half the separation factor for the column—the other half is the ratio of the same components in the bottom product. Assuming that the bottom ratio does not change between examples, the separation factor for the column itself is reduced by 18/14.14 or 21 percent by using a partial condenser. This allows a commensurate reduction in V/F in addition to that gained through the slight improvement achieved in α by lowering the pressure from 76 to 73 psia.

The reduction in V/F allowed by adding an equilibrium stage in this manner may be estimated by using the procedure in Eq. (2.32). The 0.68 power indicates that the percentage reduction is less than the percentage increase in the number of stages—not nearly so dramatic as that brought about by a change in α.

If column pressure is controlled by releasing the vapor product, the equilibrium in the accumulator will vary with condensate temperature. Therefore vapor-product composition will vary with heat load and coolant temperature as well as with feed composition. Suppose the column in Example 3.3 were controlled at 73 psia by manipulating vapor-product flow. Should the condensate temperature rise to 110°F with the same feed, more vapor must be withdrawn. Since all the propane in the feed must pass overhead, the net effect of the increase in temperature is to increase the n-butane portion of the vapor product. The equilibrium vapor composition would then be higher in n-butane and lower in the other two components to keep the same vapor pressure at the higher temperature. The preferred arrangement would have product composition controlled by its own flow while pressure is free to float with condensate temperature and feed composition.

Overhead Products in Both Phases Having examined the relationships among the variables when either a liquid or a vapor product is withdrawn, it is now possible to predict the results of withdrawing both at the same time. This practice is common when total condensation is impossible because of volatiles in the feed and the principal distillate product is a liquid. In essence, the condenser becomes a one-stage stripper.

The equilibrium between liquid- and vapor-product composition is dictated by condenser pressure and temperature as before. Secondly, their relationship to top-tray vapor composition is affected by the flow rate of the vapor product relative to top-tray vapor rate, again as seen before. But because the two products have different compositions, they tend to affect the column external balance differently. When the entire overhead product was withdrawn either as a vapor or a liquid, the mole fraction of the lightest component varied with feed composition as in Eq. (3.14). But by splitting a vapor stream away from the liquid product, the mole fraction of the lightest component in the liquid can thereby be controlled.

Ideally, the controls should be applied as follows. The concentration of heavy key in the distillate liquid should be controlled by manipulating that flow or reflux as recommended elsewhere. Then the vapor pressure of the liquid may be controlled by withdrawing vapor, thereby holding the concentration of the lightest component constant. Vapor flow rate will then vary with the relative concentration of that component in the feed.

To control the vapor pressure of the liquid product, however, the pressure on the condensate must be made to follow its temperature. Subcooling cannot take place in a partial condenser, so that condensate temperature is a true indication of composition at a given pressure and will vary with heat load and coolant temperature. If column pressure is to be controlled at a fixed point, condensate temperature must also be controlled by manipulating the rate of heat removal. Otherwise, the pressure set point must be programmed as a function of condensate temperature along a vapor-pressure curve representative of the desired product.

Unfortunately, the practice of holding pressure constant regardless of condensate temperature is almost universal. Temperature a few trays down from the top is typically controlled to hold the heavy-key–light-key ratio constant. But then distillate-liquid composition varies with heat load and coolant temperature unless vapor pressure is specifically controlled as described above.

TWO-PRODUCT COLUMNS

The bulk of multicomponent fractionation seems to fall into the category of two-product columns. Typically a feed contains two principal components with relatively smaller amounts of heavier or lighter contaminants. Or a feedstock consisting of many components in relatively equal concentrations may be split into several products in a series of two-product columns. In either case, an understanding of the two-product column is fundamental to studying more complex systems. As will be seen, most of the groundwork has already been laid in the previous chapter on binary systems.

Key Components A two-product column separates the heavy and light keys. To determine which components are the heavy and light keys in a given feed, tabulate the feed and product compositions in order of increasing boiling point as shown in Table 3.1. The horizontal line in the table indicates the split between the key components. Isobutane and propane are lighter than the light key (*n*-butane); *n*-pentane, isohexane, etc., are heavier than the heavy key (isopentane).

Not all splits are quite so definite. For example, butenes boil very closely to *n*-butane. If any were then present in this wide range of components, they would have to be lumped together with *n*-butane as the light key. Also, a component will occasionally appear in low concentration between the keys. It will tend to be split evenly between the two products if there is no side draw to remove it.

Only the key components can be controlled. In the separation described in Table 3.1, the isopentane in the distillate and the *n*-butane in the bottom product are controllable. Alternatively, the *n*-butane in the distillate and the isopentane in the bottom product could be controlled, although their limits are quite sensitive to the amount of off-key components in the feed. A third

TABLE 3.1 Compositions in a Typical Two-product Column

Component	Atm bp, °F	z, %	y, %	x, %
Propane	−43.7	1.06	4.2	
Isobutane	10.9	7.4	29.1	0.11
n-Butane	31.1	17.6	64.0	2.0
Isopentane	82.3	16.2	2.0	21.0
n-Pentane	97.3	29.3	0.73	38.9
Isohexane	140.4	14.8		19.8
Heavier		13.6		18.2

choice is to control the ratio of n-butane to isopentane in both products, which is essentially independent of off-key components.

The problem with controlling *purities* (as opposed to impurities) is the aforementioned sensitivity to off-key components. An attempt to control the light key in the distillate requires that variations in lighter components must be offset by equivalent adjustment in the heavy-key content. Since the lighter components are uncontrollable, gross changes in their percentage in the feed will have a pronounced effect on the allowable concentration of the heavy key. The same is true of the heavy key and heavier components in the bottom. The off-key components can therefore determine to a great extent how profitably a separation may be made.

If a product from one column is the feed to the next, the content of its key impurity must be controlled. This key impurity will be the lighter-than-light key or heavier-than-heavy key in the next product. Consequently, it will have a pronounced effect on separation in that column.

Separation between Components The separation factor described in Chap. 2 is also used in multicomponent systems to estimate product compositions. Equation (2.21) is repeated here with different subscripts to simplify usage with many components:

$$S_{ih} = \frac{y_i/x_i}{y_h/x_h} \tag{3.16}$$

Subscript i denotes any selected component relative to the heavy key h. Choice of the heavy key as a reference component is consistent with Hengstebeck's method [4].

Equation (3.16) may also be rearranged in a form relating the ratios of component concentrations in distillate (y) and bottom (x) products:

$$S_{ih} = \frac{y_i/y_h}{x_i/x_h} \tag{3.17}$$

The significance of this form of the separation equation lies in the fact that component *ratios* are often the variables which should be controlled in multicomponent systems.

To demonstrate this point, consider Fig. 1.2 where ethane and propane products are distilled successively from a multicomponent feed. Ethane product specifications typically call for a maximum propane content; if lighter components are insignificant, then the ratio of ethane to propane in the ethane product can be controlled. The ethane content of the propane product is also to be controlled, but this must be effected at the deethanizer since ethane is not a key in the depropanizer. As described by Eq. (1.12), the ethane content of the propane product will remain constant if the ethane-propane ratio in the deethanizer bottom product is constant. Therefore *both* deethanizer products must have fixed ethane-propane ratios. This can only be accomplished by holding deethanizer separation constant.

As in a binary system, the separation factor is a function of V/F, n, E, and so forth. First the separation factor S_{lh} between the heavy and light keys must be determined to satisfy product specifications. Then separation factors for the other components i are related by relative volatility to S_{lh}:

$$\frac{\log S_{ih}}{\log S_{lh}} = \frac{\log \alpha_{ih}}{\log \alpha_{lh}} \tag{3.18}$$

example 3.4

Check the concentration of n-pentane given for the products in Table 3.1 by using Eq. (3.18). Volatilities of n-butane (l) and n-pentane (i) relative to isopentane (h) are 2.33 and 0.79, respectively.

$$S_{lh} = \frac{64.0/2.0}{2.0/21.0} = 336$$

$$\log S_{ih} = \log 336 \frac{\log 0.79}{\log 2.33} = -0.704$$

$$S_{ih} = \log^{-1}(-0.704) = 0.198$$

$$\text{Check:} \quad S_{ih} = \frac{0.73/38.9}{2.0/21.0} = 0.197$$

Again, this method is not intended to be as accurate nor as detailed as that proposed in Ref. 4 or elsewhere in the literature. Nonetheless, its simplicity makes it extremely useful for evaluating the effects of control action and for estimating optimum operating conditions.

For a column separating a given feed mixture into products of known composition, separation factors for each component relative to the heavy key may be calculated. It is then unnecessary to know all the relative volatilities to predict the effect of a change in V/F. Most columns are operated in the linear region of Fig. 2.3. For them, $\log S_{ih}$ for all components varies directly with V/F. For those whose operating condition is expected to lie in the nonlinear region of the curve, $\bar{\alpha}_{lh}$ must be used and values of E estimated until a point is found which falls on the curve. Again, variations in V/F will affect $\log S_{ih}$ in direct proportion to $\log S_{lh}$.

The preceding method is particularly suited to modeling existing columns. Then actual composition data instead of tabulated K values may be used to determine the various separation factors. The effects of changes about the known operating point are then predictable with greater accuracy and tend to be far more representative of actual conditions.

Multicomponent Material Balances Actually, the equations used for multicomponent material balances are no different than those describing binary systems—there are only more of them. For any component i, binary Eq. (2.36) can be restated:

$$\frac{D}{F} = \frac{z_i - x_i}{y_i - x_i} \tag{3.19}$$

When stated for all components and taken together with a set of Eq. (3.16) for all components, all compositions may be found for given values of D/F and S_{lh}. However, this requires the solution of $2(k-1)$ equations for k components. Although this calculation is readily soluble with a digital computer, a simpler approximation soluble with a pocket calculator is desirable.

Quite often only one component lighter than the light key appears in significant concentration in the bottom product and one heavier than the heavy key appears to any extent in the distillate. The following approximation is derived to fit this general case. The approximation begins by formulating a binary mixture with the light key and lighter components as the light group and the rest as the heavier. A binary separation factor may be calculated by using this binary composition as in Eq. (2.22). A change in distillate or bottom binary composition may then be imposed either with the same value of separation or a new one. Then the binary composition of the remaining product may be calculated by using the appropriate form of Eq. (2.33). When both new binary compositions have been found, D/F may then be calculated by using Eq. (2.36). Up to this point, the system is handled as a simple binary (which it is not).

Those lightest components having no significant contribution to the bottom product will appear in the distillate in concentration:

$$y_i = \frac{z_i}{D/F} \tag{3.20}$$

The heaviest components having no appreciable concentration in the distillate will appear in the bottom product as

$$x_j = \frac{z_j}{1 - D/F} \tag{3.21}$$

example 3.5

The mixture given in Table 3.1 can be represented by a binary whose compositions are

$$z_b = 26.0\%$$

$$y_b = 97.3\%$$

$$x_b = 2.11\%$$

Calculate D/F and binary separation. Then determine what D/F is required to raise x_b to 3.0 percent. Finally evaluate new concentrations of those components appearing in one product only:

$$\frac{D}{F} = \frac{0.26 - 0.0211}{0.973 - 0.0211} = 0.252$$

$$S_b = \frac{0.973(0.9789)}{0.0211(0.027)} = 1{,}666$$

For $x_b = 3\%$:

$$y_b = \frac{1{,}666(0.03)}{1 + (0.03)1{,}665} = 0.981$$

$$\frac{D}{F} = \frac{0.26 - 0.03}{0.981 - 0.03} = 0.242$$

Propane: $y_i = \dfrac{1.06\%}{0.242} = 4.38\%$

Isohexane: $x_j = \dfrac{14.8\%}{1 - 0.242} = 19.5\%$

Heavier: $x_j = \dfrac{13.6\%}{1 - 0.242} = 17.9\%$

The concentrations of the keys and their adjacent components are to this point unknown, but their sums are known. Subtracting the known lightest components in the distillate from y_b yields the sum of the light key and next lighter component. The sum of these two components in the bottom product is x_b, already known. This can then be treated as a binary mixture whose separation factor may be calculated from the original set of conditions. By combining a single separation and material-balance equation for this two-component mixture, individual compositions may be found. The complete listing of compositions calculated by this method, starting with Example 3.4, is given in Table 3.2.

Note that the ratio of the lighter/light-key concentration in the bottoms and the ratio of the heavier/heavy-key concentration in the distillate tend to be nearly constant with D/F. This allows the above mixture and those like it to be treated as a binary for control purposes. The *key impurities* (light key in the bottoms and heavy key in the distillate) behave essentially like their binary counterparts in response to D/F and V/F.

MULTIPLE-COLUMN UNITS

Whenever a multicomponent feedstock must be split into several relatively pure fractions or a cut must be taken from the heart of the feed, more than one column is ordinarily required. Control strategies devised for a single column often have to be adjusted to satisfy the demands of other columns. When the product from one column is the feed to another, for example, wide fluctuations in its flow must be avoided. And as has already been demonstrated, the

TABLE 3.2 Effect of *D/F* on Product Compositions

	$D/F = 0.252$		$D/F = 0.242$	
Component	y, %	x, %	y, %	x, %
Propane	4.2		4.38	
Isobutane	29.1	0.11	29.8	0.16
n-Butane	64.0	2.0	63.9	2.84
Isopentane	2.0	21.0	1.41	20.9
n-Pentane	0.7	38.9	0.49	38.6
Isohexane		19.8		19.5
Heavier		18.2		17.9

control of the quality of certain products may have to be implemented one or more columns upstream.

Energy integration also becomes possible with multiple-column systems, adding a new dimension to the control problem. It increases the probability of interaction between columns beyond what is normally encountered in multicomponent systems. But the most interesting aspect of distillation trains is the possibility of complete unit optimization—what may be optimum policy for a single column may not be optimum for the unit as a whole.

A classic design problem exists in ordering the sequence of separations for a multicomponent feed. A feedstock consisting of k major components requires $k - 1$ columns for complete separation. But they may be arranged in a variety of ways. Reference 5 describes a computer-aided design procedure for selecting the optimum (least cost) arrangement, both with and without energy integration. This means of arriving at an optimum arrangement is beyond the scope of this text. However, common arrangements are presented so that their individual and collective control problems may be pointed out.

Light-ends Fractionation Perhaps the most common application of multiple-column fractionation is the separation of light hydrocarbons. Natural-gas liquids, virgin naphthas, and the light products from cracking reactions all contain a spectrum of hydrocarbons from ethane through gasoline fractions. The components are often removed sequentially, beginning with the lightest. This practice probably gave rise to the nomenclature common in refineries, where towers are named according to the light component removed in them, e.g., deethanizer, depropanizer, etc.

Since the lightest components are the most difficult to condense, they are usually removed first. Then downstream columns may be operated at lower pressure and without refrigeration. Isomers which are relatively difficult to separate are often split out together, however, to be separated alone. These relationships are only qualitative, since the best arrangement depends heavily on relative concentrations as well as on opportunities for energy integration.

The example of a multicomponent feed given in Table 3.1 will be developed further at this point. Already considered was the column splitting between n-butane and isopentane—the debutanizer. The two products from this column may be further fractionated in a sequence such as that shown in Fig. 3.1. The author makes no claims as to whether Fig. 3.1 represents an optimum arrangement: it is simply used to illustrate control problems associated with distillation trains.

Each product from the train may have two specifications—a higher and a lower boiling component. Thus the propane product may have specifications on ethane and butane content, the isobutane product may have specifications on propane and n-butane content, etc. For the lightest and heaviest products—in this case propane and hexane (and heavier)—only one specification is controllable, however. If the ethane limit on propane must be met, it must be met in the feed.

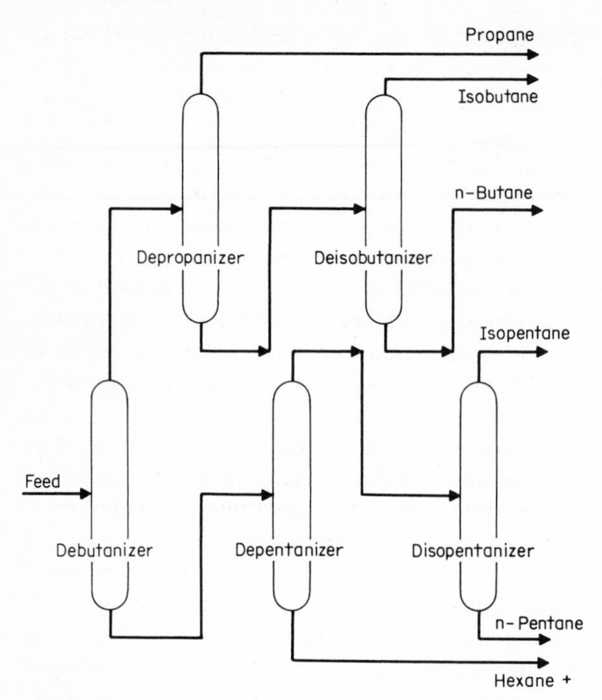

figure 3.1 *One possible configuration for separating the multicomponent mixture given in Table 3.1.*

Products from the last two columns are the most difficult to control. The amount of n-butane in the deisobutanizer distillate and isobutane in the bottom product are both controllable at the deisobutanizer. But the percentage of isopentane in the n-butane can only be adjusted at the debutanizer—two columns upstream. Closed-loop feedback control of this variable is simply not feasible—the delays distributed through the three towers are excessive. Instead, the isopentane content of the debutanizer distillate must be controlled in ratio to its n-butane content in the manner described in Eq. (1.12). In this way, the two components will be in approximately the correct ratio when they emerge from the deisobutanizer.

It may be desirable to operate below the isopentane limit. If the remaining specification on n-butane is simply a purity limit without regard to the isobutane content, reducing the isopentane level will allow the isobutane to increase. This has the benefit of reducing the separation required of the deisobutanizer, but at the same time it increases that of the debutanizer. If the isopentane and isobutane command different selling prices, their value difference also enters into the operating cost. Thus a two-column optimization problem exists with the debutanizer overhead composition as the cost-determining element.

This is not the only possibility. The amount of n-butane in the isopentane product must also be controlled at the debutanizer, and its level affects the separation of the deisopentanizer as well. In the same manner, propane in the isobutane product affects separation in both the depropanizer and the deisobutanizer.

Another problem common to multiple-column systems is that a bottleneck in any column can limit unit production. It is possible to shift the loading somewhat between columns through the same mechanism as suggested for optimization. If, for example, the deisobutanizer alone has reached an operating constraint, some of its separation load may be shifted to the debutanizer by reducing isopentane in its distillate. At the same time, some loading could be shifted to the depropanizer by reducing propane in the deisobutanizer feed. These relationships are mentioned at this point only to suggest the complexities of operating a distillation train. Actual optimization procedures are developed in Chap. 11.

Recycle Streams Occasionally a recycle stream appears in a flowsheet. The usual reason is to satisfy a reactor of some kind in which complete conversion of a feedstock to a product is not feasible. Thus the unconverted feed may be separated from the product and returned to the reactor. If there is no reaction associated with the recycle stream, its purpose is usually to improve recovery of a final product. Figure 3.2 illustrates one possibility: the light ends from a refining column are returned to recover whatever product may be there. If the final product were taken overhead, the heavy cut would be recycled to the heavy-ends column to improve recovery.

When confronted with a system like that shown in Fig. 3.2, it is not easy to arrive at a sensible quality-control philosophy. The reason for the dilemma is that there are too many columns for the number of products. A three-product system ordinarily requires only two columns. Similarly, control of two specifications on the major product can be done with only two columns. The third column and its recycle stream add a new dimension to the control

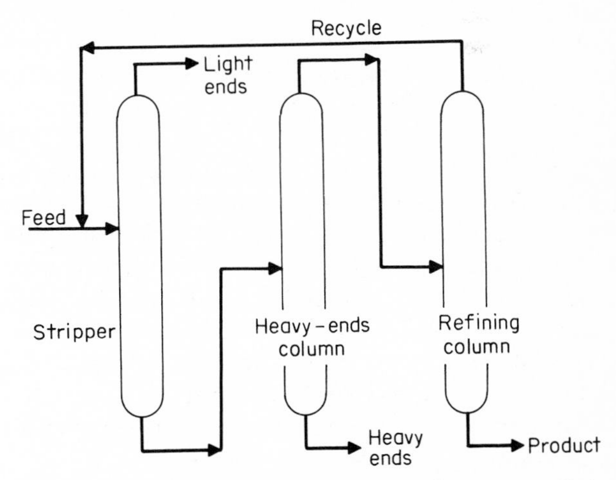

figure 3.2 *The recycle stream is introduced to recover whatever product may leave the top of the refining column.*

problem. The question of control over the light-component content must be resolved. If it is controlled at the refining column, then what specification should be met in the bottom product from the light-ends column? If that product already meets final specifications, the recycle flow can be zero. If it does not, a recycle flow will result, proportional to the excess of light component. Since the recycle returns to the light-ends column, the two act like a double column having a single purpose. It is doubtful whether an optimum recycle rate exists—shifting the load from one column to another by increasing the recycle flow is not likely to have a significant effect on unit operating cost. In effect, this type of unit is usually more efficient if designed with the minimum number of columns. Even a double column is more efficient than two separate ones since it has a single reboiler and condenser. The refining column, as a rule, is added to allow trimming of the final-product quality and may not be required if control of upstream columns is adequate.

With a reactor in the recycle loop, conditions are entirely different. Figure 3.3 shows a simplified flowsheet of a sulfuric acid alkylation unit as an illustration. The acid catalyzes the reaction of isobutane with an olefin such as isobutylene to form an isooctane:

$$\begin{array}{c} \mathrm{C} \\ | \\ \mathrm{C{-}C{-}C} \end{array} + \begin{array}{c} \mathrm{C} \\ | \\ \mathrm{C{=}C{-}C} \end{array} \rightarrow \begin{array}{c} \mathrm{C}\quad\;\;\mathrm{C} \\ |\qquad\;\; | \\ \mathrm{C{-}C{-}C{-}C{-}C} \\ | \\ \mathrm{C} \end{array} \tag{3.22}$$

Equation (3.22) is but one of many possibilities. Butenes are so difficult to separate from one another that the olefin stream may contain butene-1 and butene-2 as well. Other olefins, if present, will also participate. And as is typical of many organic reactions, side reactions also take place, yielding a mixture of highly branched paraffins, principally in the 6–9 carbon range. If the reaction is properly controlled, the alkylate is a high-octane motor fuel.

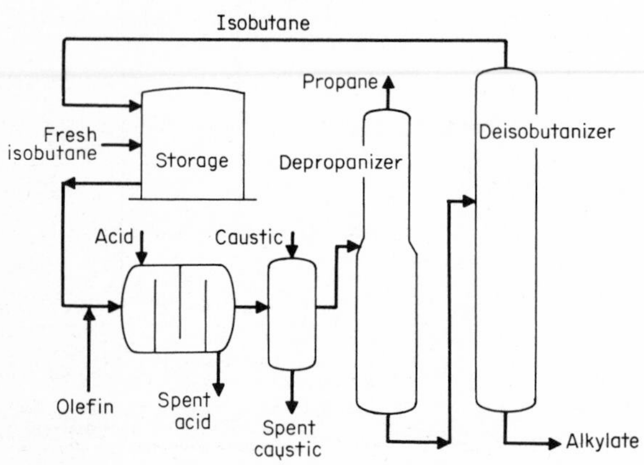

figure 3.3 *A simplified flowsheet of a sulfuric acid alkylation unit.*

One of the control problems is to keep the olefin components from reacting with each other. They have a strong tendency to polymerize, and the polymer is a poor fuel since it is unsaturated. Polymerization is minimized by diluting the olefins with isobutane, in a ratio of 1 part in 5 to 10, and by operating at low temperature with refrigeration. The reactor effluent then contains only 1 part of alkylate in 4 to 9 parts of isobutane, which must then be separated and recycled. Figure 3.3 shows a deisobutanizer making that separation.

In a recycle system, care must be taken to avoid accumulation of components not participating in the reaction. Thus a depropanizer is necessary to remove propane and lighter hydrocarbons which will build up even if present only in minute amounts in the feed. Normal butane is also unreactive and must be removed with the alkylate if no other outlet is provided. In the next section, a sidestream will be added to remove it.

Principal specifications on the isobutane recycle stream are propane and n-butane. Both take up reactor space and therefore reduce production. Isobutane losses with the alkylate are undesirable since it raises the vapor pressure excessively. More is said on this subject under multiproduct towers.

Inventory Controls Proper sizing and control of the storage tank are essential to the successful operation of any recycle system. For example, consider the alkylation unit in a steady state with the storage tank half full. To increase production by 20 percent would require a 200 percent increase in fresh isobutane if the level in the storage tank were to remain constant, assuming a 10:1 ratio of isobutane to olefin. Similarly, a 20 percent decrease in production would require a 200 percent decrease in fresh isobutane, which would clearly not be possible even if the 200 percent increase were. Eventually most of the increase or decrease in isobutane will be returned as recycle from the deisobutanizer. But this takes time because levels in the reactor and column trays must change to accommodate the new flow rate. It is therefore impossible to control the level in the storage tank in the unsteady state. Furthermore, any attempt to control it by executing the gross manipulations in fresh feed estimated above would place unrealistic demands on the feed-processing unit.

If the flow variations imposed on the fresh feed are to be no greater than the changes in reactor feed, the storage tank must be large enough to absorb the entire change in inventory of the balance of the system. Therefore the storage tank level must be low when the production rate is high and high when production is low. The level should not be tightly controlled at midscale. On the contrary, if fresh feed is set proportional to tank level, a reasonably high feed rate will be maintained when the level is low, but only enough to keep the level from falling further. It is not necessary to add enough to raise the level to midscale—this overstresses the upstream feed-processing unit.

These basic rules apply whenever a storage vessel is located between columns or processing units. The intended function is usually to absorb upstream fluctuations while providing a nearly constant feed rate to the next unit. These vessels are often called "surge" tanks, descriptive of their role in absorbing surges. In the steady state, outflow must match inflow, but rapid inflow fluctuations should not be passed on. Consequently, level in the surge tank cannot be held within narrow limits since this will shrink the effective volume of the vessel to those limits. If the level controller manipulates the outflow, the level should be high when the flow is high and low when the flow is low. In this way, the vessel will be best able to absorb upsets in the direction of more normal operation or of the opposite extreme. For further discussion on this subject, the reader is directed to Ref. 6.

Energy Integration By "energy integration" is meant interchanging of energy between process streams to minimize overall consumption. A simple form was shown in Fig. 2.10, where heat was exchanged between feed and bottom product of a single column. This practice is also common between columns, but the exchange is nearly always between a bottom product and a feed.

Much more energy can be saved if the integration is made by combining reboiling and condensing. In this way, two columns may be operated with three heat exchangers instead of four and with the energy consumption of one column. Such a scheme is shown in Fig. 3.4.

To function at all, column 1 must be operated at a much higher pressure than column 2 because the bottom of column 2 will be richer in higher-boiling components than the top of column 1. Alternatively, a compressor may be used to lower the pressure in column 1. To be successful from an energy-conservation standpoint, nearly all the vapor in column 2 must be generated from reboiler 1. The steam-heated reboiler on column 2 is intended only for trimming. While this scheme is admirable, it does present some control problems—the two columns must be reboiled at essentially the same rate. This takes away a degree of freedom by removing a variable which is ordinarily manipulated for quality control.

When two columns are operated with independently adjustable boilup, four

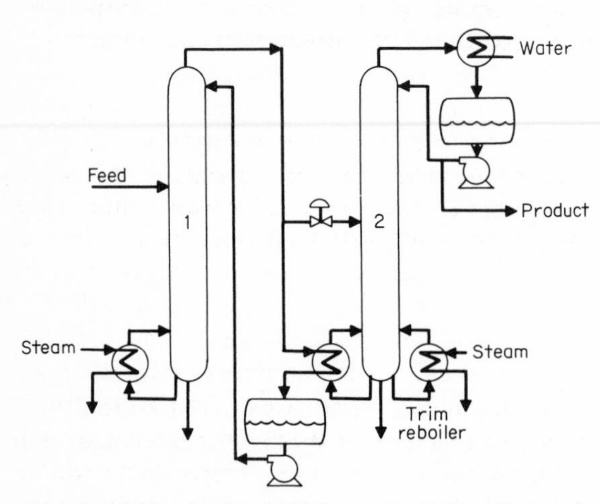

figure 3.4 *Energy integration forces the boilup of one column to be dependent on another.*

composition variables may be controlled—two in each tower. Thus the distillate leaving column 2 typically will have two components controlled. With the heat integration of Fig. 3.4, however, one of the column 2 compositions will float, but it can still be regulated. A constant V/F ratio in column 1 will allow control of both product compositions there. For a reasonably constant feed composition, a constant V/F in column 2 will also be maintained. Thus if the distillate composition of column 2 is controlled directly, the bottom composition will be maintained constant by virtue of constant separation. As separation in column 2 tends to remain constant, however, it is also very difficult to improve. So if specifications there are tightened, additional boilup must be provided either by the trim reboiler or by column 1.

The pressure in column 1 will tend to vary with column 2 bottom temperature and heat load. Pressure may be controlled by flooding the condenser-reboiler or bypassing it, but not by opening the valve feeding column 2. This is a partial condenser whose equilibrium is determined by temperature and pressure. So pressure cannot be set independent of temperature without affecting composition. Instead, feed to column 2 should be set to control composition. Pressure must be controlled elsewhere or allowed to float on the condenser-reboiler.

Heat integration may be applied to a multiple-column unit such as Fig. 3.1 by matching heat loads and temperatures. Typically, the lighter components must be condensed directly whereas heavier components may be condensed by boiling lighter ones. The depentanizer has the highest-boiling distillate, which could be used to reboil either the depropanizer or deisobutanizer. Of these two columns, the deisobutanizer reboiler will have the lower temperatures if both are condensed against the same sink. So the depentanizer condenser could conceivably be the deisobutanizer reboiler—other combinations are not likely without substantial elevation in pressures or adding compressors. Reference 5 gives an example of separating a five-component mixture with four columns wherein only one heat-integration connection was feasible.

Heat is typically recovered from final bottom products by interchanging with the feed. Typically the bottom products from both the depentanizer and the deisopentanizer in Fig. 3.1 would be used to preheat the feed to the unit. Feedback loops associated with this heat recovery tend to be slower than with a single column but can still create instability if feed temperature is not tightly controlled.

MULTIPLE-PRODUCT COLUMNS

It is entirely possible to separate multiple-component mixtures into multiple products in a single column. It is also efficient, in that a single reboiler and condenser are used. However, the mixtures which can be separated and the product specifications achievable are limited. It is not possible, for example, to extract a pure component from the heart of a complex feed mixture—it would have to be withdrawn near the feed tray where other impurities are high. Yet there are many worthwhile applications of multiproduct towers, some of which are described here.

Refining or Pasteurizing Columns Perhaps the most common tower of the refining or pasteurizing type is the ethylene fractionator, shown in context with other columns of its unit in Fig. 3.5. Its feed contains a small amount of methane along with the principal impurity, ethane. The ethylene fractionator actually has two final products: the distillate containing virtually all the methane in the feed, and the refined product, high-grade ethylene, leaving as a sidestream.

Actually, more of the methane could have been removed in the demethanizer but at the cost of higher boilup or increased ethylene losses there. Withdrawing the high-grade product as a sidestream allows its methane content to be controlled precisely at the point of withdrawal. This is an important feature in a multicolumn unit. But having a market for low-grade ethylene adds another dimension to the picture—the yield of that product is tied to the operation of the demethanizer.

To illustrate this relationship, assume that all the methane and ethylene leaving the demethanizer are recoverable. Then a methane balance yields

$$Bx_1 = Dy_1 + Pw_1 \tag{3.23}$$

where B = demethanizer bottoms flow
D = low-grade ethylene flow
P = high-grade ethylene flow
x_1, y_1, w_1 = methane contents of associated streams

A similar balance may be drawn on the ethylene, indicated by subscript 2:

$$Bx_2 = Dy_2 + Pw_2 \tag{3.24}$$

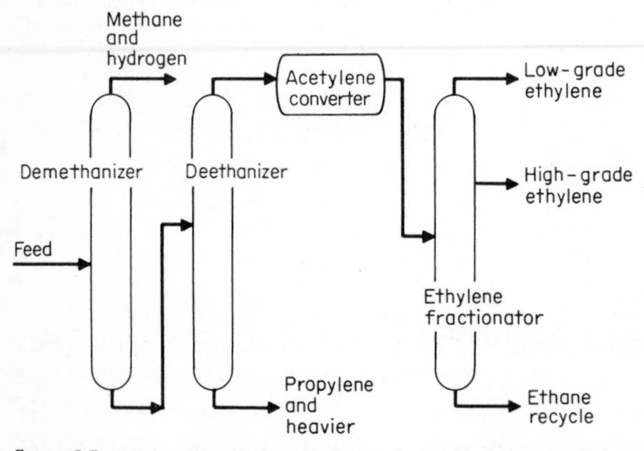

figure 3.5 *Virtually all the methane from the bottom of the demethanizer must leave with the low-grade ethylene.*

If the two equations are combined by eliminating B, the relationship of the D/P ratio to demethanizer product composition appears:

$$\frac{x_1}{x_2} = \frac{y_1 D/P + w_1}{y_2 D/P + w_2} \tag{3.25}$$

Equation (3.25) has a twofold significance. It indicates how the D/P ratio must be adjusted to control y_1 or w_1 for an imposed x_1/x_2 ratio. Secondly, it gives the x_1/x_2 ratio needed to satisfy a specification on y_1 or w_1 and to provide the desired D/P ratio to meet market demands. In actual practice, D must be manipulated to control w_1. However, the x_1/x_2 ratio should also be adjusted at the demethanizer so that the demands for both products may be satisfied.

example 3.6

Estimate the ratio of methane to ethane in the demethanizer bottom product to satisfy the following specifications:

$$y_1 = 2\% \qquad w_1 = 0.1\%$$

$$y_2 = 98\% \qquad w_2 = 99.7\%$$

$$\frac{D}{P} = 0.2$$

$$\frac{x_1}{x_2} = \frac{0.02(0.2) + 0.001}{0.98(0.2) + 0.997} = 4.19 \times 10^{-3}$$

The relationship among the compositions on the right side of (3.25) is a function of the separation in the top section of the ethylene fractionator:

$$S = \frac{y_1/w_1}{y_2/w_2} = \frac{y_1/y_2}{w_1/w_2} \tag{3.26}$$

This statement is required for a complete description of the top section.

example 3.7

Starting with the conditions stated above, determine how the D/P ratio must be adjusted to change w_1 to 0.15 and 0.05 percent, assuming a constant separation.

$$S = \frac{0.02/0.98}{0.001/0.997} = 20.35$$

Solving for y_1 for new values of w_1 requires substituting $1 - y_1$ for y_2 and $0.998 - w_1$ for w_2, along with rearranging (3.26):

$$y_1 = \frac{Sw_1}{Sw_1 + w_2} = \frac{Sw_1}{Sw_1 + (0.998 - w_1)}$$

For $w_1 = 0.15\%$:

$$y_1 = \frac{20.35(0.0015)}{20.35(0.0015) + 0.998 - 0.0015} = 0.0297$$

For $w_1 = 0.05\%$:

$$y_1 = \frac{20.35(0.0005)}{20.35(0.0005) + 0.998 - 0.0005} = 0.0101$$

Next the D/P ratio may be found by rearranging (3.25):

$$\frac{D}{P} = \frac{w_2 x_1/x_2 - w_1}{y_1 - y_2 x_1/x_2}$$

For $w_1 = 0.15\%$:

$$\frac{D}{P} = \frac{0.9965(4.19 \times 10^{-3}) - 0.0015}{0.0297 - 0.9703(4.19 \times 10^{-3})} = 0.104$$

For $w_1 = 0.05\%$:

$$\frac{D}{P} = \frac{0.9975(4.19 \times 10^{-3}) - 0.0005}{0.0101 - 0.9899(4.19 \times 10^{-3})} = 0.618$$

The results are summarized in Table 3.3.

Because the separation is constant, the y_1/w_1 ratio is essentially constant. Since separation is likely to be adjusted only for meeting the ethylene specification in the ethane product, separation between the top products is not ordinarily manipulable. This was the same problem encountered when heat integration was applied to two columns. The ethylene fractionator has essentially three manipulated variables for quality control—two product flow rates and separation. Two controlled variables must be the methane and ethane content of the high-grade product. As mentioned, the ethylene content of the recycled ethane should also be regulated.

Thus the distillate-product flow must be manipulated to control w_1—for fixed separation y_1 will follow w_1. Therefore, no direct control exists over y_1. Finally, as mentioned earlier, the D/P ratio is a function of demethanizer bottoms composition. Reference 7 describes techniques for optimizing the ethylene separation unit.

Vapor vs Liquid Sidestreams A word should be said about the differences between withdrawing vapor or liquid as a sidestream product, particularly from a refining column. Figure 3.6 describes a refining column with *vapor* sidestream. When the major product is the sidestream, variations in its flow can cause severe upsets to trays above. The author has observed a column of this type whose top and bottom streams were both manipulated for control of sidestream quality. Then the sidestream flow was used to control bottom level while heat input was fixed. A rising level caused the vapor valve to open wider, which had the capability of interrupting vapor flow into the top section. Being a sieve-tray column, the liquid in the top trays then would fall to the

TABLE 3.3 Effect of Flow Ratio on Methane Content of Ethylene Products

Flow ratio D/P	High grade w_1, %	Low grade y_1, %
0.618	0.05	1.01
0.200	0.10	2.00
0.104	0.15	2.97

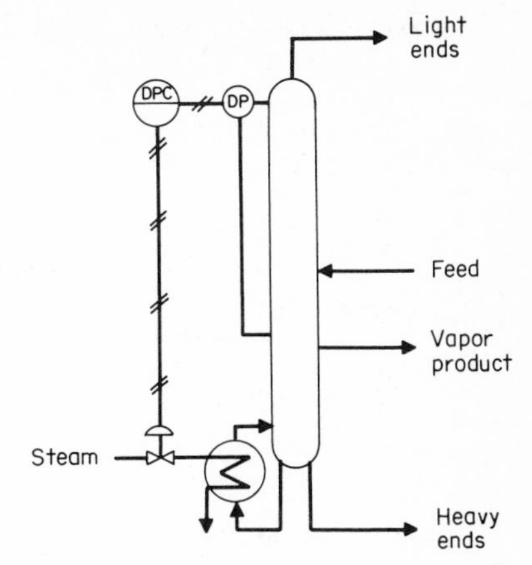

figure 3.6 *When a major product is removed as a vapor sidestream, vapor flow above the point of withdrawal must be controlled.*

bottom, further raising the level and augmenting the disturbance. This situation was corrected by using heat input to control differential pressure (DP) across the top section as shown in Fig. 3.6. Then as the sidestream flow increased, the differential pressure controller (DPC) raised the heat input to maintain top-section vapor rate. The increased boilup in turn caused bottom level to fall, and stable level control was thereby achieved.

Withdrawing a liquid sidestream may be accomplished in several ways. The column may be fitted with a "chimney" or "total trap-out" tray which is not a vapor-liquid contactor. This type of tray catches all the liquid so that reflux must be returned to a distributor by an external loop. A level transmitter or controller (LC) is often fitted to the tray to control the rate of reflux or withdrawal as shown in Fig. 3.7.

Perhaps more common is a "partial trap-out" tray which contains an overflowing chamber under the downcomer. There is no need to withdraw reflux—the product may be withdrawn under flow control with the balance of the liquid overflowing to the next tray.

Multiple Feed Streams Recent years have seen towers performing combinations of functions. A classic example is the alkylation deisobutanizer introduced about 1964. In addition to separating unreacted isobutane from alkylate as shown in Fig. 3.8, it also yields an n-butane side product. This feature allows n-butane in the feed to be increased beyond what is normally

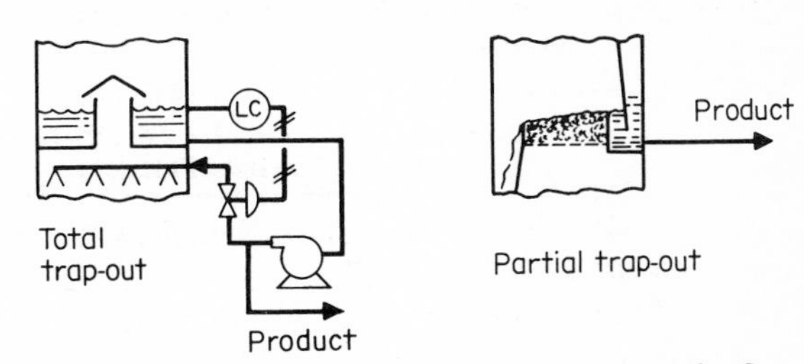

figure 3.7 *A total trap-out tray requires an external-reflux loop, whereas a partial trap-out tray simply overflows.*

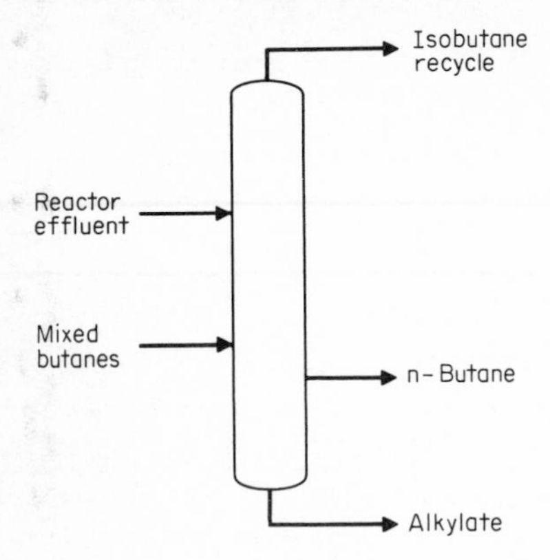

figure 3.8 *Being able to withdraw n-butane as a side product allows adding a mixed-butane feed.*

desirable either in the alkylate or recycle isobutane. Consequently, the tower is also capable of splitting a mixed-butane feedstock.

The flowsheet in Fig. 3.3 shows fresh isobutane being added to storage in an amount equal to what is consumed in the reaction. This feed had to be separated from n-butane in another column. Because the deisobutanizer with its n-butane sidestream can also split a feedstock of mixed butanes, the feed splitter can be eliminated.

The mixed-butane feed enters the deisobutanizer below the reactor effluent because it contains only 40 to 50 percent isobutane compared to 80 percent or more in the effluent. The n-butane content of the distillate is ordinarily controlled since it affects conversion in the reactor. Isobutane content of the sidestream should be controlled to maximize its recovery and to improve the value of the sidestream as a gasoline-blending component (this was discussed in Chap. 1).

Control of sidestream composition is not easy. This can be demonstrated by examining the material balance. Let the ratio of isobutane to olefin in the feed be 10:1 and the mixed-butane feed be a 50-50 mixture of the two isomers. The amount of isobutane entering the column is 10 times the amount of n-butane (9 from the reactor and 1 in the fresh feed). Therefore, to control the sidestream composition within 1 percent the isobutane material balance must be manipulated to an accuracy of 0.1 percent.

The n-butane content or the vapor pressure of the alkylate product may be controlled by manipulating the sidestream flow. But increasing its flow will tend to draw isopentane from the alkylate into the sidestream, as well as n-butane. The ratio of the ratios of n-butane to isopentane in the two streams is determined by the separation factor across the bottom section of trays. The relationship given in Eq. (3.26) for the top section of the ethylene fractionator applies here as well.

In addition to controlling compositions, the flow of mixed butanes must be established. It can best be calculated to put as much isobutane into the system as is needed to react with the olefin feed. This will vary with the purity of the mixed-butane feed. Even if this calculation is accurately performed, control over isobutane inventory will not thereby be accomplished. As described earlier, the level of the storage tank in Fig. 3.8 will tend to fall on an increase in production and to rise on a decrease as the inventory is shifted through the process. Furthermore, the rate of isobutane loss cannot be matched exactly by the calculated flow of mixed butanes. As a result, the average isobutane inventory will tend either to rise or fall gradually with time unless some regulation is applied. This can be in the form of operator adjustment to the mixed-butane feed rate based on hourly level readings or it can be done automatically. In either case, sufficient time must be allowed for the isobutane recycle stream to respond to a change in mixed-butane flow, or instability will ensue. As before, if the level is allowed to fall on a production increase and to rise on a decrease, upsets to the column will be minimized.

DISTILLATION OF CRUDE FEEDSTOCKS

The most common crude feedstock is, of course, crude oil. It belongs in a class by itself not only because of its wide boiling range but because of the admixture of components it contains. A typical crude oil is composed of nearly 50 percent aromatic hydrocarbons (many of multiple rings) along with cycloparaffins as well as straight-chain and branched paraffins. Effluents from fluid catalytic-cracking units and hydrocracking units are by comparison simpler mixtures, the former being rich in olefins while the latter are not.

Crude oil also is rich in low-volatile hydrocarbons—asphalts and waxes which require vacuum distillation. But the subject at hand is multiple-product distillation, characteristic of the pipestill or atmospheric crude column. This unit is similar in many respects to the main fractionator for catalytic-cracker and hydrocracker effluents. Consequently, the problem of crude-oil fractionation is discussed here, considering that the other separations are less difficult.

The Atmospheric Tower Figure 3.9 describes the essentials of an atmospheric tower for separating crude oil into a typical set of products. The feed is heated by the overhead condenser, two side coolers, and one or more product streams before entering the fired heater. At this point, the temperature is raised as high as possible short of cracking the feed, usually about 700°F but varying with the quality of the crude. The feed then enters the tower between 50 and perhaps 70 percent vaporized, again depending on the crude composition. There is no reboiler as such; the liquid portion falls through the few trays in the bottom section, where it is stripped of additional volatiles with steam. Because there is little vapor flow in this bottom section, the diameter is reduced.

If reflux were provided only at the top of the tower, vapor and liquid rates would vary drastically from top to bottom. The liquid flow is lower at the bottom because the sidestream products are all liquids. The vapor flow is greater at the top because the large quantity of sensible heat at the bottom is converted into latent heat at the top. In addition, the steam used for sidestream stripping also accumulates in the top.

To distribute liquid and vapor loading more equitably, heat is generally removed at two points in the column as shown. These "pump-around" loops reduce vapor flow and increase liquid flow at the same time. However, the effect is virtually a step change in distribution. Vapor and liquid flow rates are highest just below the tray where cold liquid is returned, and flooding is therefore most likely to occur at that point.

Steam stripping is used to increase the recovery of volatile components by lowering their partial pressure in the vapor at a given liquid temperature. The theory behind steam stripping is analyzed in detail in the next chapter and so will not be developed here. Van Winkle [8] indicates that up to 10 percent additional recovery is possible by steam stripping, using from perhaps 0.3 to 0.5 lb steam per gallon of product, depending on the number of plates in the stripper. In excess of 1.0 lb/gal is necessary to increase the recovery of volatiles from the topped crude by a comparable amount. Steam usage beyond these values achieves relatively little incremental recovery of volatiles.

Product Specifications A single pure component will, of course, completely vaporize at one temperature. If a binary mixture is distilled batchwise through many trays with a high reflux ratio, the boiling point will change nearly stepwise when the last of the lighter component is removed. This method of distilling results in what is known as the true boiling point (TBP) of each component. As more components are added to a mixture, the plot of temperature vs percent distilled takes on the appearance of a staircase with

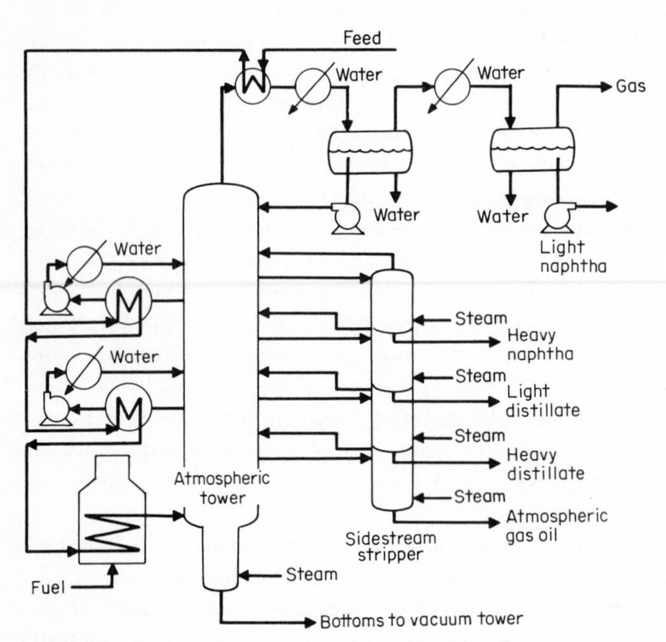

figure 3.9 *All the heat used in crude-oil distillation enters with the feed.*

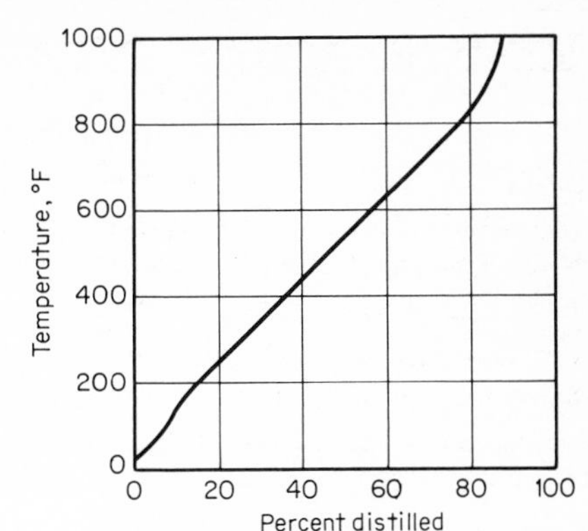

figure 3.10 *A true-boiling-point (TBP) curve for a typical crude oil.*

one step for each component. But because the reflux ratio and number of trays are less than infinite, some rounding of the steps is always evident. A crude-oil sample, or even one of its fractions, contains so many components that the TBP curve is smooth, as shown in Fig. 3.10.

A second method of evaluating a complex mixture is the American Society for Testing Materials (ASTM) method. A flask without plates is used to distill the sample under very carefully controlled conditions. The separation between components is not so great as with the TBP method, and a different curve of temperature vs percent distilled will be obtained. Van Winkle [9] offers well-documented comparisons of these methods and others used to determine the composition and behavior of complex mixtures.

The reason for describing these testing methods is that the specifications on the products fractionated from crude oil are based on their results. Samples of the crude feed and each fraction are distilled in laboratory or automated apparatus. Products are then characterized on the basis of boiling-point range. Although it would be convenient to characterize products based on the full 0 to 100 percent boiling range, it is not practical to do so. The TBP and ASTM curves are probably least reliable at the extremes, so the temperatures at which 5 and 95 percent of the sample are distilled are most often used. Although the 50 percent distilled point gives a measure of the average boiling point of a mixture, it does not indicate purity. Specifying the 5 and 95 percent ASTM or TBP temperatures is equivalent to specifying a light and heavy impurity of a three-component mixture.

Typically a crude feed will be split into about seven products in the atmospheric column. Table 3.4 lists the commonly recognized nomenclature of a set of those products along with their approximate boiling ranges and destinations.

The ultimate uses and therefore the boiling ranges of the products listed will change from one refinery to the next. They depend heavily on refinery facilities, market, and the characteristics of the crude. A refinery with a market for aromatic chemicals, for example, may carefully cut heavy naphthas from several sources for reformer feed. Another refinery without this

TABLE 3.4 Typical Products of an Atmospheric Crude-oil Column

Product	5–95% ASTM, °F	Carbon atoms per mole	Use
Gas	<100	<5	Liquefied gas and fuel gas
Light naphtha	80–180	5–6	Gasoline blending
Heavy naphtha	180–380	7–9	Reforming to aromatics or gasoline
Light distillate	380–550	10–16	Jet fuel and heating oil
Heavy distillate	550–650	16–20	Diesel fuel and heating oil
Atmospheric gas oil	650–750	20–30	Cracking to olefins and gasoline
Bottoms	>700	>30	Vacuum-distilled to cracker feed, lube oil, residual fuel, asphalt

outlet will concentrate on gasoline blending stocks. Whether a light distillate is to be used for commercial jet fuel, military jet fuel, or heating oil will determine its specifications as to boiling range and flashpoint. By the same token, diesel fuel and heating oil have different specifications.

As in other multiproduct towers, separation is not individually adjustable for each product. In fact, it is scarcely adjustable at all. Consequently only one specification can be met on each product. The chosen specification is usually the 95 percent distilled point or "end point" as it is sometimes called. The only way two specifications on a single product can be controlled is by sacrificing one on an adjacent product. For example, to control both the 5 and the 95 percent distilled temperatures for the heavy naphtha, no control can be exercised over the 95 percent point of the light naphtha. Overlap, if any, between the end point of a light product and the initial boiling point of the next heavier product is essentially fixed by tower design and percent vaporization of the feed.

Overflash The heaviest distilled product—atmospheric gas oil—may have no end-point specification. Although production of this stream should be maximized, liquid cannot be totally removed from the column. If it is, there will be no fractionation or scrubbing action below that point and nonvolatiles can be carried upward by the vapor into the gas-oil stream. Since these materials can interfere with cracking of the gas oil, some liquid must be returned to wash them down the tower. This liquid, which is flashed from the feed but not taken as a distillate, is called "overflash." It typically amounts to perhaps 2 to 4 percent of the feed.

From an economic standpoint, it is desirable to minimize the overflash, which is the same as maximizing the yield of gas oil. The best approach to this problem is to specify the amount of nonvolatiles permitted in the gas oil as measured by color. Then gas-oil flow can be raised to the level at which color bodies begin to appear, indicating insufficient overflash.

Material Balances For purposes of mathematical analysis, the product breakdown for crude distillation may be likened to a component breakdown for a simpler mixture. A given feedstock may contain—as indicated by TBP or ASTM distillation—a certain percentage of light naphtha, heavy naphtha, etc., in accordance with their specified boiling ranges. The column can do no better than to yield those same percentages as products. Poor separation will cause overlap between adjacent products. Separation can be maximized by heating the feed as high as practicable and using enough stripping steam. For a fixed separation, the flow rates of all the products are determined wholly by material balance.

In any distillation where composition control is applied to more than a single product, there is a possibility for interaction. The composition controllers tend to fight against one another trying to satisfy their individual set points. Because of the delays always present through the process, this interaction becomes manifest in sustained oscillations when more than one controller is in automatic.

Nowhere is interaction between compositions more pronounced than in an atmospheric crude-oil column, because of the multitude of products and their similar flow rates. If the end point for the heavy-naphtha product rises too high, for example, its controller will reduce the flow to force the higher-boiling components further down the column. However, the components rejected from the heavy naphtha will tend to displace heavier components in the light distillate, reducing its end point. The light-distillate end-point controller will then have to increase the flow of light distillate to return the endpoint to its previous value. Any delay in this control action could allow the disturbance to propagate downward into heavier streams. An estimate of the degree of this interaction and its resolution are presented in Chap. 10.

Heat Recovery Heat is recovered to the crude feed wherever possible. Most actual flowsheets are much more complex, in the number of heat exchangers used, than Fig. 3.9 indicates. But in every case cooling is directed first to the crude feed and then to cooling water. An important objective in operating the atmospheric column is then to reject as much heat to the feed and as little to the cooling water as possible. From this standpoint, cooling water should only be used where essential, i.e., for overhead condensing, and avoided for cooling the pump-around loops.

Flooding is most likely to occur just below the top pump-around loop, where vapor and liquid rates are highest. Therefore, the heat removed at that point must be controlled to maximize liquid flow short of flooding. The lower pump-around loop recovers heat to the crude, but at the expense of reflux higher in the column which enhances separation. In essence, the cooling duty is split among the two side loops and the reflux condenser, the latter taking whatever the side loops do not. The heat-transfer rates through both side loops then must be closely controlled in relation to feed rate if an optimum combination of heat recovery and separation is to be achieved.

To maximize heat recovery, as much heat as possible should be rejected to the lower side loop. But this will tend to reduce vapor flow to the overhead condenser, which will reduce reflux. As a consequence, separation between light and heavy naphtha would suffer. To avoid this situation, the heat removed by the lower side loop should be manipulated to maintain a desired reflux flow.

REFERENCES

1. Perry, J. H.: "Chemical Engineers' Handbook," 4th ed., p. 14-3, McGraw-Hill Book Company, New York, 1963.
2. Treybal, R. E.: "Mass-Transfer Operations," p. 205, McGraw-Hill Book Company, New York, 1955.
3. Daniels, F.: "Outlines of Physical Chemistry," p. 203, John Wiley & Sons, Inc., New York, 1948.
4. Hengstebeck, R. J.: An Improved Shortcut for Calculating Difficult Multicomponent Distillations, *Chem. Eng.* (*NY*), Jan. 13, 1969.
5. Rathore, R. N. S., K. A. Van Wormer, and G. J. Powers: "Synthesis Strategies for Multicomponent Separation Systems with Energy Integration," MIT Industrial Liaison Program, April 1974.
6. Shinskey, F. G.: Controlling Surge-tank Level, *Instrum. Control Syst.*, September 1972.
7. Shinskey, F. G.: Values of Process Control, *Oil Gas J.*, Feb. 18, 1974.
8. Van Winkle, M.: "Distillation," p. 359, McGraw-Hill Book Company, New York, 1967.
9. *Ibid.*, pp. 127ff.

PART THREE

Specialized Distillation Processes

CHAPTER FOUR

Absorption, Stripping, and Extractive Distillation

Absorption is the process of capturing or condensing certain components in a gaseous stream by a liquid absorbent. Stripping is the reverse: absorbed components are removed or desorbed from the liquid. The two processes may be conducted independently of one another, such as the absorption of sulfur dioxide from flue gas or the stripping of hydrogen sulfide from a "sour-water" stream. More often, however, they are combined to recover certain components selectively from a gas stream. In this way, natural-gas "liquids," i.e., propane, butane, and pentane, are commonly recovered from the gas by absorption in oil followed by stripping.

Some absorbents are chosen for their selectivity: for example, monoethanolamine is used to remove carbon dioxide. In some cases it is possible to affect the absorption or desorption by altering the chemical composition of the absorbent. In every case, regardless of whether a physical or chemical equilibrium is involved, the condition of the absorbent must be altered to shift the equilibrium between absorption and desorption.

These processes are very similar to distillation, as Fig. 4.1 indicates. In fact, the absorber resembles the top section of a distillation column since the feed enters the bottom and there is no reboiler; the absorbent replaces the reflux. As will be seen, the vapor-liquid equilibria are essentially the same as in a distillation column.

The stripping column, on the other hand, may take on any one of several

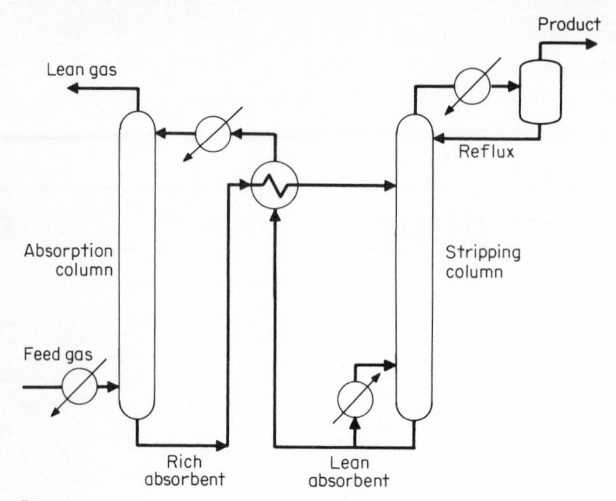

figure 4.1 *Heat must be removed from the absorption column and added to the stripping column to maintain the desired equilibria.*

configurations. The stripping column shown in Fig. 4.1 is the most complex since it has both a reboiler and condenser. This arrangement is typical of stripping in conjunction with absorption, since the condenser and reflux minimize losses of absorbent while the reboiler provides the heat necessary to drive off the volatile components. The absorbent then loops continuously between absorber and stripper, carrying product from one to the other.

The simplest stripping column is shown in Fig. 4.2. It is essentially identical in appearance to an absorber, its function being to transfer some volatile component from the liquid to the gas. This type of stripper is customarily used when stripping carbon dioxide from water with air or steam or in some similar service. Its function is to reduce the volatile content of the liquid product, with less regard for the composition of the off-gas.

When steam is used as the stripping medium, another possibility arises. If

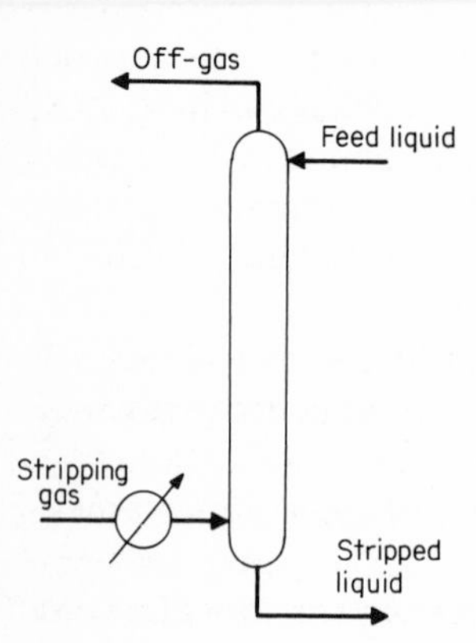

figure 4.2 *This simple stripping column is intended to remove a volatile component from the liquid to the gas stream.*

the liquid in contact with the steam is immiscible with water, an unusual vapor-liquid equilibrium appears. This is sufficiently different from those relationships already described that a separate section later in the chapter is devoted to steam stripping.

Also included in this chapter is a presentation of extractive distillation. As will be seen, the extractive-distillation column differs from a conventional unit only in the addition of an extractant stream. Since its function is similar to that of an absorbent, consideration in this chapter is natural. Furthermore, the stripping column used to recover the extractant is identical to that used in absorbent recovery.

PHYSICAL ABSORPTION AND DESORPTION

This section considers the absorption and release of vapors or gases using a nonselective absorbent, i.e., where physical rather than chemical relationships govern the equilibria. Because the absorption of light hydrocarbons in an oil is so common, this example will be used to illustrate the principles. Fortunately, most of the theory behind the absorption equilibrium was developed in the previous chapter under the discussion of third components overhead. The essence of the equilibrium is the solubility of relatively volatile vapors in a solvent of low volatility.

Absorption Equilibria The limits cited for the applicability of Henry's law in Chap. 3 seem to be satisfied by the conditions commonly encountered in absorbers. Although many absorbers operate above 1 atm, most of the components will be absorbed below their critical temperature at a partial pressure less than half their vapor pressure. In hydrocarbon systems, only methane (−117°F) and ethylene (50°F) have critical temperatures below cooling-water temperatures. For nonhydrocarbon systems, the critical points of nitrogen (−233°F) and carbon dioxide (88°F) are low whereas hydrogen sulfide (212°F) and ammonia (271°F) are above cooling-water temperature. Also, the dominance of the absorbent in the liquid phase would in most cases allow the use of Raoult's law to estimate the partial pressure of the absorbent. To this point, then, the relationships between vapor and liquid compositions can for most mixtures be modeled by Eqs. (3.1) to (3.5).

One notable point is the relationship between the equilibrium vaporization ratio K and total pressure as described by (3.3):

$$K_i = \frac{H_{ij}}{p} \tag{3.3}$$

As total pressure p increases, K_i decreases for a given value of Henry's law constant H_{ij}. This is confirmed by the existence of a "convergence pressure," where K approaches 1.0 for all components of a mixture. The convergence pressure would only correspond to the critical pressure of a component if the temperature were also its critical temperature. The convergence pressure must be found or approximated for most of the methods used in estimating

K values. It can be of significance in high-pressure absorption when a wide boiling range of components is encountered.

Fortunately, K values are available in Refs. 1 and 2 for hydrocarbons and many nonhydrocarbon gases. Therefore, equilibrium concentrations of components in each phase can readily be estimated following the procedure given in Example 3.3. However, this method is restricted to nonselective systems—it depends on solutions being reasonably ideal.

Note that substantial recovery of a volatile component by absorption depends on its K value being below 1.0. This condition can only be achieved when the system pressure exceeds the vapor pressure of the component at the solvent temperature. In essence, an absorber differs from a conventional distillation column principally in the use of a selected solvent rather than condensed overhead vapor as a reflux medium. The solvent acts to lower the mole fraction of the absorbed component in the liquid phase, so that its mole fraction in the vapor phase may also be low.

The reverse process, desorption or stripping, should be conducted under conditions of reduced pressure and/or higher temperature where the K value of the component to be recovered exceeds 1.0. Because absorption and desorption essentially involve condensing and vaporization, latent heat must be removed or supplied as shown in Fig. 4.1 if these favorable conditions are to be maintained.

As in any equilibrium process, multiple stages of vapor-liquid contact will improve the separation between the overhead and bottom products. Consequently, absorption and stripping columns are typically fitted with trays or packing to provide several equilibrium stages just like distillation columns. However, in the case of absorbers at least, all stages are not equally effective. Owens and Maddox [3] discovered that 80 percent of the mass transfer in an absorber takes place in the top and bottom stages. Although their survey indicates that most absorbers contain 6 to 10 theoretical stages, only the terminal ones are particularly effective. The ineffectiveness of interior stages is due to an unusual temperature profile. Temperatures of the terminal trays are related closely to the temperatures of the absorbent and feed gas. But the absorption which takes place in these stages and within the column causes internal temperatures to rise, so that the terminal trays are coolest. As a result, the equilibrium in the interior trays is shifted away from absorption. This situation can only be corrected with interstage cooling like that used in crude-oil distillation. If more than 6 to 10 stages are needed, this cooling must be provided.

Absorbers and most stripping columns have an overhead product in the vapor phase. Therefore, the properties of partial condensation as described in Chap. 3 apply. If the absorbent or stripper reflux temperature is variable as a function of coolant temperature or heat load, column pressure should be adjusted accordingly. The ideal program of pressure versus temperature would keep the K value of the key component constant. Then the concentration of that component in the off-gas is not likely to change with temperature.

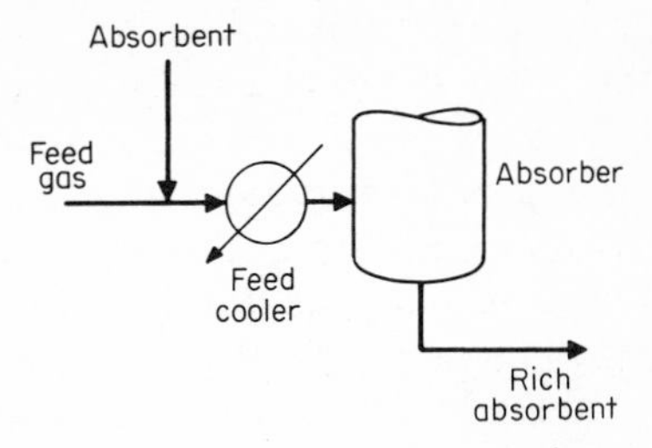

figure 4.3 *Mixing some absorbent with the feed entering the cooler can add another equilibrium stage and increase the capacity of the absorber.*

Preabsorption Another equilibrium stage can be added if part of the absorbent is mixed with the gas feed upstream of the feed cooler, as shown in Fig. 4.3. Absorbent used in this fashion is often called a "sponge oil." In absorbing some of the components from the vapor, it releases heat which is removed by the cooler. Additionally, the presence of the liquid in the cooler promotes heat transfer. Thus part of the absorption is already complete upstream of the absorber without raising feed temperature. Furthermore, both the liquid and the vapor loading in the absorber are reduced: vapor loading, in that some of the feed is already condensed; liquid loading, in that part of the absorbent bypasses the tower. This practice is especially helpful in increasing the capacity of an already heavily loaded absorber.

Properties of the Absorbent Absorbers have an additional degree of freedom which distillation columns do not—the choice of the absorbent. This choice is important even in nonselective systems. In these systems, *molar* concentrations of volatile components absorbed in a solvent are essentially independent of the properties of the solvent. However, Treybal [4] points out that their *weight* concentrations then vary with the *molecular weight* of the solvent. Consequently, heavier oils tend to recover fewer pounds or barrels of volatile hydrocarbons per pound or barrel of oil. They consequently require higher circulation rates and consume more energy than lighter solvents. The principal disadvantage of lighter absorbents is that their volatility tends to favor somewhat higher loss with the effluent vapors. Absorbents with a combination of low molecular weight and low volatility are thereby preferred. Water, for example, makes a good absorbent for some vapors.

To illustrate the effect of molecular weight, Table 4.1 gives the equilibrium concentrations of methane and *n*-butane in the vapor and liquid leaving the top tray of an absorber at 300 psia and 50°F. Weight percentages of both components in 500°F boiling-point oil (molecular weight 200) are compared with those achieved using heptane (molecular weight 100). Where heptane is used as the absorbent, the vapor leaving the top tray contains 0.2 mole percent heptane under the stated conditions. Loss of oil is much lower if entrainment is nil. However, the oil flow required to reach the equilibrium mole percent

TABLE 4.1 Equilibrium Concentrations of Vapor and Liquid at 300 psia and 50°F

		Mole percent		Weight percent	
	K	Vapor	Liquid	In oil	In heptane
Methane	8.0	97.9	12.2	1.3	2.4
n-Butane	0.1	2.1	21.0	8.2	15.0

given is almost twice as great as that required for heptane, which would result in proportionately heavier entrainment losses.

Reducing the molecular weight of the absorbent has the obvious effect of requiring a lower liquid–vapor mass ratio in the absorber. This increases the capacity of a tower for both vapor and molar liquid flow rates. Thus, towers that are capacity-limited can actually have their production rate *increased* while using a *lower* mass flow of absorbent just by reducing its molecular weight.

When absorption and stripping are combined, the stripping column also benefits by the reduced molecular weight of the absorbent. Both its higher vapor pressure and its lower flow rate reduce the energy required to heat it to the boiling point. By the same token, less sensible heat must be removed as the lean absorbent is returned to the absorber. If the absorber described in Table 4.1 operates in conjunction with a stripping column held at 100 psia, the 500°F oil would have to be heated to 650°F to boil whereas the heptane would boil at 360°F. The sensible heat added by the reboiler and removed by the cooler to return the absorbent to 50°F would be 3.5 times as great for the oil as for the heptane, assuming equal molar flow rates. More than just the sensible heat must be transferred, since the product must be first condensed and then vaporized. But the sensible heat represents a total loss, and its reduction can greatly improve the thermal efficiency of the unit.

Material Balances To examine the effect of the liquid-vapor ratio on product compositions of either an absorber or a simple stripping column, refer to Fig. 4.4. An overall mole balance is

$$V + L = G + B \tag{4.1}$$

where V and L are vapor and liquid feed rates and G and B are vapor and liquid product flows. In addition, a mole balance may be written for each component i:

$$Vz_i + Lw_i = Gy_i + Bx_i \tag{4.2}$$

where w, x, y, and z are mole fractions of component i in their respective streams. If solvent losses are negligible, then

$$B = \frac{Lw_a}{x_a} \tag{4.3}$$

where w_a and x_a are the mole fractions of absorbing component in the liquid feed and bottom product, respectively. When Eqs. (4.1) and (4.3) are substituted into (4.2), compositions of the products can be related to the L/V ratio:

$$\frac{L}{V} = \frac{x_a(z_i - y_i)}{w_a(x_i - y_i) - x_a(w_i - y_i)} \tag{4.4}$$

As may be perceived, variations in feed compositions z_i and w_i must be countered with appropriate changes in L/V if bottom composition is to remain constant.

Off-gas composition varies with K_i at the top, as well as L/V and w_i. Should K_i vary because of changes in the temperature-pressure relationship, L/V could be adjusted to control y_i. However, such an adjustment will affect x_i. In general, off-gas composition should be controlled by the top-tray temperature-pressure relationship, leaving L/V to determine bottom composition. These rules seem to apply equally to both the absorber and the gas stripping column. In the former case, absorbent flow would be manipulated to set the L/V ratio; in the latter case, the flow rate of the stripping gas would be manipulated.

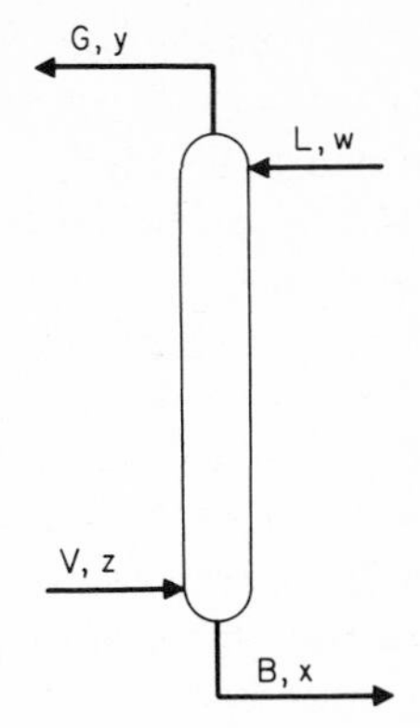

figure 4.4 *This representation may be used either for an absorber or a simple stripping column.*

Reboiled Stripping Columns The reboiled stripping column shown in Fig. 4.1 is identical in operation to a multicomponent distillation column with partial condensation as described in Chap. 3. Its function is to deliver a bottom product or lean absorbent with a controlled amount of volatiles while losing a minimum amount of absorbent overhead. Partial condensation is useful in the extra equilibrium stage it provides. In this regard, only enough vapor should be condensed to provide the reflux needed to control absorbent losses. As with the absorber, column pressure should be forced to follow reflux temperature variations to maintain a constant vapor-liquid equilibrium overhead.

Bottom temperature can be controlled to determine the initial boiling point of the bottom product. In a combined absorption-stripping system, it determines the composition of the absorbent and affects the losses of light components in the absorber off-gas. Increasing this temperature will reduce losses but will also increase the heat load on the stripper and cooling load on the absorber. If the column pressure is adjusted to follow top-temperature variations, pressure compensation must be applied to the bottom-temperature measurement to infer composition.

CHEMICAL ABSORPTION AND DESORPTION

This section considers selective absorption based on chemical as well as physical equilibria. For example, certain specific components may be removed from a gas stream by an absorbent having a chemical attraction for them. In this way a polar component may be selectively removed from a mixture of less polar gases by a polar solvent. A notable example is the selective absorption of carbon dioxide from a mixture of carbon monoxide, hydrogen, and nitrogen by a solution of an ethanolamine in water. The ethanolamines are water-soluble, volatile organic bases having an affinity for acid gases. Other acid gases such as sulfur dioxide or hydrogen sulfide would also be absorbed if present. The ethanolamine, being volatile, is recovered in a reboiled stripping column—the circuit is essentially that shown in Fig. 4.1. Reference 5 gives operating guides to setting flow rates, compositions, and temperatures.

Chemical Equilibria In the preceding example, absorption is selective whereas desorption is physical. The next group of systems to be studied are those whose equilibria are shifted between absorption and desorption by chemical means. These systems all feature acid-base reactions, where the absorbed or stripped components are acid or basic gases such as carbon dioxide or ammonia. The acid gases are absorbed by basic solutions and desorbed by the addition of a nonvolatile acid. Conversely, basic gases are absorbed by acids and stripped in the presence of a nonvolatile base.

Consider the problem of absorbing ammonia from a mixture of other gases, using water as the absorbent. The concentration of free NH_3 in the water solution is determined by the Henry's law constant at the prevailing temperature and the partial pressure of NH_3 in the gas. Although the absorption equilibrium is governed by Henry's law, a chemical equilibrium also exists between the NH_3 in solution and the NH_4^+ ion:

$$NH_3 + H_2O \overset{K_B}{\rightleftharpoons} NH_4^+ + OH^- \tag{4.5}$$

Because of this ionization, the total ammonia in the liquid phase will always exceed the free ammonia in equilibrium with the gas.

The ionization equilibrium is governed by an ionization constant K_B, defined as the ratio of the product of the ions to the free ammonia in solution:

$$K_B = \frac{[NH_4^+][OH^-]}{[NH_3]} \tag{4.6}$$

The brackets indicate an expression of concentration of the component within. The units customarily used are gram ions per liter of solution, known as normality (N).

Let the total ammonia in solution be identified as x_B:

$$x_B = [NH_3] + [NH_4^+] \tag{4.7}$$

Then, from (4.6),

$$x_B = [NH_3]\left(1 + \frac{K_B}{[OH^-]}\right) \tag{4.8}$$

In any aqueous system, the hydroxyl-ion concentration $[OH^-]$ is determinable as a function of the hydrogen-ion concentration $[H^+]$:

$$K_W = [H^+][OH^-] \tag{4.9}$$

Constant K_W is known as the ionization constant for water and is 10^{-14} at 25°C. Finally, the hydrogen-ion concentration is measurable as solution pH:

$$[H^+] = 10^{-pH} \tag{4.10}$$

Inserting the last two expressions into Eq. (4.8) yields x_B as a function of $[NH_3]$ and pH:

$$x_B = [NH_3]\,(1 + 10^{pK_W - pK_B - pH}) \tag{4.11}$$

Here pK_W and pK_B are defined as $-\log K_W$ and $-\log K_B$, respectively, to place them in the same terms as the measurable pH. For ammonia, pK_B is 4.75 and pK_W is 14 at 25°C.

pH Adjustment Note that when the exponent in Eq. (4.11) is zero, $x_B = 2[NH_3]$; that is, half the total ammonia is ionized. For ammonia at 25°C, this condition occurs at pH 9.25. Every unit increase in pH reduces the ammonium-ion concentration by one decade such that at pH 10.25, $x_B = 1.1[NH_3]$. Each unit decrease in pH increases the ion concentration one decade such that at pH 8.25, $x_B = 11[NH_3]$. Thus the equilibrium concentration of total ammonia in solution can be shifted by a factor of 10 by adjusting solution pH between 8.25 and 10.25. Because this relationship is extremely nonlinear, it is best presented in tabular form (see Table 4.2).

As a gas is absorbed, heat is evolved which raises the temperature of the liquid, thereby tending to impede further absorption. Similarly, as ammonia is absorbed the pH of the solution rises, also impeding further absorption. To maintain the pH at a favorable level, an acid reagent must be added to the solution at a rate equal to the rate of absorption.

Stripping involves shifting the equilibrium by raising temperature, lowering pressure, and, in the case of ammonia, elevating the pH with a nonvolatile base. The amount of base required to release the ammonia is equal to the amount of acid used to absorb it. Control over pH of the stripping operation is both more important and more difficult than for absorption. Table 4.2 shows that the pH during absorption, say 7 to 8, has a pronounced effect on ammonia concentration; conversely, variations in ammonia concentration and hence acid usage have relatively little effect on pH. During stripping, the pH must be controlled above 9 where the ammonia content is minimal and pH is much more sensitive to the flow of the base. Furthermore, insufficient base will leave residual ammonia and any excess is wasted.

The set of titration curves in Fig. 4.5 demonstrates this point. A 1.0 N solution of ammonia will have a pH of 11.62 at 25°C. Addition of acid to that solution will lower the pH along the $x_B = 1.0\ N$ curve by generating ammonium ions. To free the ammonia from the ionized state, caustic must be added

TABLE 4.2 Effect of pH on the Absorption of Ammonia at 25°C

pH	$x_B/[NH_3]$
7	178.8
8	18.78
9	2.79
10	1.179
11	1.018
12	1.002

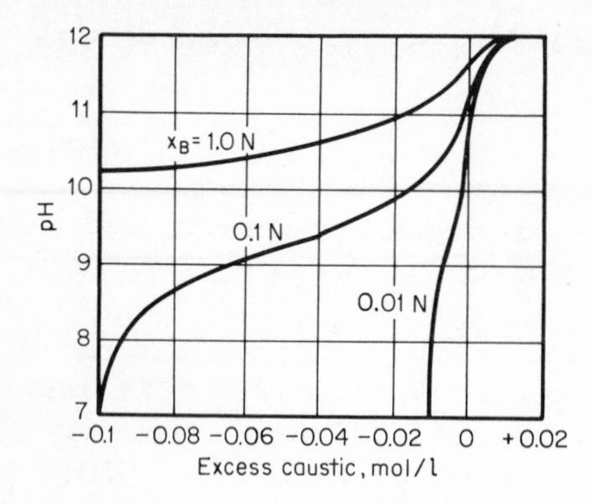

figure 4.5 *The pH should be increased above 11 to strip ammonia effectively.*

to return the pH to 11.62. Similarly, weaker solutions of ammonia absorbed in acid must have their pH adjusted to 11 or thereabouts to neutralize the acid and thereby free the ammonia. Observe that the slope of each curve is maximum at the point of zero excess caustic—the "end point." This confirms that pH is the most difficult to control at the end point. The slope is least at the half-neutralization point, where pH = 14 − pK_B.

The end point may be approximated as halfway between the half-neutralization point and the end of the pH scale. Ammonia is half-ionized at pH 9.25; the average of 9.25 and 14 is pH 11.625. Exact calculation of the end-point pH requires the trial-and-error solution of Eq. (4.12) for bases:

$$x_B = 10^{\mathrm{pH}-\mathrm{p}K_W}(1 + 10^{\mathrm{pH}-\mathrm{p}K_W+\mathrm{p}K_B}) \tag{4.12}$$

and Eq. (4.13) for acids:

$$x_A = 10^{-\mathrm{pH}}(1 + 10^{\mathrm{p}K_A-\mathrm{pH}}) \tag{4.13}$$

Note that the end point is also the pH of a solution of the gas in pure water.

The Acid Gases Absorption and stripping are commonly conducted on hydrogen chloride, chlorine, hydrogen fluoride, hydrogen sulfide, carbon dioxide, sulfur dioxide, and the nitric oxides. Of these, only hydrogen chloride completely ionizes in solution—ionization of the others varies with pH, as was the case with ammonia.

Hydrogen chloride absorbs so readily in water that a mole of water may be vaporized for every mole of HCl absorbed adiabatically. Hydrochloric acid forms a maximum-boiling azeotrope at about 20 wt percent HCl, the highest concentration obtainable adiabatically. To produce commercial 31 percent HCl, external cooling must be applied. Its great solubility is indicated by a partial pressure of HCl gas of only 0.02 psi in equilibrium with the 31 percent acid solution at 70°F. Under the same conditions, ammonia would generate a partial pressure of 5.8 psi.

Absorption of chlorine in water forms the weak hypochlorous acid as well as hydrochloric acid:

$$Cl_2 + H_2O \rightarrow HCl + HClO \tag{4.14}$$

Similarly, absorption of nitrogen dioxide yields strong nitric acid and weak nitrous acid:

$$NO_2 + H_2O \rightarrow HNO_3 + HNO_2 \tag{4.15}$$

The presence of the strong acid tends to depress the pH of these solutions and therefore reduce the solubility. Addition of caustic first neutralizes the strong acid, then the weak. Stripping requires the addition of enough acid to regenerate both the strong and the weak components of the gas. The end point is essentially the pH of the strong component.

Many of the acid gases ionize in two steps; for example,

$$SO_2 + H_2O \overset{K_1}{\rightleftharpoons} HSO_3^- + H^+ \tag{4.16}$$

$$HSO_3^- + H_2O \overset{K_2}{\rightleftharpoons} SO_3^{2-} + H^+ \tag{4.17}$$

Only the first step is important from an absorption-desorption point of view, in that it determines the concentration of gas dissolved but not ionized. A second end point appears midway between the two steps at a pH which is the average of the two pK values. At this point, half the acid is neutralized; i.e., step 1 is complete. Table 4.3 lists the pK values for the weak acids formed, along with their estimated end points. Because end-point pH varies with concentration, the values given in Table 4.3 are typical for the solubilities of the particular gas. Where a strong acid is formed, e.g., with chlorine and nitrogen dioxide, the end-point pH is determined entirely by the concentration of that acid.

Figure 4.6 describes the successive stripping of hydrogen sulfide and ammonia from a refinery sour-water stream. The water contains a nearly neutral solution of ammonium sulfide resulting from treating petroleum feedstocks. Since the effluent is aqueous, live steam may be used as a stripping medium—no reboiler is required. The pH of the stripped effluent may be high and therefore require adjustment before disposal. It cannot be recycled because it contains sodium sulfate, which will accumulate with reuse. But the sulfide and ammonia pollutants have been removed by the strippers and may be recovered for further processing. Titration curves and details on the measurement and control of pH for this application are given in Ref. 6.

Flue-gas Desulfurization Selective absorption between two acid gases is possible if their end points are far enough apart. A common problem in industry

TABLE 4.3 Ionization Constants and End Points of Various Acid Gases

Gas	Weak acid	pK	End-point pH
Cl_2	HClO	7.5	−log [HCl]
CO_2	CO_2	6.35, 10.25	4.0, 8.3
HF	HF	3.17	1.8
H_2S	H_2S	7.0, 12.9	4.0, 10.0
NO_2	HNO_2	3.2	−log [HNO_3]
SO_2	SO_2	1.8, 6.8	1.0, 4.3

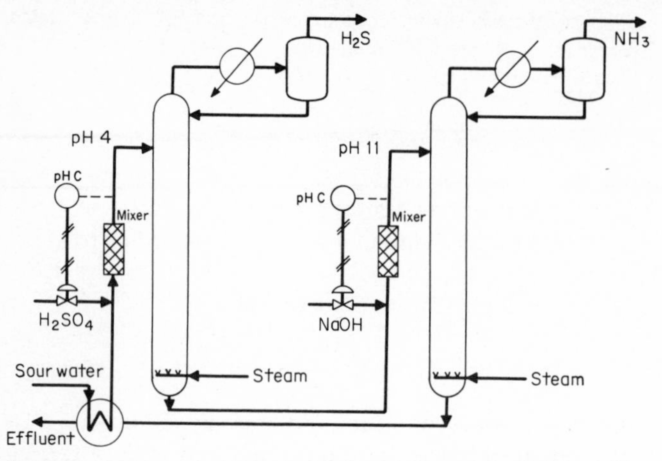

figure 4.6 *Refinery "sour water" may be stripped of hydrogen sulfide and ammonia successively by appropriate adjustment of its pH.*

is that of absorbing a few hundred ppm of SO_2 from a flue gas while not absorbing the 16 percent CO_2 present. Absorption of CO_2 consumes reagent unnecessarily, increases the sludge volume, and promotes scaling, all of which are undesirable.

Table 4.3 shows that the second end point for SO_2 and the first end point for CO_2 are nearly coincident in the pH 4 to 4.5 range. Therefore, in this range absorption of SO_2 is essentially complete while that of CO_2 is virtually nil. Figure 4.7 shows a titration curve of SO_2 with a basic solution of sodium sulfite in a partial pressure of 0.2 atm CO_2. The end point for SO_2 absorption appears to be pH 4.3. The rich absorbent in this sodium-based process is then reacted with lime to raise the pH and precipitate calcium sulfite. The clarified solution, which is principally sodium sulfite, is then recycled to the absorber.

Table 4.4 demonstrates how little CO_2 is absorbed in comparison to SO_2 at various pH values despite its high partial pressure. The flue-gas composition

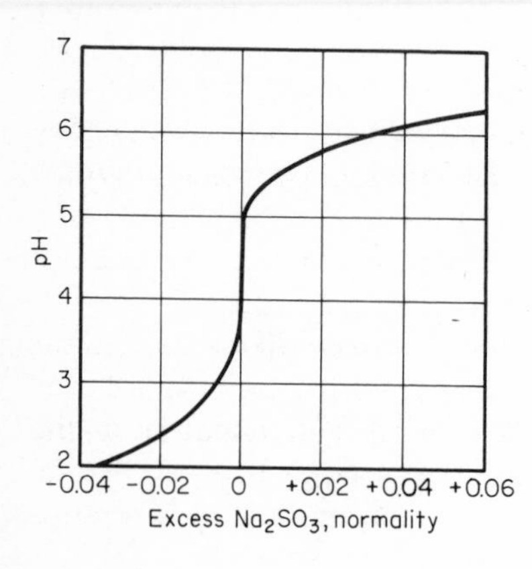

figure 4.7 *Sulfur dioxide absorption is essentially complete at pH 4.3.*

TABLE 4.4 Solution Concentrations in Equilibrium with Flue Gas Containing 200 ppm SO_2 and 16 Percent CO_2 at 120°F

pH	SO_2, mol/l	CO_2, mol/l
5	0.588	0.00328
6	6.70	0.00454
7	149.5	0.0172

is typical of scrubber effluents. Note that only a small fraction of the absorbed SO_2 (3.65×10^{-4} mol/l) exists as the dissolved gas under the above conditions—the balance is ionized. The concentration of absorbed CO_2 (3.14×10^{-3} mol/l) as the dissolved gas is actually greater, but these pH levels do not favor its ionization.

Absorbed gases can also be transferred to the solid state by precipitation as an insoluble product. If a calcium-based scrubbing medium is used (either lime or limestone), the calcium ions will combine with sulfite ions to form a precipitate:

$$Ca^{2+} + SO_3^{2-} \overset{K_s}{\rightleftharpoons} CaSO_3 \downarrow \tag{4.18}$$

Here K_s is known as the solubility-product constant and is defined as

$$K_s = [Ca^{2+}]\,[SO_3^{2-}] \tag{4.19}$$

Again, solution pH determines the absorption of SO_2 and its conversion to SO_3^{2-} ions. Consequently, this reaction is also characterized by a titration curve of pH versus lime addition similar in some respects to Fig. 4.7. However, CO_2 absorption interferes with the precipitation because it forms $CaCO_3$, whose solubility is 2.5 orders of magnitude lower than $CaSO_3$. In the presence of CO_2, or if limestone ($CaCO_3$) is used as a reagent, the solubility limit of $CaCO_3$ is reached at about pH 6.7—the pH cannot be increased above that point. Consequently it must be controlled at or below 6 in this system to avoid carbonate formation.

EXTRACTIVE DISTILLATION

Extractive distillation is a true distillation process which uses a solvent to improve the relative volatility of the components of interest. It can be used to separate homogeneous azeotropes or simply close-boiling components which do not form an azeotrope. The solvent is selected to have a particular affinity for one component or class of components. In this respect it acts much like a selective absorbent.

Two steps are required for complete separation of two components. In the the first, the extractant attaches itself to one component while the other is distilled. A second column then strips the captured component from the extractant. Consequently the extractive distillation process bears a distinct relationship to the conventional absorber-stripper combination. Extractive distillation is commonly used to separate close-boiling, nearly ideal mixtures of olefins and paraffins as well as less ideal mixtures such as acetone and methanol.

The Extractant Circuit The first column shown in Fig. 4.8 has neither sufficient trays nor reflux to make the required separation between the key components by conventional distillation. But the extractant introduced near the top preferentially lowers the vapor pressure of component B, sweeping it to the bottom of the column. The few trays above the point of extractant introduction are needed to separate it from component A. From the extractant entry point downward, the column performs essentially like a conventional unit except that the liquid rate tends to be much greater than the vapor rate.

Typically, very little of the extractant boils in the first column—most of the vapor is provided by component B. But since the vapor rate may be substantially greater than the flow of component B leaving the column, there is a sizable change in composition across the bottom tray. This can be demonstrated with a simple material balance, using Fig. 4.9. The flow of extractant L_E is essentially the same everywhere in the column. But the flow of component B in the liquid phase decreases from the first tray to the bottom-product stream by the vapor rate V. Thus the fraction of component B in the bottom product is

$$x_0 = \frac{B}{L_E + B} \tag{4.20}$$

whereas the composition leaving the first tray is

$$x_1 = \frac{B + V}{L_E + B + V} \tag{4.21}$$

Typically $B \ll V \ll L_E$, which makes $x_0 \ll x_1$. This step change in composition is indicated by a step change in temperature.

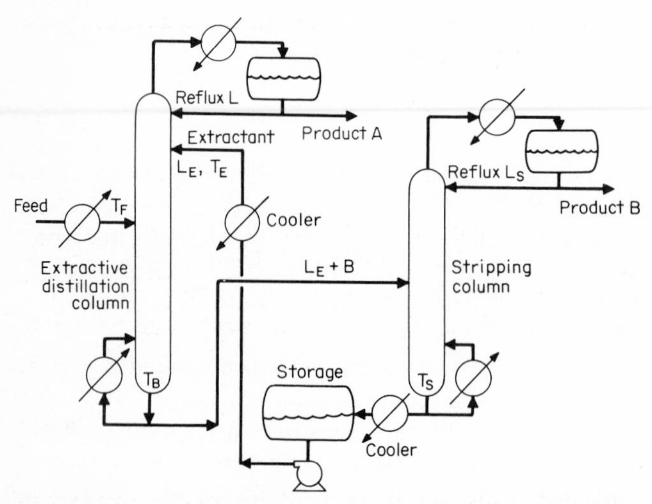

figure 4.8 *The extractant captures component B, which is then released in the stripping column.*

In its role of enhancing relative volatility, the extractant flow can be set in proportion to the feed or, more precisely, in proportion to the captured component in the feed. This will bring about a uniform concentration of extractant in the feed to the stripping column. Increasing the extractant flow has an effect similar to that of increasing reflux—by diluting component B further, it depresses its vapor pressure and improves separation.

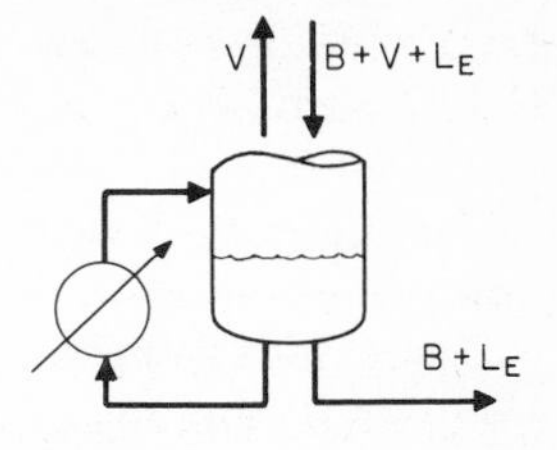

figure 4.9 *Since little of the extractant boils, the liquid composition changes dramatically across the bottom tray.*

The stripping column is essentially identical to those reboiled stripping columns operated in conjunction with absorbers. Its goal is to produce an extractant-free distillate while returning a minimum of component B to the extractive-distillation column. Relatively few trays are required when the boiling point of the extractant is much higher than the distillate. This is, in every respect, a conventional distillation process. The only unusual feature would be the sizable temperature gradient observable—bottom temperature will approach the boiling point of the extractant. The amount of component B recycled with the extractant influences the separation in the extractive column in a manner similar to variations in feed composition. Thus an increase in the concentration of component B in the extractant requires an increase in extractant flow to offset it.

Because the boiling point of the extractant tends to be considerably higher than that of the components to be separated, it is usually cooled in the recycle loop. The cooling could be eliminated if the towers were operated at different pressures. This is not usually possible, however, since the two components which are separated must both be condensed and their boiling points must be close together (or extractive distillation would not have been necessary).

The Extraction Mechanism The extractant acts to produce a nonideal liquid mixture. The relative volatility of a given liquid mixture may be expressed as a function of vapor pressures and activity coefficients γ:

$$\alpha_{AB} = \frac{\gamma_A p_A^\circ}{\gamma_B p_B^\circ} \tag{4.22}$$

For an ideal mixture, these activity coefficients are unity. When the components of a nearly ideal mixture such as n-butane and 1-butene have virtually the same vapor pressure, separation is difficult. Then an extractant is useful that can alter one activity coefficient relative to the other.

The activity coefficients for a nonideal liquid mixture can be approximated by van Laar's equation [7]:

$$\log \gamma_i = \frac{Z_{iE}}{[1 + (Z_{iE}x_i/Z_{Ei}x_E)]^2} \tag{4.23}$$

In Eq. (4.23) subscript E is used to identify the extractant and i may refer to either of the components to be separated; the coefficients Z_{iE} and Z_{Ei} are van

Laar's binary constants, some sets of which are given in Ref. 7. For the case of infinite dilution,

$$\log \frac{\gamma_A}{\gamma_B} = Z_{AE} - Z_{BE} \tag{4.24}$$

Using data from Ref. 7, activity coefficients and relative volatilities for the n-butane/1-butene system were evaluated with furfural as an extractant (see Table 4.5). Calculations were made for conditions at the extractant entry point (T = 150°F, $x_B \to 0$) and near the bottom of the column (T = 200°F, $x_A \to 0$).

The vapor pressure of 1-butene is actually greater than that of n-butane. In fact their ratio at 150°F is 0.853. But if sufficient furfural is present, it raises the activity coefficient of n-butane enough for it to become the more volatile component. The 2-butenes are slightly less volatile than n-butane and are therefore somewhat easier to separate in the extractive column. In a multicomponent mixture, n-butane and 1-butene are the keys, with isobutane lighter and the 2-butenes heavier than the keys.

Because the extractant concentration has such a strong influence over the relative volatility, a certain concentration must be reached before the separation can be achieved at all. This is particularly true of the system illustrated here in that α crosses 1.0 by the addition of the extractant. Like the V/F ratio in a conventional column, maximum separation is achievable only with an infinite ratio of extractant to product.

Selecting the Extractant Van Winkle [8] has compiled lists of extractants effective in improving the relative volatility of several systems, including aromatic-paraffin and olefin-paraffin mixtures. Most of the solvents selective for aromatics are aromatic in nature, such as nitrobenzene, but some are alcohols such as propylene glycol. Many of the solvents selective for olefins contain nitrogen, such as nitriles, amines, and nitro-compounds; oxygenated organics including ketones, ethers, and alcohols are also effective. This reference compares relative volatilities of the hydrocarbons for certain solvent-dilution factors. For example, diluting an unspecified butane-butene mixture 3:1 by volume with furfural is listed as raising the relative volatility to 1.40 at 158°F.

Naturally, the extractant must be miscible with the components to be sep-

TABLE 4.5 Estimated Relative Volatilities of *n*-Butane/1-Butene Mixtures in Furfural

Furfural-hydrocarbon mole ratio	T, °F	n-Butane		1-Butene		Relative volatility,
		x_A	γ_A	x_B	γ_B	α_{AB}
∞	150	0	10.0	0	5.75	1.48
20	150	0.05	8.15	0	5.75	1.21
	200	0	8.12	0.05	4.43	1.56
10	150	0.10	6.70	0	5.75	0.99
	200	0	8.12	0.10	3.94	1.76

arated, in the proportions that exist in the column. If it is not, the extracted component is not in fact extracted, since it is able to exert its own vapor pressure. Water, for example, is capable of extracting olefins from paraffins, but its solubility in these hydrocarbons is so low that its effectiveness is severely limited.

It is possible to use an extractant that forms a high-boiling azeotrope with the extracted component. In fact, this approach would be quite effective, although it would present a problem in recovering the extractant. An extractant which forms a low-boiling azeotrope will pass overhead—this becomes a different type of distillation, which is covered in detail in the next chapter.

The volatility of the extractant is an important consideration. A very volatile extractant will require more trays above the entry point and a greater reflux in both the extractive column and the stripping column. However, increasing the difference between its boiling point and that of the product means that more sensible heat must be removed prior to recycling. This sensible heat is a net energy loss—it contributes nothing to the separation but still must be provided by the two reboilers.

Some practitioners have found that the volatility of an otherwise effective extractant can be adjusted by adding another component. A commonly used mixture is a solution of about 4 percent by weight of water in furfural; this is enough to depress the boiling point nearly to that of water. The amount which may be added is low because of a solubility limitation—the 4 percent solution is saturated with water at about 60°F.

Provision must then be made for withdrawing any excess water which may have been added above the solubility limit. Even within this limit the effect is considerable in reducing the sensible heat lost to cooling. However, the presence of the additional component alters the temperature-composition relationship in the bottom of the stripping column. The temperature control point there must then be adjusted as a function of the water content of the furfural if the desired concentration of the extracted component is to be maintained.

If a water layer is always maintained in the furfural storage tank, and if the temperature of the tank is controlled by the stripper bottoms cooler, a constant water content in the extractant can be attained. In this case further cooling prior to extraction cannot be applied because this would allow water to separate from the mixture.

The addition of water to the furfural apparently does not detract from its effectiveness as an extractant and in fact enhances it. Reference 8 indicates that dilution of a butane-butene mixture by a factor of 3.7 by volume with furfural containing 4 wt percent water improves the relative volatility to 1.78 at 128°F. By contrast, aniline does not seem to show an improvement in selectivity when a similar amount of water is added.

Heat Balances The relative volatility of the components to be separated in an extractive-distillation column is a function of their dilution by the extractant. If the feed to the column is a liquid, dilution by the extractant is less below the feed tray than above. This discontinuity can be avoided by completely vaporizing the feed in a preheater, which is common practice in these separations.

The energy which must be supplied to the extractive-column reboiler (Q_E) serves to vaporize the reflux L and increase the sensible heat of the extractant L_E and bottom product B. Credit is given for condensing component B from the feed:

$$Q_E = (L - B)H_V + L_E C_E (T_B - T_E) + BC_B(T_B - T_F) \tag{4.25}$$

Here H_V is the heat of vaporization of the components to be separated; C_E and C_B are the specific heat of extractant and bottom product and T_E and T_B are their temperatures; T_F is feed temperature.

In the stripping column, heat Q_s must be provided to vaporize reflux L_s and distillate B and to supply the sensible heat gained in raising the extractant flow to its boiling point T_s. Some heat is recovered in cooling product B to reflux temperature T_L:

$$Q_s = (L_s + B)H_V + L_E C_E (T_s - T_B) - BC_B(T_B - T_L) \tag{4.26}$$

Combining these two equations gives the total heat Q which must be supplied to the system. If T_L is close to T_F, the last terms in the two equations above will cancel, yielding

$$Q = (L + L_s)H_V + L_E C_E (T_s - T_E) \tag{4.27}$$

Separation between components A and B in the extractive column can be improved by

1. Increasing reflux flow
2. Increasing extractant flow
3. Reducing extractant temperature

The last variable has a pronounced effect on internal reflux and its distribution. If extractant temperature T_E is below the boiling point T of the components at the point of entry, some vapors will be condensed, increasing L below that tray by ΔL:

$$\Delta L = \frac{L_E C_E (T - T_E)}{H_V} \tag{4.28}$$

If extractant temperature exceeds the component boiling point, ΔL will be negative, indicating a *lower* reflux below the extractant feed tray. Since reducing T_E increases internal reflux, it tends to increase separation below that tray for a given external reflux.

The relationship between separation and heat input is different for each of the three variables listed above. The effect of the reflux–component B ratio on separation between the components is some function of the extractant concentration:

$$\frac{\partial S}{\partial L/B} = f\left(\frac{L_E}{L}\right) \tag{4.29}$$

The effect of extractant temperature is essentially the same in that it adjusts internal reflux:

$$\frac{\partial S}{\partial T_E} = -\frac{L_E C_E}{BH_V} \frac{\partial S}{\partial L/B} \tag{4.30}$$

The effect of extractant flow is conversely expected to depend on the L/B ratio:

$$\frac{\partial S}{\partial L_E/B} = f\left(\frac{L}{B}\right) \tag{4.31}$$

Note that these functions are highly nonlinear and interacting. From Table 4.5 it can be seen that no amount of reflux can make the required separation if insufficient extractant is present.

A choice must be made regarding which variable should be adjusted to control separation. It is not possible to base the selection on the partial derivatives given in the preceding equations because there is no common denominator. The common denominator that applies most universally is the energy input to the system. The expense in terms of energy requirements must be evaluated for the three manipulations. From (4.27),

$$\frac{\partial Q}{\partial L} = H_V \tag{4.32}$$

$$\frac{\partial Q}{\partial T_E} = -L_E C_E \tag{4.33}$$

$$\frac{\partial Q}{\partial L_E} = C_E(T_s - T_E) \tag{4.34}$$

The value of the three manipulations can then be determined by dividing the set of (4.29) to (4.31) by (4.32) to (4.34). As it happens, the improvement in separation attained for a given increment in heat input is the same when manipulating either L or T_E:

$$\left.\frac{\partial S}{\partial Q/B}\right|_{L_E} = \frac{\partial S}{\partial L/B} \frac{1}{H_V} \tag{4.35}$$

Actually, manipulating L would be slightly more beneficial than T_E in that it increases reflux above the point of extractant entry as well as below. For manipulation of extractant flow,

$$\left.\frac{\partial S}{\partial Q/B}\right|_{L,T_E} = \frac{\partial S}{\partial L_E/B} \frac{1}{C_E(T_s - T_E)} \tag{4.36}$$

The choice of the variable to manipulate for control of separation would fall on the one with the greater effect on separation per unit heat input as evaluated in (4.35) and (4.36).

If reflux manipulation is more efficient, reflux flow should be set at maximum and extractant temperature at minimum so that extractant flow is no greater than required to achieve the desired separation. Should extractant flow have the greater effect on separation per Btu, its flow and temperature should be high, with just enough reflux provided to keep the extractant out of the overhead product.

STRIPPING WITH STEAM

Open steam is used as a stripping or reboiling medium in two principal situations:

1. When the bottom product is water
2. When the volatiles are immiscible with water and a minimum stripping temperature is desired

The first situation was demonstrated in Fig. 4.6. Other familiar applications are the stripping of relatively small quantities of methanol, ethanol, or acetone from water. The principal advantage in using open steam is to eliminate the reboiler. Its only drawback is a dilution of the bottom product. If a feed is rich in methanol, for example, a proportionally large amount of steam would be necessary to distill it overhead. Then the small quantity of water in the feed would be augmented by the larger quantity of steam condensate. Since equilibrium in the column is on a mole-fraction basis, increasing the water flow in the bottom product would increase the rate of methanol loss.

A Single Liquid Phase Steam stripping of a water-immiscible liquid may be conducted with or without a liquid aqueous phase present. If no aqueous phase is present, the steam acts just like any stripping medium to reduce the partial pressure of the other components and thereby promote their vaporization. Following condensation, the water may be readily decanted from the other liquid phase. This is why steam is injected into the bottom of the atmospheric crude-oil tower. The crude oil has already been heated to its highest allowable temperature (~700°F). By diluting the vapors with steam, further vaporization is possible. No aqueous phase can exist since the oil temperature is well above the boiling point of water. In fact, the steam as introduced is below the oil temperature and this produces some cooling. The additional vaporization it brings also causes a reduction in temperature.

The vaporization achievable by steam injection is limited by two effects. First, cooling by the steam and the additional vaporization reduce the temperature of the liquid. Second, the loss of volatile components from the liquid lowers their partial pressure in the vapor. As a result, the amount of additional hydrocarbon vaporized per pound of steam falls off sharply as the steam rate is increased. Van Winkle [9] has plotted this relationship for several sets of conditions common in crude-oil fractionation.

Steam stripping will be more effective in recovering additional volatiles when its temperature is higher relative to the boiling point of the oil and when the boiling range of the oil is reduced. This is borne out by graphs in Ref. 9, which show more than twice the recovery for kerosene as for atmospheric bottoms, when stripped with 0.5 lb steam per gallon.

The molar ratio of steam w to oil o in the vapor is the ratio of their partial pressures:

$$\frac{y_w}{y_o} = \frac{p_w}{p_o} \tag{4.37}$$

Since the oil vapor is in equilibrium with its liquid, its partial pressure is also its vapor pressure:

$$\frac{y_w}{y_o} = \frac{p_w}{p_o^\circ} \tag{4.38}$$

Two Liquid Phases The partial pressure of the steam cannot be increased indefinitely. When it reaches its vapor pressure at the temperature of the system, water will condense, forming a second liquid phase. This will happen in the overhead condenser for the crude-oil column, and it could happen on some of the top trays as well.

When two liquid phases exist in equilibrium with a vapor, an entirely new relationship appears: *each liquid phase exerts its own vapor pressure independent of the other*. This situation can develop whenever steam is used to distill an immiscible liquid, depending on how much steam is used in comparison to the externally applied heat. It is also a property of heterogeneous azeotropes, which are discussed in detail in the next chapter. Water is not an essential ingredient in these systems—any two volatile immiscible liquids will do. However, water is the most common liquid and the condensate of our most universal heating medium, steam.

Open steam is used to distill certain heat-sensitive products at temperatures well below their normal boiling points without using vacuum. For example, turpentine, whose atmospheric boiling point is 309°F, may be distilled at 204°F by open steam.

The theoretical limits of the molar ratio of steam to oil in the vapor is reached when an aqueous phase exists. The partial pressure of the steam cannot exceed the vapor pressure of water at the existing temperature. Then

$$\frac{y_w}{y_o} = \frac{p_w}{p_o^\circ} \leqslant \frac{p_w^\circ}{p_o^\circ} \tag{4.39}$$

Nonvolatiles in the oil phase dilute the volatile component, thereby reducing its vapor pressure. If the concentration of the volatile component in the oil phase is x_o, then from Raoult's law

$$p_o = x_o p_o^\circ \tag{4.40}$$

This alters (4.39) to read:

$$\frac{y_w}{y_o} = \frac{p_w}{x_o p_o^\circ} \leqslant \frac{p_w^\circ}{x_o p_o^\circ} \tag{4.41}$$

The mole ratios described by Eqs. (4.39) and (4.41) represent the minimum amount of steam capable of vaporizing an amount of oil. Ellerbe [10] notes that the vaporization efficiency of open steam is much less than 100 percent. The efficiency limit is essentially the sort that impedes establishment of vapor-liquid equilibria in general—insufficient vapor-liquid contact. Consequently only 60 to 90 percent of the oil theoretically capable of being distilled per pound of steam may in fact be distilled. Fortunately, the low molecular weight of water compared with most distilled oils still results in a reasonably high yield in pounds of product per pound of steam.

The phenomenon of both liquid phases exerting their own vapor pressure has caused strange and surprising behavior in some columns. Occasionally, a component boiling between the light and heavy keys in a column will accumulate to a concentration sufficient to exceed its liquid solubility in the other components. When the second liquid phase forms, the vapor pressure of the mixture virtually doubles. At this point the rate of boiling increases sharply, which may cause the second phase to overflow to the next lower tray where the incident is repeated. The energy spent in this additional boiling lowers the temperatures of the affected trays substantially. If a temperature control point is located there, the controller may increase reboiler heat input drastically in an effort to restore the original conditions. In so doing, the heavy components in *both* liquid phases will be carried overhead, thereby contaminating the distillate product. Eventually the mid-boiling component will be stripped overhead and normal conditions may return—until it once again accumulates to the solubility limit. The only solution to this problem is to provide a side draw for removing the troublesome component.

Heat Balances Steam distillation will always use more energy than conventional distillation by the amount of steam which must be condensed with the distillate. As more external heat is applied, less open steam is needed. The minimum vaporization temperature will be maintained as long as an aqueous liquid phase exists. Above this point no open steam is condensed. Therefore, at this point all the heat required to vaporize the oil must come from the external heat source and whatever superheat may be available in the open steam. The superheat in the open steam may be considerable if the boiling point of the oil is low. But in situations where the oil is composed of high-boiling components such as fatty acids, high-pressure steam is necessary and superheat may be an unaffordable luxury.

Note that the higher the boiling point of the oil, the more steam will be required to distill it. In the case of turpentine, the maximum amount that may be distilled per pound of water is 1.33 lb if two liquid phases exist [10]. Roughly 0.17 lb additional steam will condense in vaporizing the turpentine. So in this example more than *six* times as much steam is used as is actually necessary to distill the turpentine. If the 0.17 lb of steam were used in a reboiler, the total heat load would not change, because an aqueous phase would still exist. As the reboil heat increases, so that the boiling point rises above that of water, there will be a single liquid phase. Then the boiling point will vary with the amount of open steam used per pound of turpentine vaporized by the reboiler.

Since under these conditions no open steam condenses, the weight of oil vaporized W_o is determined solely by reboiler steam W_r:

$$W_o = \frac{W_r H_w}{H_V} \tag{4.42}$$

Here H_w is the heat given up per pound of reboiler steam. The mole ratio of water to oil in the vapor is then related to the weight ratio of stripping to reboiler steam:

$$\frac{y_w}{y_o} = \frac{W_w/M_w}{W_o/M_o} = \frac{W_w}{W_r}\frac{M_o}{M_w}\frac{H_w}{H_V} \tag{4.43}$$

Here M represents the molecular weight of the species indicated by the subscript. There is really no optimum ratio in this problem. As with the crude tower, as much reboil heat should be applied as possible—to whatever temperature limit applies to the product. Then vaporization at that temperature is made possible by injecting the appropriate amount of open steam.

REFERENCES

1. Van Winkle, M.: "Distillation," pp. 46–126, McGraw-Hill Book Company, New York, 1976.
2. "Technical Data Book—Petroleum Refining," 2d ed., chap. 8, American Petroleum Institute, Washington, D.C., 1970.
3. Owens, W. R., and R. N. Maddox, Short-cut Absorber Calculations, *Ind. Eng. Chem.*, December 1968.
4. Treybal, R. E.: "Mass-Transfer Operations," p. 207, McGraw-Hill Book Company, New York, 1955.
5. Sisson, B.: How to Determine MEA Circulation Rates, *Chem, Eng. (N.Y.)*, Nov. 26, 1973.
6. Shinskey, F. G.: "pH and pIon Control in Process and Waste Streams," pp. 67–69, Interscience-Wiley, New York, 1973.
7. Perry, J. H.: "Chemical Engineers' Handbook," 4th ed., pp. 13-6–13-7, McGraw-Hill Book Company, New York, 1963.
8. Van Winkle, M.: *op. cit.*, pp. 464–469.
9. *Ibid.*, p. 359.
10. Ellerbe, R. W.: Steam-Distillation Basics, *Chem. Eng. (N.Y.)*, Mar. 4, 1974.

CHAPTER FIVE

Azeotropic Distillation

An azeotrope is a mixture of two or more volatile components having identical vapor and liquid compositions at equilibrium. Its composition, therefore, does not change as distillation proceeds—it is a constant-boiling-point mixture. This property precludes its being separated by simple distillation. The azeotropic mixture could boil at a higher or lower temperature than its components, depending on the nature of the system. Many acids such as hydrochloric and nitric form maximum-boiling-point mixtures with water. However, minimum-boiling azeotropes are far more common and are used as examples throughout this chapter.

If one attempts to distill a mixture of two components forming an azeotrope, only one of the components can be removed, the azeotrope becoming the other product. Since the azeotrope cannot be separated by conventional distillation, other methods, such as extraction, must be combined with distillation. In every case, a column is required for *each* pure component to be recovered. Thus to separate a binary azeotrope requires two columns.

In addition to the boiling-point classification, azeotropes may be either homogeneous or heterogeneous. Heterogeneous azeotropes separate into two liquid phases when condensed from the vapor. Homogeneous azeotropes do not, which makes their separation much more difficult. Nonetheless, several methods have been developed to bring about their separation. One method for separating homogeneous azeotropes has already been covered—extractive distillation. These specialized distillation processes are not reserved for azeotropes: they may be applied to any mixture that is close boiling. Whether a mixture is simply extremely difficult to fractionate (as butane-butene) or impossible (as the methanol-acetone azeotrope), extractive distillation is effective. Another classic is the acetic acid–water system—it does not form an azeotrope but is so difficult to separate that it is treated as such. And, as will be seen, a true azeotrope is used to force the separation.

There seem to be so many ways to separate these nonideal mixtures that not all can be covered in a single chapter. But the more common methods are described, each illustrated by an example from industry.

BINARY HETEROGENEOUS AZEOTROPES

The heterogeneous azeotrope can actually be easier to separate than many more ideal mixtures. The very immiscibility of the liquid phases condensed from the vapor effectively breaks the azeotrope. This allows complete separation of the components in two columns or partial separation in one column. Quite pure products may be made with comparatively few trays and little energy, owing to the limited solubility of the two liquid phases. This property is even used to advantage to assist the separation of other close-boiling mixtures and homogeneous azeotropes. The approach used is to add a third component forming a low-boiling heterogeneous azeotrope with one of the other components. But before ternary systems like this can be presented, a simple binary heterogeneous system must be examined.

The Furfural-Water System Furfural and water form a binary heterogeneous azeotrope described by the phase diagram given in Fig. 5.1. Any liquid mixture in the center of the diagram will separate into two phases whose compositions are represented by the nearly vertical curves. Reducing the temperature decreases the mutual solubility of the components, enhancing their separation.

The lighter (aqueous) layer may be separated by fractional distillation into water and vapor of the azeotropic composition. Although the difference between the boiling points of these two products is small (212 versus 208°F), their relative volatility is substantial. Vapor-liquid equilibrium data taken from Ref. 1 indicate that the relative volatility between furfural and water changes from 7.5 at 0 percent furfural to about 2.5 at the lower solubility limit (18.4 wt percent at 208°F). At the upper solubility limit (84.1 wt percent at 208°F), the relative volatility of furfural to water drops to about 0.1. Here furfural becomes less volatile and is even easier to separate from the azeotrope than water is.

If the pressure is elevated sufficiently to raise the boiling point of the azeotrope above 250°F, a single phase will condense from the vapor. At this point, the azeotrope becomes homogeneous. To take advantage of the separation afforded by the limited solubility of the two liquids, the condensate should be cooled to as low a temperature as is reasonable.

Stripping with Decanter Feed A single column is used when only one pure product is required. This would be the case when excess water is to be removed from a lean mixture while minimizing the loss of furfural. The water product should be virtually pure; the furfural product can be no more concentrated than the upper solubility limit at whatever solution temperature exists. Figure 5.2 illustrates the stripping system with decanter and column feed points indicated.

Consider first the lean furfural feeding the decanter as an aqueous phase saturated with furfural at or above the decanter temperature. In passing through the decanter, it will give up some furfural, leaving it saturated at the decanter temperature. The saturated aqueous phase entering the top of the column will yield a vapor close to the azeotropic composition. Upon condensing, a furfural-rich layer will form, to be withdrawn as product, along with a water-rich layer recycled with the lean feed. Bottom product from the stripping column will be water that is virtually free of furfural. Open steam is used for stripping.

This system is interesting to study because of the intertwining of the energy and material balances. This property appears in all examples of azeotropic distillation, but furfural stripping is perhaps the simplest illustration. The symbols given in Fig. 5.2 will be used for flow rates and weight fractions of furfural in the following analysis.

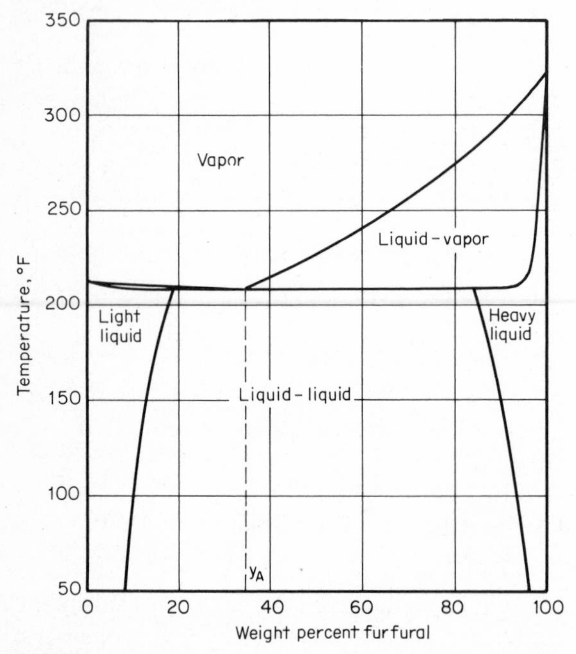

figure 5.1 *The phase diagram for the furfural-water system at atmospheric pressure.* (From G. H. Mains, The System Furfural-Water, Chem. Metall. Eng., April 26, 1922. By permission.)

Consider first the mass balance for the decanter:

$$F + V = D + L \tag{5.1}$$

Next consider the system balance for furfural only, assuming that x_B approaches zero and is therefore negligible:

$$Fz = Dx_D \tag{5.2}$$

A similar furfural balance may be written for the stripper only, using the same assumption:

$$Lw = Vy \tag{5.3}$$

Solving Eq. (5.2) for D and (5.3) for L, and substituting for D and L in (5.1), yields a solution for the boilup-feed ratio:

$$\frac{V}{F} = \frac{w(x_D - z)}{x_D(y - w)} \tag{5.4}$$

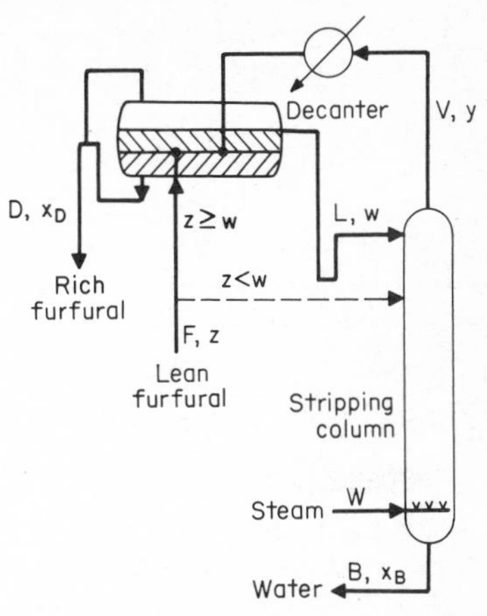

figure 5.2 *A stripping system for concentrating furfural may be fed either at the decanter or at the column, depending on feed composition.*

For control purposes, it is desirable to set boilup in ratio to the feed. However, Eq. (5.4) indicates that this ratio varies with feed composition z, which may not be measurable. Instead, boilup could be set proportional to reflux L, eliminating this problem:

$$\frac{V}{L} = \frac{w}{y} \tag{5.5}$$

Reflux composition w can be controlled by decanter temperature, or inferred from it, if a minimum rather than a controlled temperature is desirable.

Solving (5.4) for y indicates how it is affected by variations in the V/F ratio:

$$y = w + \frac{w(x_D - z)}{x_D(V/F)} \tag{5.6}$$

Increasing V/F reduces y through the action of the material balance. Yet increasing V/F for a column is known to improve separation, driving y and x_B farther apart. The combination of these two effects is to reduce x_B more than y. Although x_B was neglected as being insignificant compared to the other terms in the material balance, it is nonetheless the primary controlled variable. Although small, x_B is still a finite, measurable number (typically about 0.02 wt percent) and is controlled by manipulation of boilup. By contrast, y does not need to be controlled since it is not a final product, and any value between y_A and w will produce a furfural-rich product. Product concentration x_D is determined solely by decanter temperature.

The relationship between L/V and V/F can be seen by solving Eq. (5.2) for D and substituting it into (5.1):

$$\frac{L}{F} = 1 - \frac{z}{x_D} + \frac{V}{F} \tag{5.7}$$

As can be seen, L/F varies with V/F on a 1:1 basis but is higher for all reasonable operating conditions. Thus increasing V/F will cause L/V to decrease:

$$\frac{L}{V} = 1 + \frac{1 - z/x_D}{V/F} \tag{5.8}$$

If an operator wishes to decrease the L/V ratio by increasing V, he will find that L will subsequently increase if it is manipulated by a level controller or simply overflows from the decanter. When a new steady state is reached, L/V will be less than during the original steady state but not as low as immediately following the increase in V.

example 5.1

For the furfural stripping column, calculate V/F, D/F, L/F, and L/V on a mass basis for $z = w = 0.09$ wt frac. (at 100°F), $x_D = 0.94$, and $y = 0.25$ wt frac.

$$\frac{V}{F} = \frac{0.09(0.94 - 0.09)}{0.94(0.25 - 0.09)} = 0.509$$

$$\frac{D}{F} = \frac{0.09}{0.94} = 0.096$$

$$\frac{L}{F} = 1 - \frac{0.09}{0.94} + 0.509 = 1.413$$

$$\frac{L}{V} = \frac{1.413}{0.509} = 2.775$$

In the case of the furfural stripping column, bottom temperature is not a satisfactory measure of composition: there is at most 4°F difference in boiling points between the overhead and bottom product. But if L/V is held constant and decanter temperature is controlled, x_B should not vary appreciably.

Stripping with Column Feed Figure 5.2 indicates that the lean furfural may be fed either to the column or the decanter. In conventional distillation, feed is not blended with the reflux because it will dilute the reflux and thereby the overhead product as well. In this heterogeneous system, feeding the decanter has no effect on reflux composition (which is determined by temperature). Instead, feed rate and its composition affect only reflux flow. A leaner feed will increase reflux flow, requiring more boilup to hold L/V constant.

A feed which is significantly leaner than the reflux ought to be introduced at the appropriate tray in the column. To illustrate this point, consider a decanter mass balance with feed entering the column:

$$V = D + L \tag{5.9}$$

Next consider a furfural balance on the column only, assuming that x_B approaches zero as before:

$$Vy = Fz + Lw \tag{5.10}$$

Solving (5.9) for L and substituting into (5.10), while substituting for D from (5.2), yields

$$\frac{V}{F} = \frac{z(x_D - w)}{x_D(y - w)} \tag{5.11}$$

Note that this expression is almost (but not quite) identical to (5.4). The difference is the transposed locations of z and w in the numerator. For the case of $z = w$ as in Example 5.1, both will give the same results. But as z approaches zero, Eq. (5.4) approaches a maximum while (5.11) approaches zero.

The explanation for this behavior is related to the reflux-dilution problem. Feed entering the decanter augments the reflux, whose fixed composition increases the furfural entering the column. As the feed grows leaner, reflux flow increases, carrying more furfural to the column. All this furfural must be driven overhead to maintain a uniform bottom product.

When a lean feed enters the column directly, only enough boilup must be provided to drive overhead the furfural it contains, with a proportionate amount for that in the reflux. A leaner feed requires less boilup and hence less reflux. To demonstrate, Eqs. (5.9), (5.10), and (5.2) may be solved for L/F:

$$\frac{L}{F} = \frac{z(x_D - y)}{x_D(y - w)} \tag{5.12}$$

As z approaches zero, L/F approaches zero. But when feed is introduced to the decanter, L/F approaches maximum when z approaches zero.

example 5.2

Compare V/F for decanter feed and column feed when $z = 0.05$, $w = 0.09$, $x_D = 0.94$, and $y = 0.25$.

For decanter feed:

$$\frac{V}{F} = \frac{0.09(0.94 - 0.05)}{0.94(0.25 - 0.09)} = 0.532$$

For column feed:

$$\frac{V}{F} = \frac{0.05(0.94 - 0.09)}{0.94(0.25 - 0.09)} = 0.283$$

Controlling the stripping column when feed is introduced to it rather than to the decanter presents more of a problem because boilup cannot be set in ratio to reflux. Without boilup, there is no reflux. Therefore, boilup must be set in proportion to feed rate and composition as (5.11) indicates. This requires either a feed analyzer to measure z or a bottom-product analyzer to indicate when the V/F ratio is incorrect.

Two-column Operation To make a water-free furfural product requires a second column called a "dehydrator." The stripping column operates essentially as described except that its reflux is augmented by the aqueous portion of the dehydrator overhead stream. As Fig. 5.3 indicates, the two columns may use a common decanter and in fact a common condenser. The dehydrating column requires a reboiler to produce an essentially water-free product.

The material balances are easy to solve if the impurities in the two products are neglected. For feed introduced at the stripping column, a furfural mass balance yields

$$Fz + L_1 w_1 = V_1 y_1 \tag{5.13}$$

In the decanter, there is a net transfer of all the furfural in the feed from column 1 to column 2. Then

$$V_1 - L_1 = Fz \tag{5.14}$$

and

$$L_2 - V_2 = Fz \tag{5.15}$$

Finally, a furfural mass balance on the dehydrating column yields the bottom-product flow Fz as a function of boilup and reflux:

$$Fz = L_2 w_2 - V_2 y_2 \tag{5.16}$$

Equations (5.13) and (5.14) may be combined to give the V/F ratio for the stripping column:

$$\frac{V_1}{F} = z \frac{1 - w_1}{y_1 - w_1} \tag{5.17}$$

Observe that this equation is the same as (5.10) for $x_D = 1$. Combining (5.15) and (5.16) gives the V/F ratio for the dehydrating column:

$$\frac{V_2}{F} = z \frac{1 - w_2}{w_2 - y_2} \tag{5.18}$$

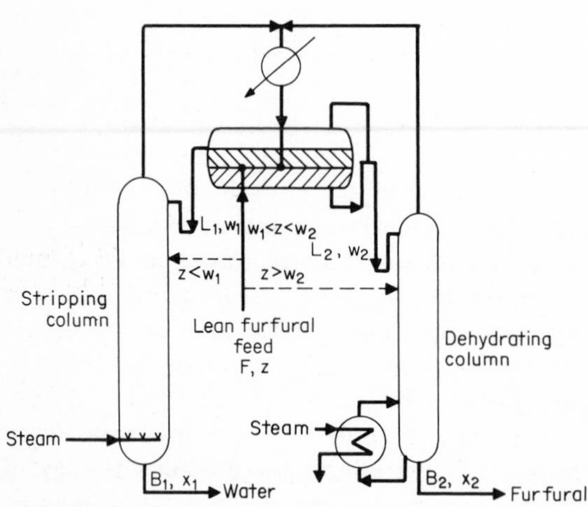

figure 5.3 *Two columns are required to separate a furfural-water mixture into its pure components.*

Because w_2 is much larger than w_1, V_2 can be much smaller than V_1.

example 5.3

Estimate the V/F ratios for a two-column system separating a feed of 5 wt percent furfural into two essentially pure products. Let $w_1 = 0.09$, $w_2 = 0.94$, $y_1 = 0.25$, and $y_2 = 0.60$.

$$\frac{V_1}{F} = 0.05 \frac{1 - 0.09}{0.25 - 0.09} = 0.284$$

$$\frac{V_2}{F} = 0.05 \frac{1 - 0.94}{0.94 - 0.60} = 0.00882$$

Note that V_2/F appears unreasonably small. Actually, however, the feed to the dehydrator is Fz rather than F, so that V_2/Fz would be a more appropriate index of heat duty. For this cited example, $V_2/Fz = 0.176$, which is comparable to V_1/F.

Whether the low V_2/Fz estimated for the dehydrator is sufficient to produce an acceptably pure bottom stream depends on the relative volatility and the number of trays. As with the furfural stripping column described earlier, increasing boilup will drive y_2 and x_{B2} apart by improving separation and will raise y_2 through the action of the material balance. The latter effect can be seen by rearranging (5.18):

$$y_2 = w_2 - \frac{z(1 - w_2)}{V_2/F} \tag{5.19}$$

As a consequence, increasing V_2/F has more net effect on bottom-product quality than on overhead composition. Furthermore, the actual value of overhead composition achieved at a given boilup rate is inconsequential since it is not a product.

The phase diagram indicated a wide boiling-point range for furfural-rich mixtures. Consequently, column temperature is a sensitive and reliable index of composition and is used successfully to control furfural quality by manipulating heat input.

Energy Balances The separation in the decanter is not achieved without cost. Sensible heat must be removed from the condensate to enhance separation and must in turn be supplied by the heat input. In the case of the stripping column, condensing the azeotrope without subcooling (at atmospheric pressure) will produce liquid-phase compositions of about 20 and 80 wt percent furfural. Purities may be doubled by cooling to 140°F; in fact further cooling is usually carried out simply because of availability of cooling media in the ambient range.

An estimate using compositions given in Example 5.1 indicates that subcooling the condensate to 100°F increases the energy required per pound of feed by nearly 40 percent. Because of the relatively large energy loss associated with this subcooling, its value in reducing the solubility of the components is questionable except for the case where a specified furfural purity (for example, 90 percent) is required with a single column. Subcooling may not be justified at all in two-column operation.

If no subcooling is provided in the two-column system, w_1 will increase, raising y_1. The material balance (5.17) shows that V_1/F will have to increase as w_1 approaches y_1 in the denominator. In addition, V_1/F must increase to maintain control of x_{B1} in the face of a rising y_1. These increases may offset the loss in sensible heat incurred by subcooling. Each particular heterogeneous azeotropic separation should be examined, either in the design stage or in actual operation, to determine whether subcooling raises or lowers the energy required to make acceptable products. An optimum decanter temperature may exist in many systems.

PRESSURE-SENSITIVE HOMOGENEOUS AZEOTROPES

Homogeneous azeotropes lack the fortunate property of liquid-phase separation. If they are to be separated, then, a method must be used to alter somehow the properties of the azeotrope. There are four principal methods presented in this book:

1. Extractive distillation
2. Pressure adjustment
3. Formation of a heterogeneous ternary azeotrope
4. Formation of a heterogeneous binary azeotrope

Of these, extractive distillation was described in the previous chapter. The balance of this chapter is devoted to the other three methods.

A pressure-sensitive azeotrope exhibits a shift in composition of its constant-boiling-point mixture as total pressure is changed. Therefore, the azeotrope formed by distillation at one pressure can be further fractionated at another pressure.

Operation at Two Pressures Tetrahydrofuran (THF) forms a homogeneous, minimum-boiling-point azeotrope with water. Lean mixtures may be separated by distillation at atmospheric pressure into water and the azeotrope, approximately 95 wt percent THF. This mixture may then be fractionated in a second column at 150 psig into essentially pure THF bottom product and an overhead vapor approaching the azeotropic mixture of approximately 88 wt percent THF. As with other azeotropic separations, this mixture may be recycled to the first column for reconcentration to 95 percent. The flowsheet is shown in Fig. 5.4.

The material balances for the two columns are very sensitive to the actual overhead compositions of each. The equations developed below assume that virtually pure products leave the bottom of each column. Symbols for the streams are given in Fig. 5.4, with all flow rates in mass units and compositions in weight fraction THF.

All the THF leaving the first column comes from the feed and the recycle:

$$D_1 y_1 = Fz + D_2 y_2 \tag{5.20}$$

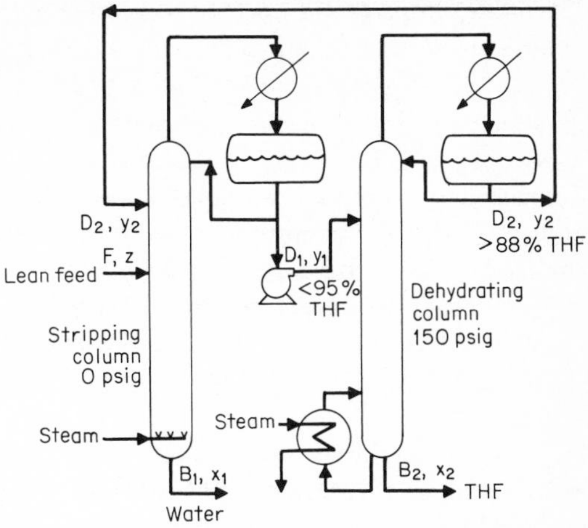

figure 5.4 *Separating pressure-sensitive azeotropes requires two columns with a recycle stream.*

The second column discharges a bottom product at a rate Fz as the difference between feed D_1 and distillate D_2:

$$Fz = D_1 - D_2 \tag{5.21}$$

Substituting (5.21) for Fz in (5.20) yields the ratio of the two distillate flows:

$$\frac{D_2}{D_1} = \frac{1 - y_1}{1 - y_2} \tag{5.22}$$

In the same manner, either distillate flow may be eliminated to yield a solution for the other:

$$\frac{D_1}{F} = z \frac{1 - y_2}{y_1 - y_2} \tag{5.23}$$

$$\frac{D_2}{F} = z \frac{1 - y_1}{y_1 - y_2} \tag{5.24}$$

Observe how the recycle flow increases as y_2 approaches y_1.

example 5.4

Given a feed containing 60 wt percent THF, with overhead compositions of 95 and 88 percent, calculate the three flow ratios listed above.

$$\frac{D_2}{D_1} = \frac{1 - 0.95}{1 - 0.88} = 0.417$$

$$\frac{D_1}{F} = 0.6 \frac{1 - 0.88}{0.95 - 0.88} = 1.029$$

$$\frac{D_2}{F} = 0.6 \frac{1 - 0.95}{0.95 - 0.88} = 0.429$$

Energy Requirements Each column is essentially a binary distillation where the relationships given in Chap. 2 apply. For example, the feed to column 2,

D_1, must command a certain energy input to achieve separation into a controlled bottom product and the recycled distillate. Decreasing column 2 boilup will tend to reduce the purity of the THF product and increase y_2. For an increase of y_2 from 88 to 90 percent, D_1/F will increase from 1.029 to 1.20 according to Eq. (5.23). This increase in column 2 feed will tend further to degrade separation by reducing the boilup-feed ratio. The recycle loop causes column 2 to be especially sensitive to variations in boilup by reinforcing disturbances through positive feedback.

Feed to column 1 is the sum of $D_2 + F$:

$$D_2 + F = F\left(1 + z\,\frac{1 - y_1}{y_1 - y_2}\right) \tag{5.25}$$

In this column also, separation determines the relative purities of product water and distillate as a function of boilup-feed ratio. An increase in y_2 caused by a reduction in column 2 boilup increases the feed to column 1 and therefore affects the separation there.

example 5.5

Estimate the effect of an increase in y_2 from 88 to 90 percent on total feed to column 1.

For $y_2 = 0.88$:

$$D_2 + F = 0.429F + F = 1.429F$$

For $y_2 = 0.90$:

$$\frac{D_2}{F} = 0.06\frac{1 - 0.95}{0.95 - 0.90} = 0.6$$

$$D_2 + F = 0.6\,F + F = 1.6\,F$$

The increase in D_2 caused by reducing y_1 can further increase D_1 and D_2. So the positive feedback is seen to penetrate both columns. Regulation can be provided, however, if both heat inputs are set in proportion to their respective feed rates. Then an increase in D_2 will not tend to degrade y_1, preventing the upset from continuing. Similarly an increase in D_1 can be countered by an equivalent increase in column 2 boilup, tending to regulate both y_2 and THF purity.

Energy Integration Figure 5.4 shows open steam being used to boil up the atmospheric column. As mentioned before, this is convenient when the bottom product is water since a reboiler is eliminated. But operating the second column at an elevated pressure presents the option of using its condenser to reboil the first column. This uses the energy fed to column 2 twice and eliminates control problems associated with trying to maintain the 150 psig pressure by conventional condenser cooling.

However, any energy integration adds a few difficulties, principal among them being limited flexibility. If column 2 boilup must be greater than column 1, additional condensation must be provided, preferably using a waste-heat boiler. Insufficient boilup for column 1 can be made up with open steam from a low-pressure source.

If column 2 pressure is to be constant even with variations in vapor loading, control must be provided. The cooling-source temperature (column 1 bottom product) is fixed at the atmospheric boiling point of water. As vapor loading increases, the temperature difference across the condenser-reboiler must increase proportionately. Pressure control can be achieved by throttling either the vapor entering or condensate leaving the condenser-reboiler.

The way energy integration couples the two columns may actually be beneficial in simplifying the controls required for this system. If the flow of high-pressure steam is set proportional to feed F, both columns will respond together to throughput variations. Overhead products do not need to meet any specifications, only the bottom products, so only two manipulated variables are required for their control.

TERNARY HETEROGENEOUS AZEOTROPES

Ternary azeotropes don't happen—they are made. The ternary systems most commonly encountered in industry are those formed to allow separation of binary homogeneous azeotropes. A volatile substance known as an "entrainer" is injected into the column containing the binary azeotrope to form a lower-boiling ternary mixture. Its principal effect is to remove one of the binary components overhead, allowing the other to be concentrated in the bottom of the column. However, the ternary azeotrope must be heterogeneous if the valuable product and the entrainer in the overhead vapor are to be recovered. As with the other applications, an example is used to illustrate the principle.

Separating Ethanol from Water with Benzene Ethanol and water form a binary homogeneous azeotrope consisting of 96 wt percent ethanol with an atmospheric boiling point of 173°F. When benzene is added to this mixture, a ternary azeotrope forms, containing about 74 wt percent benzene, 18.5 wt percent ethanol, and 7.5 wt percent water, boiling around 149°F, 29°F less than benzene. Because the ternary azeotrope is richer in water than the binary (on a benzene-free basis), all the water may be distilled overhead, leaving a dry ethanol product. A similar technique is used to manufacture propanols.

Figure 5.5 shows how two columns are combined to concentrate the ethanol in a lean feed. The stripping column yields a homogeneous distillate approaching the binary azeotrope in composition. This stream encounters a benzene-rich reflux in the dehydrating column, where the low-boiling ternary azeotrope is formed. The condensed vapor separates into two liquid phases, the lighter one rich in benzene, the heavy layer much less so. All the benzene-rich layer is returned as reflux. The heavy layer is a lean solution of ethanol in water, which is returned to the stripping column for reconcentration.

Whatever benzene is in this stream will be distilled overhead in the stripping column and thence returned to the dehydrator.

Two Liquid Phases on the Trays Note the substantial difference in water content between the dehydrator overhead vapor and reflux. This not only demonstrates the separation achieved in the decanter but also suggests the possibility of two liquid phases on the top trays of the column. This property did not appear with the binary heterogeneous azeotrope. Figure 5.3 shows the light and heavy layers being sent to different columns where they are further concentrated in their richer component. The liquid compositions in those two columns are always outside the two-phase range. Subcooling in the decanter ensures that condition by reducing the mutual solubility of the two reflux streams; their solubility then increases once equilibrium is restored in the column, so that no phase separation follows.

In the absence of benzene, the dehydrator would contain only an aqueous phase on all its trays. As benzene is added, the vapor pressure of the liquid on the trays will increase and the vapor composition will move in the direction of the ternary azeotrope. It is doubtful whether the ternary azeotropic composition can be reached with a single liquid phase on the top tray—two liquid phases must appear.

Separation will be maximized between the bottom product and the overhead vapor as the azeotropic composition is approached. It is therefore profitable to approach that composition insofar as practicable. However, the development of the two liquid phases does not seem to be stable. The formation of a second liquid phase on any given tray raises the vapor pressure, tending to boil away that phase. Additionally, conventional trays are designed for overflowing a single phase only. Should a heavier liquid phase form on a given tray, it tends to accumulate without overflowing until most of the lighter phase is displaced. A lighter phase forming is almost certain to be vaporized or overflow without accumulating. For a tray to behave as a true equilibrium stage

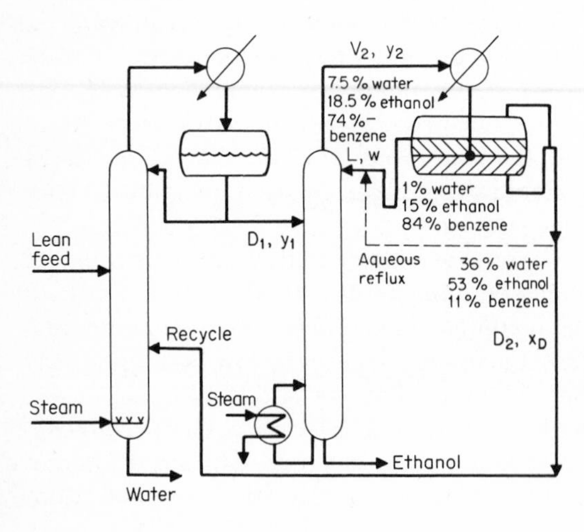

figure 5.5 *Benzene is used to entrain water, yielding an absolute alcohol product.*

with two liquid phases in contact with vapor, it must provide both overflow and underflow like a decanter.

Without the application of some external force, the trays tend to slip toward one liquid phase or the other, not both. If the inventory of benzene in the column is low, the aqueous phase will predominate and the relative volatility between overhead vapor and bottom product will be unfavorable. As additional benzene is introduced, the relative volatility improves. But benzene, being the lighter phase, tends to overflow down the column. If carried too far, it can again reduce the relative volatility and even contaminate the bottom product.

Provision is often made to reflux some of the aqueous distillate. Then two liquid phases can be ensured on the top tray at least. The control problem is to be able to maintain two phases over a certain group of trays. Temperature measurements are not particularly helpful in this regard because either single phase tends to boil at a higher temperature than the two-phase mixture. Liquid analyzers cannot be used for two-phase mixtures, and vapor analyzers like chromatographs cannot separate the azeotrope. Not even computer models can be relied upon to simulate column behavior in the region where two liquid phases exist. The only sure way to evaluate tray composition is to condense a representative sample of vapor at the tray temperature and measure the relative proportions of the resulting liquid phases.

In the ethanol dehydrating column, up to 10 percent of the reflux may come from the aqueous layer while the aqueous reflux-distillate ratio may be as high as 4.

Material and Energy Balances The two-column system shown in Fig. 5.5 is a combination of the heterogeneous two-column system with a recycle stream. Thus the material-balance equations derived for it are sensitive to the separation achieved in the decanter, but additional consideration must be given to the recycle as well.

Assuming that the amount of water leaving the bottom of the dehydrating column is negligible, all the water entering in the stripper feed must leave as the bottom product. Then all the water leaving the top of the stripping column comes from the recycle stream:

$$D_1 y_1 = D_2 x_D \tag{5.26}$$

Here y_1 and x_D are the water fractions of the two distillate streams.

An overall balance on the stripping column shows a net transfer of ethanol from the feed to the distillate:

$$D_1 - D_2 = F(1 - z) \tag{5.27}$$

Combining these two expressions yields the recycle requirements related to concentrations:

$$\frac{D_2}{F} = \frac{1 - z}{x_D/y_1 - 1} \tag{5.28}$$

A water balance may be written for the decanter,

$$V_2y_2 - L_2w = D_2x_D \tag{5.29}$$

as well as an overall balance,

$$V_2 - L_2 = D_2 \tag{5.30}$$

Combining these gives the vapor-distillate ratio for the dehydrator:

$$\frac{V_2}{D_2} = \frac{x_D - w}{y_2 - w} \tag{5.31}$$

Finally, (5.31) may be multiplied by (5.28) to eliminate D_2:

$$\frac{V_2}{F} = \frac{(1 - z)(x_D - w)}{(x_D/y_1 - 1)(y_2 - w)} \tag{5.32}$$

For a feed composed of 70 percent water, distillation using the compositions given in Fig. 5.5 requires a V_2/F ratio of 0.205. Again, the separation in the decanter is greatly beneficial. This process can be optimized by shifting the heating load between the columns until the total is minimized. There is no purity specification on the stripper distillate—reducing its water content will reduce the load on the dehydrating column and the recycle flow as well. But as y_1 approaches the azeotropic composition this benefit reaches its limit, while the V/F ratio in the stripper keeps increasing. In essence, the separation achievable in the dehydrating column is so much more efficient than that in the stripping column that y_1 should be considerably short of the azeotropic composition.

Optimization of the dehydrator involves finding the appropriate reflux composition which will minimize V_2/F by maximizing relative volatility. In the absence of reliable analyses of tray compositions, this too requires a trial-and-error approach. A certain ratio of aqueous to organic reflux will result in minimum V_2/F for a given decanter temperature. A more detailed analysis of this optimum-reflux-composition problem is developed for the acetic acid dehydrator in the next section.

TERNARY SYSTEMS CONTAINING BINARY AZEOTROPES

Actually, ternary mixtures containing binary azeotropes are much more common than ternary azeotropes. Several combinations are possible—homogeneous systems with one azeotrope, systems with one homogeneous and one heterogeneous azeotrope, etc. But the system described here is most common throughout the chemical, pharmaceutical, and synthetic fiber industries, where it is used in the concentration of organic acids.

These processes are typically carried out in three stages: liquid-liquid extraction, followed by azeotropic distillation, followed by solvent recovery. The process having the most data available is the production of glacial acetic acid using an acetate solvent. Actually, three different solvents are in common use—ethyl, isopropyl, and n-butyl acetate. Of the three, butyl acetate boils over both acetic acid and water and therefore presents some special problems. The other solvents boil lower and are in much more common use.

The Acetic Acid–Water–Ethyl Acetate System Acetic acid does not form an azeotrope with water, but their equilibria depart so much from ideality that the components are very difficult to separate by conventional distillation. The phase diagram for this system is shown in Fig. 5.6. Notice that the bubble-point and dew-point lines curve in the same direction—a most unusual characteristic. Superimposed on the same figure is a phase diagram for the heterogeneous ethyl acetate–water system. The two diagrams have been combined to describe more easily the events taking place in the distillation of this ternary mixture.

The third binary combination in this system, i.e., acetic acid and ethyl acetate, form a somewhat more ideal mixture than either of the other two binaries. Figure 5.6 is actually a projection of the prismatic ternary phase diagram, in which the acid-acetate surface is not visible. The acetate-water

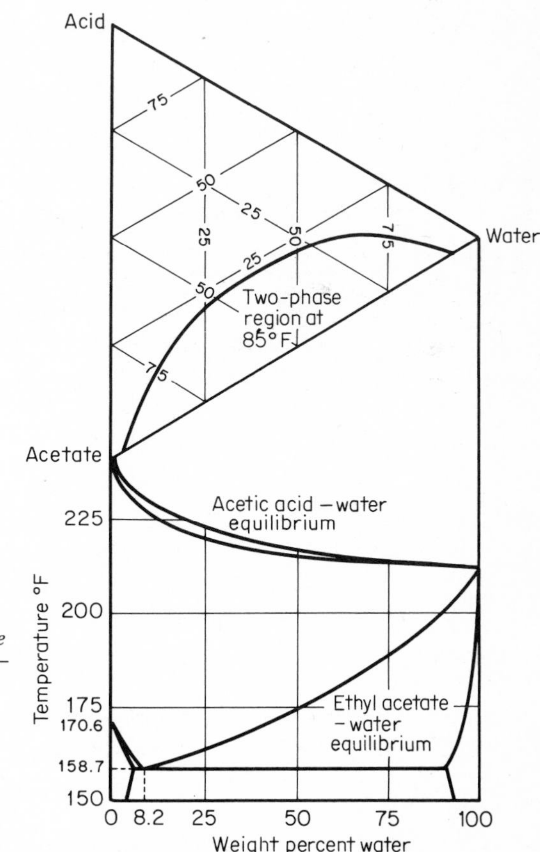

figure 5.6 *Projections of the phase diagram for the acetic acid–ethyl acetate–water system.*

phase diagram appears in the foreground while the acid-water diagram is in the background. These two planes meet at the right vertical axis (100 percent water). Looking down on the prism, the viewer will see the three-component liquid-phase diagram projected at the top of Fig. 5.6. As acid concentration is increased, the mutual solubility of acetate and water improves until eventually a single liquid phase is formed. Increasing temperature also improves their mutual solubility, shrinking the two-phase region. Hoffman [2] goes into much detail describing the variable characteristics of these phase diagrams.

The minimum boiling point for the system is that of the acetate-water heterogeneous azeotrope. Acid added to that mixture will distribute itself between the two liquid phases and raise the boiling point of the mixture. However, a minimum boiling point exists for any given concentration of acid as long as there are two liquid phases in equilibrium with the vapor. The combination of increasing acid concentration and boiling point eventually results in a single liquid phase, and the valley, characteristic of lower acid concentrations, disappears. Ellis and Pearce [3] indicate that the azeotrope shifts toward higher acetate concentrations as acid is added, until above 50 wt percent acid it disappears altogether. The valley then apparently curves toward higher acetate concentration until it flattens into a smooth plane approaching the acid-water equilibrium curve.

The Extraction Process The first stage of acid concentration is accomplished by liquid-liquid extraction of weak acid with acetate. As in other countercurrent mass-transfer processes, the water leaving the bottom of the extractor approaches equilibrium with the acetate feed. Any acid leaving the bottom of the extractor is lost and also pollutes, so the acetate extractant should be kept free of acid to avoid this loss. Acetate leaving in the aqueous effluent is recovered in a decanter-fed stripping column.

In a similar manner, the concentration of acid leaving the top of the extractor varies with the concentration of the weak-acid feed but also with the solvent-feed ratio. This relationship may be derived from a set of material-balance equations. The overall balance is

$$A + E = F + W \tag{5.33}$$

where the symbols represent mass-flow rates of weak acid, solvent, column feed, and waste, as indicated in Fig. 5.7. An acid balance on the extractor assumes no losses to waste:

$$Ax_a = Fz_a \tag{5.34}$$

where both compositions are in weight fraction of acid in the associated stream. A solvent balance may also be written:

$$Ex_e = Wx_w + Fz_e \tag{5.35}$$

with all compositions in weight fraction of acetate in their respective streams.

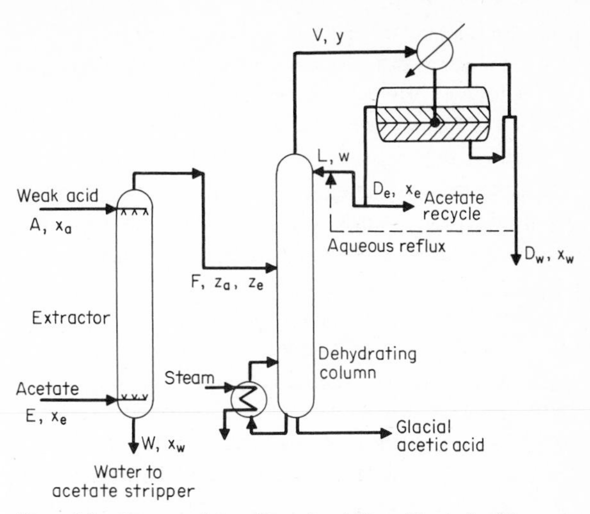

figure 5.7 *Acetate-rich reflux should be adjusted with aqueous reflux to optimize separation in the dehydrator.*

When the three balances are combined, a solution may be obtained in terms of the composition of the column feed:

$$\frac{z_e - x_w}{z_a} = \frac{(x_e - x_w)\,E/A - x_w}{x_a} \tag{5.36}$$

Of the variables given in this equation, x_w is the saturation level of ethyl acetate in water and should only be a function of temperature. If the extractant is taken from a decanter, then its saturation level of acetate x_e is fixed by decanter temperature. Therefore, these two parameters ought to be constant. Then control over the relative proportions of acetate and acid in the column feed can be achieved by adjusting the extractant–weak-acid ratio E/A as a function of weak-acid concentration x_a. If E/A is fixed, however, z_e and z_a will vary with x_a.

Equation (5.36) is a single expression with two unknowns, z_e and z_a. Another relationship is required to determine their absolute concentrations: it is the phase diagram for the ternary system. Solubility limits dictate that the liquid-phase compositions in equilibrium in the extractor lie on the curve in the triangular diagram of Fig. 5.6. Concentrations can then be taken from selected points on the curve until a fit to Eq. (5.36) is found. (Because the stream is acetate-rich, column-feed composition will lie somewhere along the left side of the curve.)

Dehydrating the Acid The distillation process has three objectives:

1. To make a glacial acetic acid product
2. To minimize acid and solvent losses
3. To minimize energy consumption

Conventional columns have but two degrees of freedom which determine overhead and bottom-product quality—they are typically the material balance and the energy input. This column has a third degree of freedom, however, the ratio of acetate to water.

The lowest boiling point of any mixture in this ternary system belongs to the acetate-water azeotrope; the highest boiling point is that of pure acetic acid. Therefore, one might assume that the average relative volatility in the column would be maximized by adjusting the acetate-water ratio to produce the azeotrope overhead. The merits of this assumption can be evaluated in light of data presented in Ref. 3. Ellis and Pearce have derived values of volatility for ethyl acetate relative to acetic acid, and for water relative to acetic acid, at various concentrations of the ternary mixture. Selected points from this source, converted into molar-average relative volatilities between the acetate-water combination and the acid, are presented in Table 5.1.

Weight percent water on an acid-free basis was chosen as a parameter to illustrate the effect of adjusting the acetate-water ratio: 8.16 percent corresponds to the acid-free azeotrope. Increasing acid content reduces the relative volatility of the 8.16 percent mixture until at 100 percent acid it no longer is maximum. The most effective fractionation would seem to be achieved when the acetate-water ratio is held at the acid-free azeotropic composition, the reduction in volatility at 100 percent acid being slight.

These columns are typically refluxed with only the organic phase from the decanter, however. Since this reflux contains only about 3 wt percent water, the relative volatility is not as high as it could be and substantially more energy is required to make the separation. As the liquid proceeds down the column, its acid content increases and so does temperature, both of which improve the solubility of the solvent and water. The liquid may become richer in water if the feed is rich in water, however. The tray composition and hence the relative volatility is then sensitive to feed composition as well as reflux composition and flow. The relationship is best borne out by the material balance.

Material-balance Considerations For simplicity, the overhead vapor is assumed to contain no acid and the bottom product no solvent or water. Then all the acetate and water in the feed leave as distillate, giving a decanter balance of

$$V = L + F(z_e + z_w) \tag{5.37}$$

where V and L are mass-flow rates of overhead vapor and reflux and where z_e

TABLE 5.1 Molar-average Volatility of Ethyl Acetate and Water Relative to Acetic Acid

Wt percent water in liquid (acid-free basis)	Wt percent acid in liquid			
	0	10	50	100
0.0	6.8	5.2	3.0	2.8
3.0	11	8.0	3.3	2.8
8.16	18	10.2	3.7	2.7
100.0	1.2	1.3	1.6	2.6

and z_w are weight fractions of acetate and water in the feed. An acetate balance gives

$$Vy = Lw + Fz_e \tag{5.38}$$

where y and w are weight fractions of acetate in the vapor and reflux.

To meet certain specifications of distillate and bottom-product purity, a certain V/F is required whose value naturally depends on the relative volatility achievable in the column. Since the relative volatility varies with the acetate-water ratio in the column, one factor of substantial significance is how overhead composition is affected by the V/F ratio. To determine this relationship, Eqs. (5.37) and (5.38) are combined by eliminating L and solving for y in terms of V/F:

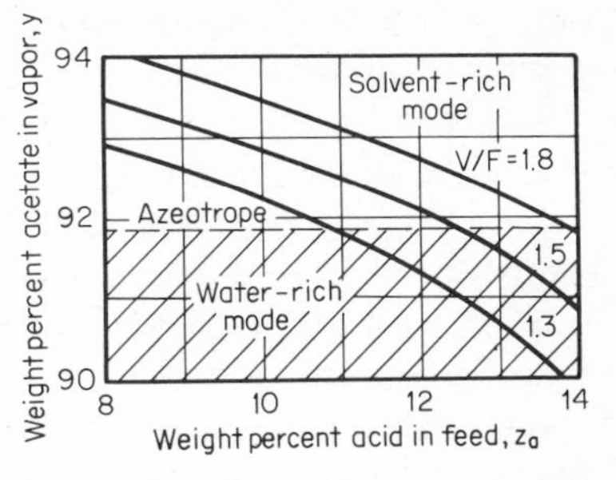

figure 5.8 *Overhead vapor composition as a function of feed composition for reflux containing 97 percent acetate.*

$$y = w - \frac{w(z_e + z_w) - z_e}{V/F} \tag{5.39}$$

For a given V/F ratio, y varies with feed composition. Because of the effect of y on α, this relationship is worth examining in detail. Accordingly, Eq. (5.39) was solved for several values of V/F at 97 wt percent acetate in the reflux; the results appear in Fig. 5.8.

Feed-composition values were taken from the solubility curve at the top of Fig. 5.6 As the acid content of the feed increases, its water content also increases, causing a drop in acetate level. Increasing acid concentration thus tends to lower the acetate-water ratio in the column. To demonstrate its effect on column operation, consider an initial condition at $V/F = 1.5$ and $z_a = 12.5$ percent, producing an overhead vapor at the azeotropic composition. An increase in z_a at that V/F will reduce y below the azeotropic composition. The liquid in the column must then become richer in water than the vapor, causing α to fall. If bottom-product composition is controlled by manipulating boilup, the controller will react to deteriorating purity by raising V/F. This action will increase y toward the azeotropic composition, thereby increasing α and returning product purity to its former value. The net result of an increase in z_a is a corresponding increase in V/F, not a normal condition by conventional standards but at least stable.

However, a *decrease* in acid in the feed generates an *unstable* situation. At a given V/F ratio, y increases above the azeotropic composition, developing a liquid phase still richer in acetate. Again α decreases, but deteriorating product quality cannot be corrected by increasing V/F. As Fig. 5.8 indicates, increasing V/F increases y and thereby further lowers α. Some columns have a history of operation in the solvent-rich mode but are prone to be upset easily. In addition, they typically allow significant losses of solvent in the acid product, and use much more energy than should be required to make the separation. Because of their high and variable V/F ratio, their feed-processing capacity also tends to be low.

Solvent-rich operation is characterized by higher acid concentration all across the column, although tray temperatures are lower. A shift from water-rich to solvent-rich operation is detectable as a sharp drop in mid-column temperature from 200 to perhaps 175°F.

Adjusting the Solvent-Water Ratio The only way out of the solvent-rich mode is to add water to the column, either in the reflux or the feed. If the solvent-water ratio in the feed is not the same as in the reflux, a composition gradient (acid-free basis) will develop across the column. In fact, the only way a gradient can be avoided is if feed, reflux, and vapor all contain the azeotropic mix of acetate and water. This would seem to provide the maximum overall relative volatility or at least very close to it. By comparison, the initial conditions described earlier had a 97 percent acetate reflux and the feed of 12.5 percent acid contained 88 percent acetate (acid-free basis). A feed this far from azeotropic proportions cannot produce an optimum volatility. The only feed composition on the solubility curve which corresponds to the azeotropic mixture of water and acetate occurs at about 7.5 percent acid, 7.5 percent water, and 85 percent acetate. This particular mix could be obtained from the extractor by increasing its extractant–weak-acid ratio, although adjustment would have to be made for variations in weak-acid concentration.

Regardless of whether feed composition is specifically adjusted to an optimum, reflux composition must be adjustable to avoid slipping into the solvent-rich mode. This is best achieved by adding some of the aqueous layer from the decanter to the reflux as shown in Fig. 5.7. Reflux composition w required to develop a specific overhead-vapor composition with varying feed can be calculated by rearranging (5.39):

$$w = \frac{yV/F - z_e}{V/F - (z_e + z_w)} \tag{5.40}$$

Equation (5.40) has been solved for $y = 0.9184$ and three values of V/F, with the results plotted in Fig. 5.9. As observed earlier, when the feed contains the azeotropic ratio, the reflux should also, regardless of the value of V/F. This would seem to be the optimum set of conditions.

Increasing the acid (and water) content of the feed requires that w increase to the point of saturation, i.e., requiring no aqueous reflux. Operation beyond this point is possible at the cost of increasing the V/F ratio. However, adding more solvent to the feed may be more economical.

The amount of aqueous reflux L_w required to adjust w to the desired point is actually quite small. It is easily calculated from the compositions of the two phases leaving the decanter:

$$\frac{L_w}{L} = \frac{x_e - w}{x_e - x_w} \tag{5.41}$$

where x_e and x_w represent the weight fractions of acetate in the organic and

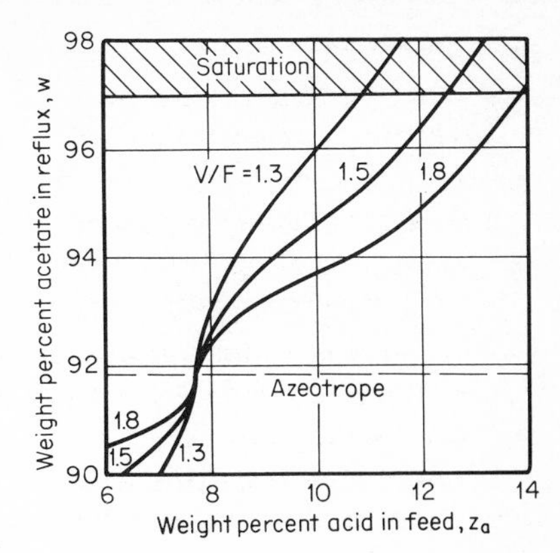

figure 5.9 *Reflux composition must be adjusted as shown here to hold the overhead vapor at the azeotropic composition.*

aqueous phases, respectively. For the case where w is desired at 0.918, given $x_e = 0.97$ and $x_w = 0.07$, $L_w/L = 0.058$. The aqueous stream is seen to contribute relatively little to the reflux flow while at the same time greatly affecting its composition.

Having a two-phase reflux ensures that two liquid phases also exist on some of the trays. But this is a natural consequence of operation in either the water-rich mode or precisely at the azeotrope throughout. Only in the solvent-rich mode may a single liquid phase exist, and this is not only suboptimum but demonstrably unstable. Therefore, the acetic acid column must be designed for operation with two liquid phases.

Not being able to analyze a heterogeneous azeotrope creates a control problem. Since reflux composition must be adjusted as a function of feed composition or to hold a certain vapor composition, some sort of analysis is necessary. One suggestion is to measure the flow of each phase leaving the decanter and calculate their ratio. If V_w is that portion of the condensed vapor leaving the decanter in the aqueous phase and V_e is the portion leaving in the organic phase, then

$$\frac{V_w}{V_e} = \frac{x_e - y}{y - x_w} \tag{5.42}$$

To hold y at 0.9184, given that x_e and x_w are 0.97 and 0.07, respectively, V_w/V_e should be 0.061. The calibration of such an "analyzer" appears in Fig. 5.10 as the solution to Eq. (5.42).

The liquid-phase flow rates will lag behind changes in vapor composition because of the large capacity of the decanter. This is to be expected and is in fact typical of product analyses made on reflux or distillate liquid streams. So the flow-ratio calculation would seem to have several of the properties of an analyzer, yet without the sampling and calibration problems. It will be temperature-sensitive, however, so decanter temperature must be controlled. In any case, it may be the only means of determining the overhead-vapor composition of a heterogeneous azeotrope.

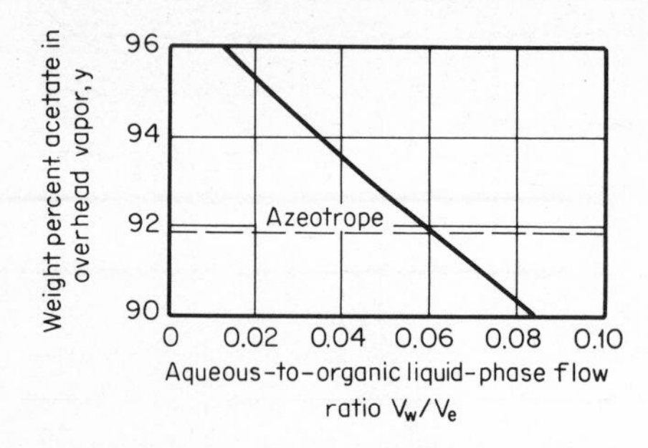

figure 5.10 *Overhead vapor composition may be inferred from the flow ratio of the two liquid phases if their composition is constant.*

Although temperature measurements are sensitive to composition variations, they can give some misleading information in this ternary system. For example, an excess of acid rising up the tower in the solvent-rich mode could give temperatures which are considered normal in the water-rich mode. Temperature can be used to indicate acid content in the top section of the column, however, if the overhead vapor is controlled at the azeotrope.

Inferring Acid Quality A temperature measurement at the bottom of the column is beset with similar problems. A reduction in temperature could be due to an accumulation of either acetate or water, depending on the mode of column operation. If a water-rich or azeotropic mode is maintained, a reproducible correlation of bottom temperature with composition can be achieved.

A greater sensitivity without error induced by pressure variations has been obtained by using a differential-vapor-pressure measurement [4]. A temperature bulb filled with a reference fluid is connected to one side of a differential-pressure measuring device as shown in Fig. 5.11. The bulb is inserted into the column at a selected point three to six trays above the bottom. The other side of the differential-pressure transmitter is also connected to the column at that point.

If the reference fluid is pure acetic acid and the tray contains pure acetic acid in equilibrium with its vapor, then the vapor pressure generated in the bulb will be the same as the pressure at the tray. Accordingly, an increase in vapor pressure in the column due to the presence of water or acetate will be sensed as a differential pressure between the column and the reference fluid. Furthermore, variations in column pressure will change the boiling point of

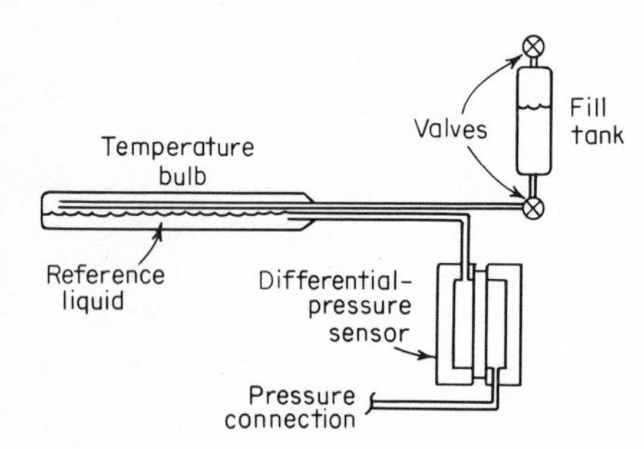

figure 5.11 *The Foxboro* dVp *cell transmitter compares the vapor pressure of a reference fluid against the pressure in the column.*

the liquid on the tray, so that the net effect on differential pressure is nil if tray composition is constant. There may be a transient effect, however, if the pressure change is faster than the thermal bulb can follow.

Sensitivity is also an important consideration. The vapor pressure of acetic acid changes about 7.3 in. of water for each degree Fahrenheit. Fortunately, the boiling point of acetic acid is sensitive to contamination by water: only 0.08 wt percent water reduces it by 1°F, as estimated from the data in Ref. 5. But differential pressure may be measured more accurately over a 7.3-in. span than temperature over a 1°F span. Consequently, the differential-vapor-pressure measurement can be employed quite successfully as a binary analyzer if its reference fluid is stable.

This turns out to be one of the problems in working with acetic acid. Apparently even the purest grades contain trace quantities of formic or other acids which attack stainless steel. Over a period of time, reaction products can build up, which even in minute amounts affect the vapor pressure of the fiuid. As a result, acetic acid has not been used satisfactorily as a reference fluid. Fortunately, methyl cellosolve has a very similar vapor-pressure characteristic and, being stable, serves admirably as a reference fluid for acetic acid distillation.

REFERENCES

1. Mains, G. H.: The System Furfural-Water, *Chem. Metall. Eng.*, Apr. 26, 1922.
2. Hoffmann, E. J.: "Azeotropic and Extractive Distillation," chaps. 5–8, Interscience-Wiley, New York, 1964.
3. Ellis, S. R. M., and C. J. Pearce: Designing Azeotropic Distillation Columns, *Br. Chem. Eng.*, December 1957.
4. The Foxboro Company: Differential Vapor Pressure Cell Transmitter, Model 13VA, *Technical Information Sheet* 37-91a, Foxboro, Mass., April 1965.
5. Ju Chin Chu: "Distillation Equilibrium Data," p. 208, Reinhold Publishing Corp., New York, 1950.

PART FOUR

Control

CHAPTER SIX

Control of Energy Inflow

Automatic control of energy inflow and outflow has actually contributed in some measure to our excessive consumption of this commodity. For example, a room thermostat responds to an open window or a blocked air duct by increasing the fuel to the furnace. A simple feedback controller such as this cannot react in any other way, since its intelligence is limited to the temperature it observes and the heating source it manipulates. Without this control, the discomfort due to the open window or blocked duct would force the intelligent operator to respond by correcting the cause.

Similarly, control of a column reboiler temperature by manipulating heat input can and does result in excessive energy consumption. The controller has no knowledge of what causes its temperature to fall: it could be low pressure, excessive reflux, increased feed, etc. Energy input should not respond equally to these disturbances, yet the simple controller has neither a means of distinguishing between them nor moderating its action accordingly. Consequently, distillation control as described in this book goes well beyond simple feedback control to relate the energy flow to the needs of the separation process. To this point, the needs of the process have been established. In this chapter the manipulation of energy inflow is presented.

The scope of this text does not allow a detailed development of the basic functions of feedback or even feedforward control. For this background the reader is directed to Ref. 1. Instead, the author assumes this basic knowledge on the part of the reader and directs his attention to the application of those concepts. Where certain characteristics peculiar to distillation merit detailed investigation, they are treated in the methodology followed in Ref. 1.

Energy conservation was listed in Chap. 1 as one of the important objectives of distillation control. And this goal was seen to be achievable by avoiding overpurification of products, by maximizing relative volatility, etc. Many additional factors, specifically relating to the means of applying and removing heat, also affect the efficient utilization of energy. Because some of these means are thermodynamic in nature, involving the interchange of heat and work, a general introduction into thermodynamic relationships seems appropriate.

THERMODYNAMICS AND ENERGY CONSERVATION

Most steam-electric generating stations derive their power from the combustion of a fuel. Heat is transferred from the flame and hot products of combustion to the evaporating and superheating sections of the boiler. The superheated steam is admitted through a throttling valve to a series of turbine stages and delivers useful work to the shaft, which turns the generator. Steam exhausting from the last stage of the turbine is still a vapor, but at a pressure and temperature (state) from which no more useful work may be extracted.

If the steam is simply exhausted to the atmosphere, it carries away with it some of the heat of the fuel and in addition represents a loss of conditioned feedwater. Therefore the steam is condensed against available cooling water and is returned to the heating cycle. In being cooled close to the water temperature, a substantial vacuum is developed on the turbine exhaust, allowing more work to be withdrawn than with atmospheric discharge.

However, the bulk of the heat contained in the steam leaving the superheater is removed by the condenser. The "heat rate" values achieved by central-station generating units range between about 8800 to 11,000 Btu/kWh. When compared to the conversion factor of 3413 Btu/kWh, the efficiencies of these plants lie in the 30 to 40 percent range. That is to say, 60 to 70 percent of the energy released by combustion of the fuel simply increases the temperature of the environment.

Tremendous economies can be realized if the heat rejected following electric power generation can be used for process heating. Unfortunately, most electric utilities have no immediate market for their low-level heat, and few process plants generate all their own electric power. But the functions of power generation and heating must be combined more extensively in the future if we are to conserve our dwindling reserves of fuel.

Irreversible Processes A reversible process is one through which the state of a system can be changed by doing work and whose original state may be restored by the same amount of work. Frictionless adiabatic compression is a reversible process through which work is done to compress a gas. In the reversal of this process—frictionless adiabatic expansion—work is done by the gas. If the identical amount were performed as was used in compression, then pressure, volume, and temperature would all return to their original states.

Unfortunately, real processes cannot be frictionless, so that some of the work done in compressing a gas is lost as heat flow. Also, some of the energy given up by the gas in expanding through an engine is lost in the same way. Consequently the work done by compression can never be completely recovered by expansion: friction takes its toll in both parts of the cycle. Hence the gas cannot be returned exactly to its original state. The degree of reversibility depends on the design and fabrication of the equipment, its thermal insulation, lubrication, etc. But the compression-expansion cycle is basically reversible by nature, whereas some processes are basically irreversible.

There are two important reasons for investigating the property of reversibility:

1. Distillation is the opposite of the irreversible process of blending.
2. Energy is supplied to and removed from distillation columns in an irreversible manner.

As reversibility is approached, processes become more efficient in extracting useful work from energy.

Entropy is the measure of irreversibility. A reversible process is isentropic; i.e., there is no change in entropy during a reversible change in state. Irreversible processes always result in an increase in entropy. Entropy has been defined [2] as the difference between energy and the available work of any state. In a reversible process, the conversion between work and energy is complete; therefore, no change in entropy takes place.

Denbigh [3] describes the maximum useful work of a process whose medium undergoes a change in state as

$$w_{max} = T_0 \Delta s - \Delta H \tag{6.1}$$

where T_0 = absolute sink temperature, °R
s = entropy, Btu/lb-°R
H = enthalpy, Btu/lb

Rather than enter into detailed calculations using numerical values of entropy, examples of various irreversible processes will be given and their effect on energy utilization estimated.

An elevator counterbalanced with a weight approaches reversibility, as opposed to a freely falling body, which does not. Similarly, charging a battery can be reversible. So is the generation of electricity by rotating a conductor through a magnetic field—the same device can be used as a motor or generator. In the battery, work is converted reversibly between chemical and electrical energy; in the generator, the conversion is between mechanical and electrical energy.

These processes are highly efficient in that they conserve *entropy*. By contrast, *energy* is, by nature, always conserved. The commonly used term "energy conservation" really means the conservation of available work. Keenan et al. [2] claim that the world energy crisis is actually an entropy crisis and that our processes should therefore be designed to conserve entropy so we may obtain the maximum useful work from our fuels.

With a little practice, irreversible processes become easy to recognize, although in many cases they are difficult to avoid. They are characterized by the downhill slide, the escape from containment, the slip from order into disorder, destruction, and waste. Specifically with regard to distillation, irreversibility is found in heat transfer (downhill slide), resistance to flow (escape), and blending (disorder). Each of these processes is investigated below.

Blending The argument against blending to control the quality of a product was given in Chap. 1 and briefly illustrated with a numerical example. At this point, the mathematical basis for that argument will be examined in light of some of the relationships developed in the intervening chapters.

The entropy of a mixture of ideal gases is given as [4]

$$\bar{s}_D = \Sigma D_i(C_p \ln T - R \ln p_i) \tag{6.2}$$

where $\bar{s}_D$ = entropy of system D, Btu/°R
D_i = number of pound moles of each component
C_p = heat capacity at constant pressure, Btu/lb-mol-°R
T = absolute temperature, °R
R = gas constant, 1.99 Btu/lb-mol-°R
p_i = partial pressure of each component, atm

Since partial pressure is the product of mole fraction y_i and total pressure p,

$$\bar{s}_D = \Sigma D_i(C_p \ln T - R \ln y_i - R \ln p) \tag{6.3}$$

Furthermore, the number of moles of each component D_i is the mole fraction of each multiplied by the total moles D in the system:

$$\bar{s}_D = D\Sigma y_i(C_p \ln T - R \ln y_i - R \ln p) \tag{6.4}$$

Consider blending D with B moles of another mixture having a set of mole fractions x_i to form F total moles having a composition described by a set of z_i:

$$F = D + B \tag{6.5}$$

$$Fz_i = Dy_i + Bx_i \tag{6.6}$$

The increase in entropy brought about by blending, $\Delta\bar{s}_b$, will be that of the blend less the sum of the entropies of the original systems:

$$\Delta\bar{s}_b = \bar{s}_F - (\bar{s}_D + \bar{s}_B) \tag{6.7}$$

If the temperature and total pressure of D and B are identical, then those terms in Eq. (6.4) will drop out when combined as above. The terms remaining are

$$\Delta\bar{s}_b = -F\Sigma z_i R \ln z_i + D\Sigma y_i R \ln y_i + B\Sigma x_i R \ln x_i \tag{6.8}$$

The change in entropy (Δs_b) per pound mole of blend may be obtained by dividing through by F:

$$\Delta s_b = R\left(-\Sigma z_i \ln z_i + \frac{D}{F}\Sigma y_i \ln y_i + \frac{B}{F}\Sigma x_i \ln x_i\right) \tag{6.9}$$

Since mole fractions are always less than unity, their logarithms are always negative. Consequently, the first term in (6.9) is positive and the others are negative. Each individual term will be maximum when the concentrations of all components in that mixture are identical. As an example, consider D to be a binary mixture. The function $y_i \ln y_i$ was calculated for both components for various values of y_i and the functions summed; the results appear in Table 6.1.

If D and B are very pure streams, then $\Sigma y_i \ln y_i$ and $\Sigma x_i \ln x_i$ will approach zero. For this limiting case,

$$\Delta s_b = -R\Sigma z_i \ln z_i \tag{6.10}$$

Blending two pure streams to produce a mixture in the 20 to 80 percent composition range is seen from Table 6.1 to increase the entropy of the system between roughly 0.5 and 0.7 Btu/lb-mol-°R.

The loss in available work caused by blending is determined by Eq. (6.1). For conditions of constant enthalpy,

$$w_{\min} = T\,\Delta s_b \tag{6.11}$$

This is the minimum amount of work which must be supplied to separate the mixture back into its components at temperature T. To separate a 50-50 mixture into its components at 100°F then requires

$$w_{\min} = 560 \times 1.99 \times 0.693 = 772 \text{ Btu/lb-mol}$$

Naturally it is impossible to separate any mixture into its absolutely pure components in a finite number of stages. Therefore, this minimum-work expression must apply to an infinite-stage process with a completely reversible energy cycle.

Perhaps of more practical significance is the work requirement as a function of feed and product composition. Table 6.1 shows how the minimum-work

TABLE 6.1 The Function $y_i \ln y_i$ for a Binary Mixture

y_1	$y_1 \ln y_1$	$y_2 \ln y_2$	$\Sigma y_i \ln y_i$
0	0	0	0
0.01	−0.046	−0.0099	−0.056
0.1	−0.230	−0.095	−0.325
0.2	−0.320	−0.179	−0.499
0.5	−0.346	−0.346	−0.693
0.8	−0.179	−0.320	−0.499
0.9	−0.095	−0.230	−0.325
0.99	−0.0099	−0.046	−0.056
1.0	0	0	0

function decreases as feed composition approaches product composition. This relationship was deficient in the binary-distillation model given in Chap. 2. Example 6.1 shows how the minimum-work relationship associates with the deviations in that model.

example 6.1

Calculate the minimum work required to make the separations given in Table 2.3. Assuming that the overhead product is propane, having a latent heat of vaporization of 137 Btu/lb at 100°F and a molecular weight of 44, the energy used to generate the V/F of 3 is

$$\frac{Q}{F} = 3(137 \text{ Btu/lb})(44 \text{ lb/mol}) = 18{,}084 \text{ Btu/lb-mol}$$

z	y	x	D/F	Y	Δs_b	$w_{\min}$, Btu/lb-mol
0.5	0.920	0.082	0.499	0.600	0.820	459
0.3	0.930	0.062	0.274	0.650	0.742	416
0.5	0.975	0.181	0.402	0.638	0.722	404
0.6	0.950	0.145	0.565	0.582	0.758	425

The figures in the $w_{\min}$ column of Example 6.1 are 40 times smaller than Q/F, indicating the energy conservation possible if the work could be applied reversibly. The product $Y\,\Delta s_b$ is nearly constant except for the last row, indicating that separation and Δs_b are related, but not uniquely.

Another example of blending that results in a loss of available work appears in Fig. 6.1. The temperature of a fluid heated by direct combustion of a fuel is often slow to respond to fuel-flow changes. Some systems are designed to bypass part of the cold fluid around the heater, blending it with the heated portion for more responsive temperature control. This is the same arrangement used to provide adjustable water temperature at various locations in a home. The process is irreversible because, after blending, the two streams cannot be separated again and returned to their original temperatures. The blending process is adiabatic since there is no heat transfer from an outside source; but it is not isentropic, and available work is lost.

Rather than estimate the loss in available work based on entropy considerations, an examination of heat losses will be used. To heat the fluid to 500°F requires a certain stack temperature. If instead of bypassing and blending, the entire stream were heated directly to 400°F, a lower stack temperature

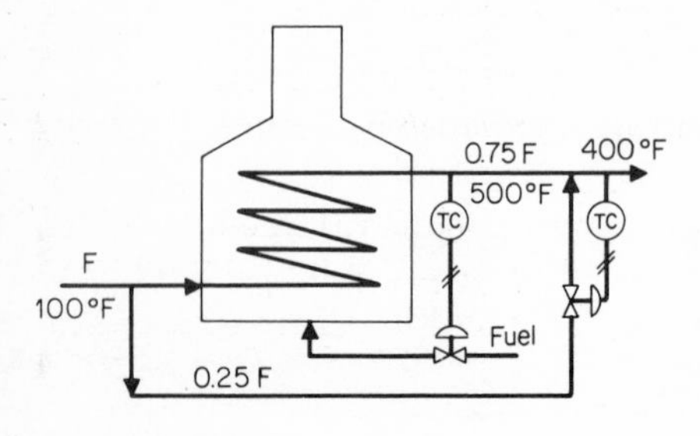

figure 6.1 *Mixing the heated and unheated fluid is an irreversible process.*

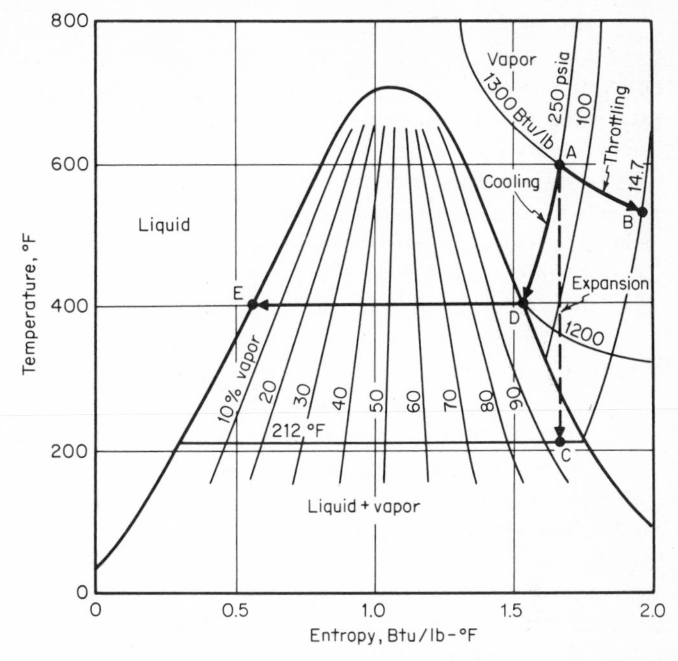

figure 6.2 *Throttling steam through a restrictor follows a line of constant enthalpy, whereas expansion through an engine is istentropic.*

would be realized. The rate of heat transfer to the fluid would be the same in both cases, but stack loss would be higher at the higher outlet temperature and hence more fuel must be consumed. The actual increase in stack loss depends on stack temperature itself. But the case against blending hot and cold is clear—it is always less efficient than heating (or cooling) directly to the controlled temperature.

Recall also how injection of nitrogen into the condenser of a benzene-toluene column depressed the dew point from 178 to 80°F. Without this blending, the level of the coolant could have been raised 98°F, thereby increasing its available work. The amount of heat transferred to it would remain the same, but its net energy level would be higher, allowing it to be used for heating other processes.

Control Valves Expansion of a gas through an engine was described as a reversible process—the reverse of compression. But expansion of a gas through a restriction accomplishes no useful work and is therefore irreversible. Expansion through a perfect engine is isentropic; expansion through a restriction is isenthalpic.

Figure 6.2 compares the two processes. It is a temperature-entropy diagram for water. The area underneath the "bell" consists of mixed liquid and vapor, with all liquid on the extreme left and all vapor on the extreme right. At the crest is the critical point.

Let the superheated vapor at point A be throttled through a restrictor from p_1 to p_2 at point B. The path follows a contour of constant enthalpy but increasing entropy. If ΔH in Eq. (6.1) is then constant, available work is lost in proportion to the increase in entropy.

But if expansion takes place through an engine such as a turbine, path AC is followed, yielding work equal to ΔH. In controlling a steam turbine, a throttle valve is normally adjusted to hold a desirable level of either electrical output or boiler pressure. So expansion takes place partly through the throttle and partly through the turbine. Pressure dropped across the throttle is lost work. Therefore, maximum conversion of energy into work can only be achieved when the throttle valve is fully open. As a result, steam turbines should be operated at full rated power for maximum efficiency.

Unfortunately, at the present state of the art, control over processes is almost exclusively exercised by throttling. Although this practice is admittedly wasteful, there seems to be little alternative. The most efficient control mechanism available is the variable-speed drive for rotating machinery. Steam-turbine-driven pumps and compressors cannot be adjusted without loss, however, since throttling is performed on the steam to the turbine just as in a power plant. Until controllable expansion turbines become available for a broad range of fluids and capacities, control valves will continue to be used. Some research [5] is moving in that direction, however.

Control valves should be sized for the minimum pressure drop which will still allow control. At maximum flow the head available from centrifugal pumps and compressors is minimum, and the pressure loss through fixed restrictions such as piping and heat exchangers is at its highest level. This tends to leave very little across the control valve, as it should be. However, when the valve is throttled to reduce flow, pressure losses elsewhere in the system decrease and the drop across the valve rises.

For liquids, the flow through a control valve is given by

$$F = aC_v\sqrt{\frac{\Delta p}{\rho}} \tag{6.12}$$

where F = flow, gpm
a = fractional valve opening
C_v = valve flow coefficient
Δp = pressure drop, psi
ρ = fluid density, g/ml

It can be seen that if Δp decreases as the valve opens, the change in flow per unit valve opening decreases. The result of this relationship is that fractional flow f (that is, F/F_{max}) varies nonlinearly with fractional valve opening as shown in Fig. 6.3.

One of the most important factors affecting control-loop performance is the uniformity of the response of the controlled variable to valve position. If the controlled variable is a linear flow measurement or is linearly related to flow,

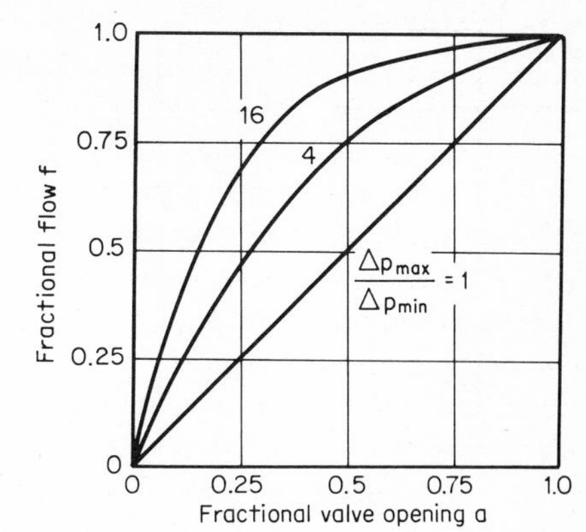

figure 6.3 *Variable pressure drop causes flow to respond nonlinearly to valve opening.*

then the nonlinear characteristics shown in Fig. 6.3 can cause trouble. Consider the example of fuel gas from a constant-pressure source throttled through a valve and a burner and then used to heat a constant stream of air. At low flow rates, when the pressure drop across the burner is nil, air temperature will respond sharply to valve position. But as full flow is approached, the sensitivity of temperature to valve position deteriorates since most of the supply pressure is dropped across the burner. If the temperature controller is adjusted for tight control at low temperatures, response will be sluggish at high temperatures (high flows). Or if adjusted for tight control at high temperatures, the increased sensitivity at low temperatures will result in oscillation there.

This undesirable relationship can be altered by using a valve with a suitably nonlinear relationship between valve opening and stem position. An "equal-percentage" characteristic exists in which the valve opening varies exponentially with the fractional stem position m:

$$a = r^{(m-1)} \tag{6.13}$$

Here r is the "rangeability" of the valve, i.e., its maximum flow divided by its minimum controllable flow under constant pressure drop. The name of the valve characteristic springs from the response of flow caused by a given incremental adjustment in stem position being an "equal percentage" of the *actual* flow for any position of the stem.

If an equal-percentage valve with a rangeability of 50 is operated under constant pressure drop, its fractional flow relates to stem position as the lowest curve in Fig. 6.4. But if it is used in a system that creates a variable pressure drop, a more linear relationship is achieved. If linearity is the only criterion of concern, then the equal-percentage valve performs admirably when the available pressure drop varies over 16 to 1.

However, variable pressure drop also reduces the effective rangeability of a valve. The maximum-flow capability of the valve will be reduced by the lower pressure drop available at that flow whereas the minimum-flow capability will

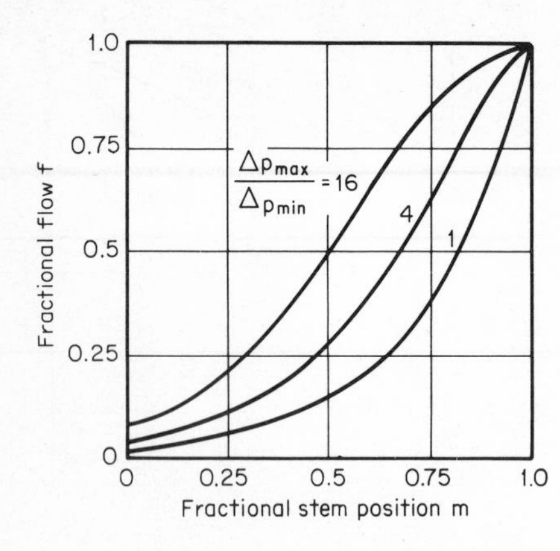

figure 6.4 *The equal-percentage characteristic compensates for variations in pressure drop.*

not. Then the effective rangeability r_{eff} will always be smaller than the rated rangeability r by the pressure-drop ratio:

$$r_{eff} = r\sqrt{\frac{\Delta p_{min}}{\Delta p_{max}}} \tag{6.14}$$

The rangeability of the equal-percentage valve described in Fig. 6.4 is reduced from 50 to 12.5 when the pressure drop varies by a factor of 16 between zero and full flow. Although this loss in rangeability can create a problem in some control loops, most of those on distillation columns will not be affected by it.

Some control loops contain nonlinearities other than the valve characteristic and can use a nonlinear valve to compensate for them. A head-type flowmeter such as an orifice, nozzle, etc., produces a differential-pressure signal h related to the square of flow:

$$h = kF^2 \tag{6.15}$$

The sensitivity of the differential-pressure signal is low at low flows, so the sensitivity of the valve should be high in that range. This characteristic corresponds to what was achieved with a linear valve having a variable pressure drop as in Fig. 6.3.

Most temperature-control loops require the opposite characteristic, even with constant pressure drop. This peculiarity of temperature loops, along with the need to compensate for variable pressure drop, explains why most control valves used in the process industries are of the equal-percentage variety. Each heat-transfer process encountered in the pages that follow will be examined to determine the valve characteristic which will provide the best control.

The power lost across a control valve flowing liquid is the product of flow and permanent pressure loss:

$$Q = F\,\Delta p \tag{6.16}$$

with Q in ft-lb/min, F in ft^3/min, and Δp in lb/ft^2. In fact this is the hydraulic power which must be delivered by the pump to force the flow through the valve. The proper conversion factors must be used for the hydraulic power to be in units of horsepower when flow and pressure drop are expressed in familiar gpm and psi:

$$Q = \frac{(F\ \text{gpm})(\Delta p\ \text{psi})(144\ \text{in}^2/\text{ft}^2)}{(7.48\ \text{gal/ft}^3)(33{,}000\ \text{ft-lb/min/hp})} = 5.83 \times 10^{-4} F\ \Delta p \tag{6.17}$$

Most of this power is recoverable if a hydraulic turbine rather than a valve is used to let the liquid down to a lower pressure. Like a centrifugal pump, its efficiency is limited, falling off at extremely high or low flow rates. Similarly, the power lost through a control valve falls to zero when the valve is shut and is low when fully open, due to the Δp being minimum. Figure 6.5 compares, for two valve sizes, the hydraulic power lost through the control valve relative to that lost through fixed resistances in the loop. The curve for the 16:1 Δp variation uses a valve twice as large as the 4:1 curve. The increased capital cost for the larger valve is offset by the reduced power loss, particularly above 80 percent flow. These curves also demonstrate why plants tend to be most efficient when operated near full capacity.

Heat Transfer Like the fluid flowing from a higher to a lower pressure through a restriction, heat flows from a higher to a lower temperature through a barrier. A hot fluid has a capacity to do useful work in proportion to its temperature above the heat sink and the quantity of energy q capable of being transferred between source and sink:

$$w_{\max} = \frac{T - T^\circ}{T^\circ} q \tag{6.18}$$

This is the familiar Carnot expression for the maximum amount of work obtainable between two reservoirs at T and T° absolute [6].

The driving force for a steam turbine is energy as opposed to simply pressure. Isentropic expansion through the turbine causes a reduction in superheat and even the condensation of liquid, as indicated in Fig. 6.2. If, instead

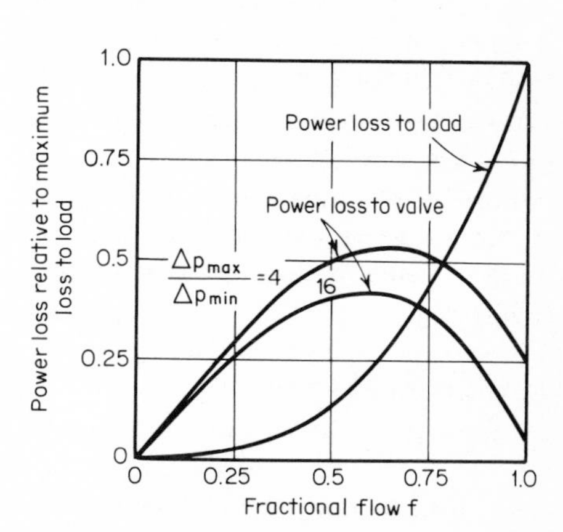

figure 6.5 *Power loss due to throttling is maximum near mid-range.*

of driving a turbine, the steam simply transfers heat at constant pressure, its available work reduces to that of the steam condensate. The cooling path follows the constant-pressure contour in Fig. 6.2 from point A through D to E. Although entropy has decreased in the cooling process, enthalpy has also decreased. However, the fluid receiving the heat gains more entropy than the hot fluid loses, so that the entropy of the system rises.

Consider the example of saturated steam at 250 psia (401°F) transferring its latent heat to boil water at 100 psia (328°F). In condensing, the 250-psia steam gives up 825 Btu/lb of enthalpy and 0.959 Btu/lb-°F of entropy. Because 889 Btu/lb is required to boil water at 100 psia, only 0.929 lb is evaporated per pound of steam condensed. But the water gains 1.129 Btu/lb-°F of entropy in evaporating. The net entropy increase for the system for every pound of steam condensed is

$$\Delta s = 1.129(0.929) - 0.959 = 0.089 \text{ Btu/lb-°F}$$

Again, an increase in entropy is a measure of the loss in available work; this loss may be calculated as $T\ \Delta s$:

$$\Delta w = (401 + 460)(0.089) = 76.6 \text{ Btu/lb}$$

The increase in entropy, and hence the loss in available work, is directly proportional to the temperature difference between the two fluids. To demonstrate, Eq. (6.18) may be used to estimate the loss in available work as a function of steam temperatures and the amount of heat transferred:

$$\Delta w = \frac{401 - 328}{328 + 460} 825 = 76.4 \text{ Btu/lb}$$

The lesson to be learned from Eq. (6.18) is to design for minimum temperature difference in order to minimize the loss in available work. This is certainly borne out in multiple-effect evaporators where the number of effects may be increased if the temperature differences across each can be reduced. Thus the thermal efficiency of the system is maximized by minimizing temperature differences across heat-transfer surfaces.

A corollary to this relationship is the desirability of transferring heat between fluids with small rather than large temperature differences. Temperature control is also more stable when the temperature difference across the heat-transfer surface is low. References 7 and 8 describe how exothermic chemical reactions become unstable as the temperature difference between the reaction mass and the coolant is increased above 60°F.

The rate of heat flow into or out of a stream is proportional to its mass-flow rate W, specific heat C, and its rise (or fall) in temperature:

$$Q = WC(T_1 - T_2) \tag{6.19}$$

A given rate of heat transfer into a fluid may be achieved by an infinite number of sets of flow and temperature rise (or fall). But the minimum loss in available work is coincident with the minimum temperature difference and therefore the maximum flow. Higher flow rates are also helpful in reducing resistance to heat transfer and fouling of surfaces. An optimum flow exists where the improvements are balanced by increased pumping cost.

Consider the example of a fired heater transferring heat to an oil used by several reboilers. A low oil flow requires a higher reboiler inlet temperature to transfer a given flow of heat to the column. And the higher inlet temperature in turn requires a higher flue-gas temperature in the heater and therefore causes higher stack losses. Maximum efficiency will be realized when oil flow is always maximum and oil temperature is minimum.

The control system should be arranged so that the reboiler demanding the most heat will receive full oil flow, with oil temperature set to deliver that heat. Flow to the other reboilers may then be throttled to match their requirements. Such a system is shown in Fig. 6.6. Each column has its own heat-input controls manipulating hot-oil flow. The three valve-position signals are compared in the high selector, and the highest is sent to the valve-position controller. This device adjusts oil temperature until the highest valve signal is at or near full opening. Then the oil temperature will be at its minimum acceptable value, as will the hydraulic power loss through the control valves. The valves are free to be manipulated by the individual column controls for fast response in the short term while the slower-acting valve-position controller minimizes energy loss in the long term. The valve-position controller will come to see considerable service in energy-conservation schemes throughout the rest of this book.

Conventional heater controls always include a bypass, recirculating hot oil back to the cold-oil line, to protect against loss in flow through the heater. The bypass valve is usually manipulated to control differential pressure between hot- and cold-oil lines. As less heat is required by the reboilers, the bypass valve opens to maintain a constant flow through the heater. However, bypassed hot oil represents a loss in available work since it is blended with the cold oil. In the scheme shown in Fig. 6.6, bypassing is not normally required since one of the load control valves is always nearly full open. However, a bypass valve that fails open should still be used to protect the heater against a control failure.

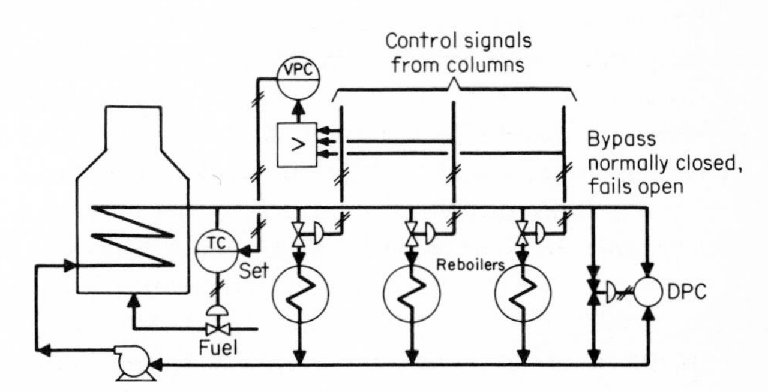

figure 6.6 *The valve-position controller holds one valve fully open by adjusting oil temperature to its minimum acceptable value.*

The practice of heating oil or any process fluid by direct combustion of fuel results in loss of available work in proportion to the temperature difference between the flame and the fluid per Eq. (6.18). If the products of combustion are first used to drive a gas turbine that generates electricity, and then to provide process heat, the maximum amount of work may be extracted from the fuel. The ideal plant would have all its electricity generated by gas turbines and all its process heat extracted from the turbine exhausts, either directly or through heat-recovery boilers. A gas turbine can operate at much higher temperatures than a steam turbine. In this way it can take advantage of the available work between flame temperature (>2000°F) and the highest steam temperatures in common use (1050°F). This advantage is vividly demonstrated by the unusually low heat rates attained when the exhaust of a gas turbine is used to generate steam for a steam turbine. These combined-cycle units easily achieve 8000 Btu/kWh without high pressures or temperatures in the steam cycle.

REBOILERS

Chapter 1 included discussions on both the cost of heat and the limitations of reboilers, so those aspects of heat-input controls need not be repeated here. Instead, this chapter looks into the characteristics of the media commonly used and the equipment through which the heat is transferred. Those features most directly related to control are examined in detail: linearity, speed of response, and sensitivity to disturbance. Each heating medium and each type of reboiler has its own special needs which must be met if column performance is to be maximized. Not infrequently an engineer is surprised by a completely unexpected and apparently inexplicable response of a reboiler to control action. Enough of these characteristics are documented here to minimize the number of surprises encountered in the plant.

Boiling When the vapor pressure of a liquid exceeds the static pressure imposed on it, the liquid is said to boil. However, to bring about boiling by transferring heat through a surface to that liquid is a little more complex. Boiling is characterized by formation of vapor bubbles at the heat-transfer surface which break away and travel to the vapor phase. The bubbles form at specific sites or nuclei on the surface; hence this type of boiling is called "nucleate" boiling.

Nucleate boiling will not begin until the heat-transfer surface is raised above the boiling point of the liquid, perhaps by as much as 50°F. But once boiling has begun, it may continue as the surface temperature is reduced as low as only 20°F above the boiling point, before subsiding. Below this temperature difference, heat is transferred simply by convection and no boiling takes place.

In the nucleate boiling regime, heat-transfer rate is directly proportional to the temperature difference between the surface and the fluid until a maximum is reached. At this point a film of vapor begins to form at the heat-transfer surface. A further increase in temperature difference actually results in a decrease in heat transfer, as shown in Fig. 6.7. This figure, called the "pool-boiling curve," describes how heat transfer varies with the temperature difference between a pool of liquid and a hot surface. Its construction is described in Ref. 9.

After passing through a minimum, the heat-transfer rate again rises with temperature difference as a stable film is formed. Of the two stable regions on the curve, nucleate boiling is far more efficient, being able to transfer much more heat with a smaller temperature difference. Considering the discussion of temperature difference and available work just concluded, film boiling ought to be avoided.

From a control standpoint, the two unstable regions of the curve also must be avoided. Once boiling has begun, the nucleate regime is stable. However, the initial rate of heat transfer achieved could, in some instances, exceed that desired. This problem area seems reserved to start-up situations and need not be a concern in continuous control.

The transitional regime is quite another case. The negative response of heat transfer to the temperature of the heating medium virtually eliminates the possibility of operating in this region. As the heat-transfer rate begins to fall, its cooling effect on the heating medium will decrease, actually causing its temperature to rise further. This naturally reduces heat transfer even more. An attempt to control the rate of boiling in the transitional region will result in constant-amplitude oscillation approaching the maximum and minimum points on the curve. Normally, reboilers are designed to operate in the stable nucleate regime because of its superior efficiency. But a severe upset or abnormal operating condition could cause the reboiler to slip into the transitional regime, and the engineer should be able to recognize its symptoms.

In the nucleate-boiling regime, the heat-transfer rate is linear with temperature difference, according to Adiutori [9]:

$$Q = UA(\Delta T - \Delta T_0) \tag{6.20}$$

where Q = heat-transfer rate, Btu/h
A = surface area, ft^2
U = proportionality constant, Btu/h-ft^2-°F
ΔT = temperature difference, °F
ΔT_0 = ΔT at which boiling subsides, °F

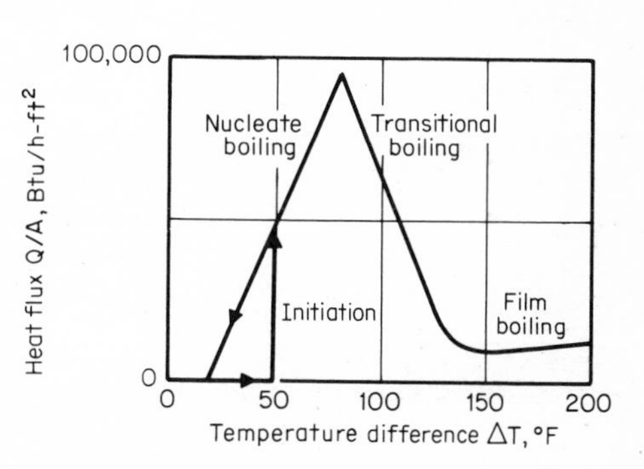

figure 6.7 *A typical pool-boiling curve indicates two unstable regions: the initiation of nucleate boiling and the transition to film boiling.*

This expression will be useful in analyzing the response of boilup to variations in flow of the heating medium.

For organic liquids in a vertical thermosiphon reboiler, Frank and Prickett [10] present curves of Q/A versus ΔT as a function of the reduced temperature T_r of the liquid. Reduced temperature is defined as

$$T_r \equiv \frac{T}{T_c} \tag{6.21}$$

where T is the liquid temperature and T_c is its critical temperature in absolute units. Equation (6.22) gives a reasonable fit to those curves in the form of Eq. (6.20):

$$Q = \frac{A(\Delta T - 11)}{0.012 - 0.011T_r} \tag{6.22}$$

For aqueous solutions, the denominator becomes simply 0.002.

Reboiler Characteristics Reboilers may be classified into three groups on the basis of the manner in which vapor is moved from the heat-transfer surface to the column:

1. Immersed
2. Natural circulation
3. Forced circulation

Each set includes subsets differing in physical construction or heat source. And each of the subsets has some outstanding physical characteristics directly affecting its control.

A reboiler tube bundle may be immersed in a "kettle" exterior to the column or internally in some designs. Figure 6.8 compares the two installations. Either may be heated with steam or liquid within the tubes. Both achieve 100 percent vaporization; i.e., there is no liquid discharged along with the vapor.

The weir in the kettle reboiler is intended to protect the tubes against loss of liquid in the event of a failure of the level controller. This does not necessarily ensure that the tubes are always covered, however, for a sudden increase in boilup (due to a loss in pressure, for example) could vaporize liquid faster than it is returned. The column base is normally empty of liquid, and the only controllable quantity lies below the weir. Since this reservoir is small, its liquid level can be difficult to control and subject to rapid fluctuation. To gain

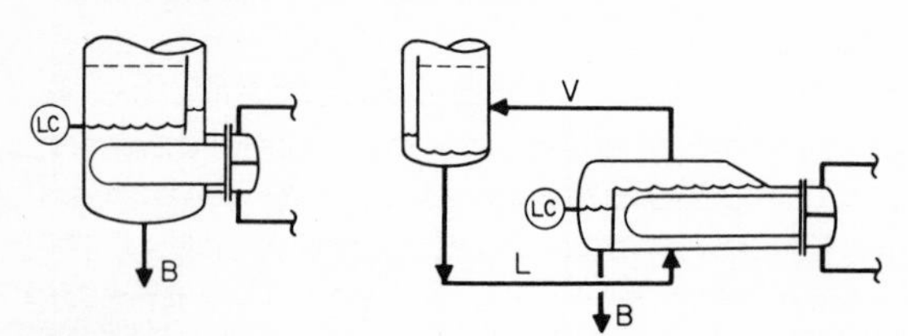

figure 6.8 *A tube bundle may be inserted either directly in the column base or in an externally mounted kettle.*

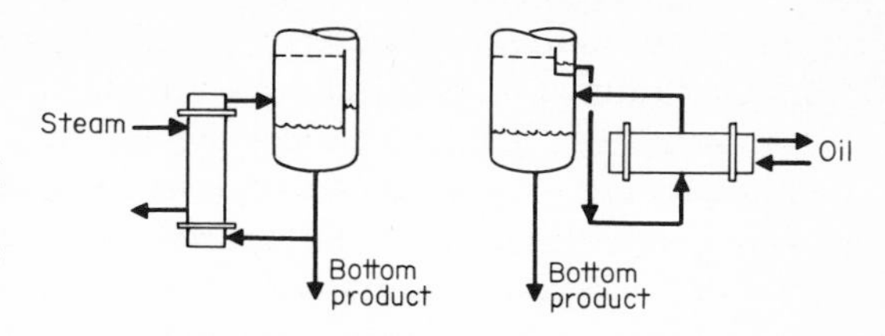

figure 6.9 *A vertical thermosiphon reboiler (left) usually recirculates bottom product, whereas a horizontal unit (right) takes liquid from the bottom tray.*

more capacity at the point of level measurement, Lupfer [11] recommends setting the control point above the weir or removing the weir altogether.

Natural-circulation, or "thermosiphon," reboilers may be mounted either horizontally or vertically as shown in Fig. 6.9. Vertical thermosiphon reboilers are used primarily in the chemical industry—their single-pass construction and high tube-side velocity minimize fouling and facilitate cleaning. By comparison, a horizontal thermosiphon reboiler usually has the process in the shell and heating medium in the tubes. Its process-side velocity is lower, and additional liquid head is usually employed, often by trapping the liquid off the bottom tray as shown. Horizontal thermosiphon reboilers are more common in petroleum refining and are usually heated with circulating oil.

The high velocities encountered within the vertical thermosiphon reboilers are attained with as little as 5 wt percent vaporization. This is an important consideration since circulation then depends on that concentration of volatile components—otherwise vaporization will be lost. Columns containing substantial concentrations of only slightly volatile components may exhibit unstable operation for this reason. This problem can be more severe in horizontal thermosiphon reboilers where the percent vaporization is higher. Trapping the liquid from the first tray then becomes essential for the reboiler to function.

Note that the liquid circuit between column base and thermosiphon reboiler forms a "U tube." Liquid contained in a U tube has a capability of resonant oscillation like a pendulum. Its period is derived in Ref. 12 as

$$\tau_o = 2\pi\sqrt{\frac{l}{2g}} = 0.78\sqrt{l} \tag{6.23}$$

where τ_o = natural period, s
l = total length of U tube, ft
g = acceleration of gravity, 32.2 ft/s^2

The length of the U tube in most columns is about 15 ft, giving a 3-s period. In absence of friction, oscillation, once begun, would continue indefinitely. However, normal frictional resistance of the piping provides some damping. Nonetheless, bubble formation in the reboiler produces enough random disturbances or "noise" to induce oscillation at the resonant period. Buckley [13] describes unstable reboiler operation having a period of 3 to 4 s coincident with high heat-flux rates. A restriction in the liquid line can correct this problem by providing damping.

Forced circulation is used with vacuum distillation or when heat is obtained from the combustion of a fuel. Vacuum operation is usually intended to minimize the boiling point and therefore the rate of thermal degradation of the bottom product. However, the gravity head of liquid above the reboiler heat-transfer surface and ΔT_o both contribute to raising liquid temperature in immersed or natural-circulation reboilers. Forced circulation may be applied to either a horizontal or a vertical reboiler for a vacuum tower, but the flow in the vertical unit must be downward. At a particular flow rate, the static head and velocity head can offset one another so that relatively little net pressure drop exists through the tubes. Forced circulation also stimulates nucleation and reduces ΔT_o. Some forced-circulation systems use narrow tubes or a fixed restriction downstream of the reboiler to maintain a backpressure within the tubes. This prevents boiling in the tubes, allowing the product to flash as it leaves. Although this may be desirable from the standpoint of heat transfer, pumping costs are increased by the restriction.

Direct firing may be used to heat an oil without boiling, as shown in Fig. 6.6. But a fired heater may also be used as a reboiler, as in the fractionation of crude oil (Fig. 3.9) and other products having boiling points in excess of 400°F. Vaporization ranges from a few percent up to 40 to 50 percent. The flow of the liquid to the heater is nearly always controlled to maintain efficient heat transfer. Reduction in flow tends to cause surface overheating within the tubes and subsequent deposition of polymers, tar, and coke. These deposits not only interfere with heat transfer but also tend to restrict flow, further aggravating the problem.

In addition, the flow is usually conducted through several parallel passes. Should the flow through one of these passes be somewhat lower than through the others, its percent vaporization will increase because essentially the same heat is being absorbed in all passes. The resulting expansion in volume increases the velocity through that pass, increasing the frictional and velocity-head loss. But if the pressure drop through all passes is the same due to manifolding, the tube generating more vapor will have its mass flow reduced even further and its percent vaporization increased more. In the absence of selective control, some tubes will be carrying high flow rates with little vaporization while others carry low flow rates with much vaporization. This condition has a detrimental long-term effect on the heater because of uneven temperature distribution. Tubes with lower flows will foul first, causing further reduction in flow. Ultimately these tubes will plug or even burn through.

Figure 6.10 shows a control system designed to provide equal flow through all passes. The highest valve signal is selected as the input to the valve-position controller. It, in turn, adjusts the set points of all flow controllers to hold that valve exactly at the fully open position. This arrangement results in the minimum attainable pressure loss through the valves. Note that the actual flow will change as tube resistance changes, but flow will always be maximum. No additional control over the total stream flow would seem necessary. If the circulating pump is oversized, flow may be reduced and horsepower saved more effectively by trimming the pump impeller than by throttling its discharge.

For bottom products with a wide boiling range, percent vaporization can be indicated by temperature. However, temperature is also a function of composition and pressure. Consequently, to use it as a measurement of percent vaporization requires that temperature or composition be controlled in the column, too, with pressure compensation provided if necessary. Percent vaporization can also be inferred from the ratio of heater outflow to inflow using orifice meters. If both orifices are identical, the same differential pressure will be developed across each when there is no vaporization. The increased velocity brought about by expansion of liquid into vapor will increase the outflow differential with percent vaporization. Because of the two-phase outflow, an eccentric orifice must be used if the line is horizontal. Even then, calibration is not predictable and must be verified by actual field tests.

Inverse Response The fractional vaporization in some reboilers poses a potential problem if heat input is to be used to control column base level. The percent vaporization increases with heat input such that the vapor volume in the reboiler also increases. Therefore, a sudden increase in heat input must move some liquid from the reboiler to the column base. This causes the base level to rise, but only temporarily, because the increased rate of boiling will drive the level down in the long term. The response of base level l_b to a step change in vapor flow V is shown in Fig. 6.11. This characteristic is known as "inverse response."

The base-level controller must be arranged for the long-term response, in which level falls on increased heat input. But the short-term or transient response is in the opposite direction, which causes a severe stability problem. To determine the magnitude of the transient, consider that the mass of liquid W_r in the reboiler changes with the boilup rate V:

$$W_r = W_0 \left(1 - \frac{kV}{V_M}\right) \tag{6.24}$$

where W_0 = liquid mass at zero boilup
V_M = maximum boilup rate
k = a proportionality constant

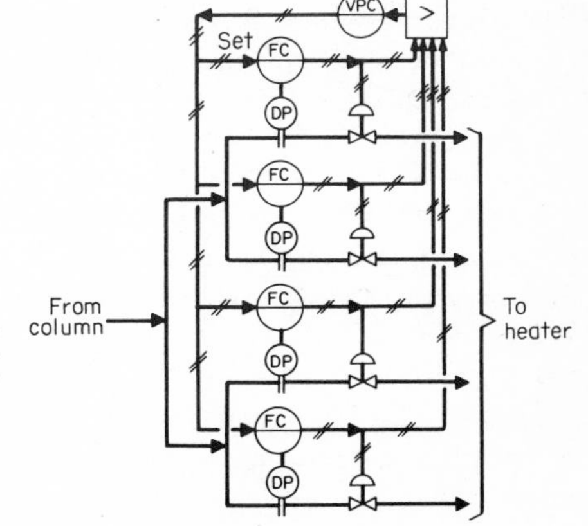

figure 6.10 *The valve-position controller sets all flows to whatever point will maintain one valve fully open.*

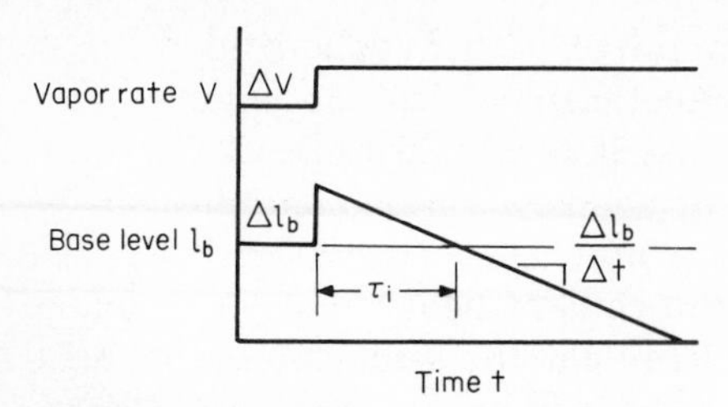

figure 6.11 *A step increase in vapor flow can displace a proportionate volume of liquid and actually cause base level to rise before it falls.*

The total mass of liquid in reboiler W_r and column base W_b varies with the difference between inflow and outflow:

$$W_r + W_b = \int (L - V - B)\, dt \tag{6.25}$$

where L = liquid flow into base
B = bottom-product flow
t = time

The liquid-level transmitter will be sensitive to base contents and will produce a signal l_b:

$$l_b = \frac{W_b}{W_M} \tag{6.26}$$

where W_M = maximum base capacity.

If the preceding three equations are combined and solved for l_b in terms of V/V_M, we have

$$l_b = \frac{V_M}{W_M} \int \left(\frac{L - B}{V_M} - \frac{V}{V_M} \right) dt - \frac{W_0}{W_M} \left(1 - k \frac{V}{V_M} \right) \tag{6.27}$$

The step response of Fig. 6.11 displays Δl_b as a function of ΔV:

$$\Delta l_b = \frac{\Delta V}{V_M} \left(k \frac{W_0}{W_M} - \frac{V_M}{W_M} \Delta t \right) \tag{6.28}$$

The time at which l_b crosses its original value is significant—it is identified as the inversion time τ_i in Fig. 6.11. It may be found by solving for Δt in (6.28) at $\Delta l_b = 0$:

$$\tau_i = \frac{kW_0}{V_M} \tag{6.29}$$

Substituting for kW_0/V_M in (6.24) gives the inversion time in terms of reboiler contents and boilup rate:

$$\tau_i = \frac{W_0 - W_r}{V} \tag{6.30}$$

The other important parameter in this system is the time constant τ of the column base, i.e., the time required to change the level full-scale when L and B are zero and V is V_M:

$$\tau = \frac{W_M}{V_M} \tag{6.31}$$

Substituting fractional boilup v for V/V_M, and τ_i and τ into Eq. (6.27), yields a general expression for systems with inverse response:

$$dl_b = \frac{\tau_i}{\tau}\, dv - \frac{1}{\tau}\int dv\, dt \tag{6.32}$$

The sinusoidal response of level to boilup as derived in Appendix D is

$$\frac{dl_b(\tau_o)}{dv(\tau_o)} = \frac{1}{\tau}\sqrt{\tau_i^2 + \left(\frac{\tau_o}{2\pi}\right)^2}\ \angle -180^\circ + \tan^{-1}\frac{\tau_o}{2\pi\tau_i} \tag{6.33}$$

where τ_o is the period of the sinusoidal oscillation. Undamped oscillations will persist when the product of controller gain and process gain dl_b/dv is 1.0 and the sum of their phase angles is -180°.

A proportional controller is described by

$$m = \frac{100}{P}\, e + b \tag{6.34}$$

where m is its output, e the deviation from set point, P its proportional band in percent, and b an output bias. Its gain is $100/P$ and phase is zero. When this controller is used in a loop with the inverse-responding element alone, undamped oscillations will persist at $\tau_o = 0$ when $P = 100\tau_i/\tau$. If the loop is to be stable, the loop gain should be 0.5 such that

$$P = 200\,\frac{\tau_i}{\tau} \tag{6.35}$$

A steady state will be reached when $V + B = L$. Therefore, any change in L or B must be corrected by an equal change in V. The sensitivity of liquid level to changes in L or B may be estimated by determining how great a change in v would be necessary to restore the steady state. The magnitude of the deviation in level (dl_b) due to an upset in mass flow (dv) into or out of the bottom of the column can then be estimated by substituting dl_b for e and dv for $m - b$ in (6.34):

$$dl_b = \frac{P}{100}\, dv = 2\,\frac{\tau_i}{\tau}\, dv \tag{6.36}$$

This allows a prediction of the size of upset which could be sustained without losing liquid level. For example, if $\tau_i = \tau$, a change of 10 percent in B/V_M or L/V_M would cause a 20 percent change in level.

Proportional-plus-reset control is used when a steady-state deviation dl_b is undesirable. The reset action adjusts the manipulated variable continuously, until the deviation is corrected, at a rate inversely proportional to its reset time constant R:

$$m = \frac{100}{P}\left(e + \frac{1}{R}\int e\,dt\right) \tag{6.37}$$

The gain and phase of the controller are described by

$$G_{PR} = \frac{100}{P}\sqrt{1 + \left(\frac{\tau_o}{2\pi R}\right)^2} \qquad \phi_{PR} = -\tan^{-1}\frac{\tau_o}{2\pi R} \tag{6.38}$$

When this controller is used on a process with inverse response, setting $R = \tau_i$ will cause $-180°$ loop phase to develop for all values of τ_o. Then the period of oscillation of the loop varies with the controller gain—a very unusual situation. For most loops, the period is virtually constant while the damping varies with the gain. As a consequence, the controller on an inverse-responding process is very difficult to adjust, and no combination of proportional and reset may give optimum settings. However, R must exceed τ_i for stability to be achieved at all.

The sensitivity of level to sudden upsets is as described by (6.36) except that the addition of reset requires the proportional band nearly to double to achieve the same damping [14]. Sensitivity to a ramp change in B or L can be estimated by substituting dl_b for e and v for m in (6.37) and differentiating with respect to time. The disturbance is evaluated as dv/dt, the output rate necessary to hold level constant:

$$dl_b = \frac{PR}{100}\frac{dv}{dt} > 4\frac{\tau_i^2}{\tau}\frac{dv}{dt} \tag{6.39}$$

The factor of 4 takes into account the doubling of the proportional band; the inequality indicates that $R > \tau_i$. If, for example, $\tau_i = \tau = 2$ min, a 2 percent/min rate of change of flow into or out of the column base would cause at least a 16 percent change in level.

In actual practice, the boilup rate cannot change stepwise as Fig. 6.11 indicates. Stepwise motion of the heat-input valve will cause an exponential (first-order lag) response in boilup following a short delay. The delay is due mostly to the response of the valve and thermal lags in the heat-transfer surface, while the first-order lag represents the heat capacity of the fluids. In most cases the delay is but 2 to 5 s and the first-order lag perhaps 10 to 20 s. If the inversion time is shorter than this lag, it should not have much effect on the level-control loop.

Usually, fired reboilers are located some distance from their columns. The reboiler return line is partly filled with vapor whose portion increases with boilup rate. When the entire line and reboiler are full of liquid, their contents are W_0. At a given boilup rate V, their contents are reduced to W_r by the percent vaporization. From these three parameters, τ_i can be estimated from (6.30). It is not unusual to encounter fired reboilers whose volume compares with that of the column base. This precludes base-level control by heat-input manipulation.

Buckley et al. [15] also describe inverse response arising from displacement of liquid by vapor within the trays of the column itself. An increase in vapor flow can lower the froth density, lifting more liquid over the weirs and into the column base. In essence, the column liquid holdup varies with vapor flow. Again, if the change in holdup is comparable to the capacity of the column base, level control by heat-input manipulation may be unsuccessful.

The inverse response centers not so much on the actual froth density as on how much it changes with boilup. Van Winkle [16] indicates that it changes more sharply with boilup at low rates than high, so inverse response observed at low boilup rates may disappear at higher rates. Also, perforated trays are more likely to give inverse response than bubble-cap trays.

The column described in Ref. 15 had 100 valve trays, which gave an inversion time of 11 min in response of base level to heat input. A forced-circulation reboiler was used. Level control by heat-input manipulation was impossible, and bottom flow was too low to be used. Although 3 min dead time elapsed in the effect of reflux on base level, reflux was selected as the best variable for its control. The authors determined that valve trays give inverse response at all vapor rates, whereas perforated trays give inverse response at low rates, direct response at high rates, and effectively dead time at intermediate rates.

HEATING MEDIA

There are three basic types of heating media used in distillation: vapors, liquids, and gases from the combustion of fuel. Their heat-transfer characteristics, sensitivity to disturbances, and control response differ markedly. Therefore, it is essential that each medium be evaluated in detail to arrive at the control system which will be most effective in manipulating it. Although steam is the most universally used medium in chemical plants, hot-oil and direct-fired reboilers are quite common in petroleum refineries. Furthermore, the problems presented by heating with liquid media will also appear in preheaters and air- or water-cooled condensers. Consequently, the description of heat-transfer mechanisms given in this section serves as a foundation for related discussions later in the book.

Steam and Other Vapors Steam is not the only vapor used for heating, but it is by far the most common. When temperatures higher than can be obtained with saturated steam are required, another vapor such as one of the Dowtherms is often used. In certain low-temperature applications, a refrigerant is used; the reboiler condenses the compressed vapor and the condenser uses the liquified refrigerant for cooling. All vapors are basically the same in the manner in which they transfer heat.

Superheated steam is a relatively poor heat-transfer medium, as are most gases, from the standpoint of the rate of heat transfer per unit temperature difference. Saturated steam, on the other hand, is an excellent medium because of low resistance to heat transfer and a high latent heat of vaporization. Yet superheated steam is nearly always supplied to reboilers. Even if the steam is saturated at the supply pressure, the pressure loss across the steam control valve develops superheat. This can be seen on the temperature-entropy diagram of Fig. 6.2 by following a contour of constant enthalpy from the saturation pressure to a lower pressure. Throttling drops the pressure, keeps enthalpy the same, and has very little effect on temperature.

With the control valve on the inlet to the reboiler as shown in Fig. 6.12, the saturation pressure in the shell varies with heat load. Since the heat is being transferred between a condensing and a boiling fluid, neither changes temperature greatly in the process. If the relationship between heat-transfer rate and temperature difference is linear and the boiling point of the process fluid is constant, then steam temperature will vary directly with boilup rate. A "steam trap" or similar condensate seal is necessary to drain condensate without releasing steam. If the trap cannot remove the steam as fast as it condenses, shell pressure will rise but steam flow will fall as the shell fills with condensate.

When a steam valve is being used, boilup rate may be controlled either by steam flow or shell pressure. The steam-flow measurement is usually made upstream of the control valve where pressure is ordinarily constant, rather than downstream where it most certainly varies. The calibration of the orifice flowmeter commonly used is sensitive to variations in pressure and temperature. Compensation may be applied for both if a mass-flow measurement is desirable. However, a heat-flow measurement is more meaningful and requires only pressure compensation. Variations in steam temperature affect density and enthalpy in opposite directions so that there is little ultimate effect on heat-flow rate.

If steam pressure rather than flow is controlled, boilup will automatically increase should the boiling point of the process fluid fall. This action compensates correctly for changes in bottom-product composition but not for changes in column pressure. Fouling of the heat-transfer surface will also affect the relationship between steam pressure and boilup. Furthermore, the rate of boilup is not linear with steam pressure nor is the steam pressure zero at zero boilup. Consequently, steam-flow control is preferred to pressure control to establish and maintain a boilup rate, prevent flooding, and control product quality.

There are certain advantages to placing the control valve on the condensate

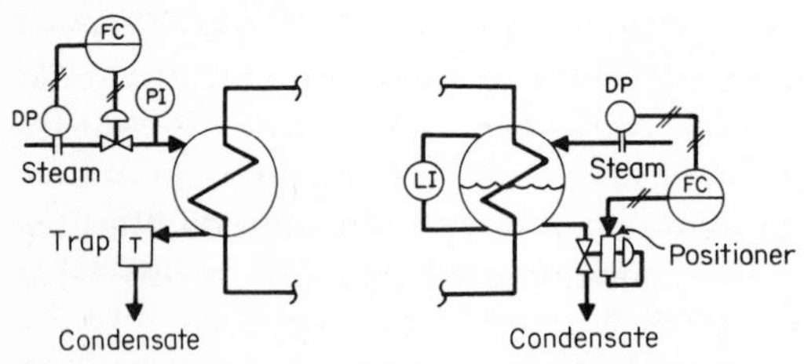

figure 6.12 *Steam flow may be controlled with a valve on either the steam or the condensate.*

side of the reboiler. In the first place, the valve can be smaller. The density of steam at 250 psia, for example, is one-hundredth that of condensate at that pressure. Because of the square-root relationship between valve size C_v and density ρ in Eq. (6.12), the condensate valve need have only one-tenth the capacity of the steam valve for the same pressure drop. But since capacity C_v varies with the square of pipe diameter, the condensate valve will be typically one-third the line size of a steam valve used for the same service. At lower pressures, the ratio is even more favorable.

Secondly, the steam reaching the reboiler is saturated (if the supply is saturated) and at a higher pressure than with a valve on the steam line. Hence the maximum heat-transfer rate is higher, and in addition there is no trap to limit the rate of condensate removal.

Steam flow is the rate of condensation, which in this case is controlled by adjusting the heat-transfer area exposed to condensation. As condensate flow is reduced, the level of condensate in the shell will rise, exposing less surface to the steam and more to the condensate. The condensate tends to be colder and a less efficient heat-transfer medium than steam. Therefore, a rising level results in a lower heat-transfer rate, a lower condensing rate, and a lower steam flow. Because of the flooding of the shell with condensate from which heat continues to be withdrawn, the condensate leaving tends to be subcooled. This may offer no advantage over a trap, however, since condensate discharged from a trap tends to be subcooled with respect to supply pressure when the shell pressure is low.

If the condensate valve cannot carry away the condensate as fast as the reboiler can produce it, a maximum steam flow will be reached with the shell still partially flooded. But if the reboiler cannot condense the steam as fast as the valve is directed to remove condensate, some steam will pass through the valve. The flow at which this limit is reached will decrease as the heat-transfer surface fouls.

The dynamic-response characteristics of the two schemes shown in Fig. 6.12 differ markedly. Manipulating the control valve in the steam line causes steam flow to change immediately. Shell pressure and hence the rate of heat transfer will lag behind a few seconds, but indicated steam flow responds directly. By constrast, the condensate valve has no direct effect on steam flow. Condensate level determines steam flow, and level takes time to change. With the valve in the steam line, the flow controller requires a wide proportional band (>100 percent) and fast reset time (1 to 2 s) typical of most flow controllers. But when a condensate valve is used, the flow-controller settings are more representative of a level controller—a narrower proportional band (25 percent) and longer reset time (30 s). (For an explanation of how control-mode settings relate to the process, see Ref. 17.) In addition, the condensate valve should be equipped with a positioner whereas the steam valve does not need one. Reference 18 describes how a valve positioner can destabilize a flow loop but stabilize a level loop.

Although installing the control valve on the condensate line has several advantages, its slow response in changing heat-transfer rate can be detrimental to base-level control. The base of the column usually has little capacity, and the level measurement tends to be noisy and exhibits the hydraulic resonance associated with a U tube. These effects combine to make base level difficult to control stably when manipulating boilup rate. If the control valve is in the steam-condensate line, the relatively large capacity of the reboiler to store condensate is interposed between the column and the valve, and stable level control may no longer be achievable.

The steam valve is not altogether free of dynamic irregularities either. Buckley [13] notes that the sensitivity of steam flow to the steam valve increases markedly when sonic velocity is reached in the valve. This occurs when the absolute downstream pressure is less than half the absolute upstream pressure. Because downstream pressure then has no effect on flow, the influence of the reboiler capacity apparently is lost and control destabilizes. This condition is most likely to develop at low flow rates and when steam supply pressure is much higher than necessary.

Heating with Liquids Whenever a liquid stream is used to boil up a column, there arises the problem of how to measure and control the heat input. With steam heating, the rate of condensation is directly proportional to the steam flow. But as will be seen, with liquid media the relationship is very nonlinear and boilup is subject to change even with flow held constant. To demonstrate this point, the relationships governing heat transfer will be developed.

A liquid medium gives up heat flow Q proportional to its mass-flow rate W and temperature loss in passing through the reboiler:

$$Q = WC(T_1 - T_2) \qquad (6.19)$$

The rate of heat transfer to a boiling liquid has already been described:

$$Q = UA(\Delta T - \Delta T_0) \qquad (6.20)$$

The temperature difference ΔT between the heating medium and the boiling liquid varies as the medium passes through the exchanger. Let the temperature of the heating medium at any point be T, with the temperature of the boiling liquid designated T_b. Then for an increment of heat-transfer surface dA, (6.19) and (6.20) may be written as

$$dQ = -WC\,dT = U\,dA(T - T_b - \Delta T_0) \qquad (6.40)$$

Next, let (6.40) be solved for dA in terms of dT and integrated from T_1 to T_2 to yield A:

$$A = \int_0^A dA = -\frac{WC}{U}\int_{T_1}^{T_2} \frac{dT}{T - T_b - \Delta T_0}$$

$$= \frac{WC}{U} \ln \frac{T_1 - T_b - \Delta T_0}{T_2 - T_b - \Delta T_0} \qquad (6.41)$$

Then outlet temperature may be related to inlet temperature and flow by converting (6.41) into an exponential form:

$$T_2 - T_b - \Delta T_0 = (T_1 - T_b - \Delta T_0)e^{-UA/WC} \tag{6.42}$$

The heat-transfer rate may be related to T_1 and W by expanding Eq. (6.19) and substituting (6.42) into it:

$$Q = WC[(T_1 - T_b - \Delta T_0) - (T_2 - T_b - \Delta T_0)]$$

$$Q = WC(T_1 - T_b - \Delta T_0)(1 - e^{-UA/WC}) \tag{6.43}$$

From (6.43) it may be seen that heat transfer is linear with temperature difference $T_1 - T_b$ but decidedly nonlinear with flow W. To assess this nonlinearity, divide (6.43) by Q_M, the maximum heat flow corresponding to an infinite flow of heating medium, i.e., when $T_2 = T_1$:

$$\frac{Q}{Q_M} = \frac{WC(T_1 - T_b - \Delta T_0)(1 - e^{-UA/WC})}{UA(T_1 - T_b - \Delta T_0)}$$

$$\frac{Q}{Q_M} = \frac{WC}{UA}(1 - e^{-UA/WC}) \tag{6.44}$$

The solution to Eq. (6.44) is plotted in Fig. 6.13. The actual operating level of WC/UA for a given reboiler may be found by solving (6.41). An equal-percentage control valve should be used to compensate this nonlinearity.

In actual practice, U is not expected to remain constant but will tend to vary in somewhat direct proportion to liquid flow W. If U varied linearly with W, then the exponent in (6.43) would be constant and Q would vary linearly with W. Heat transfer tends to vary with the 0.8 power of velocity, however. If the

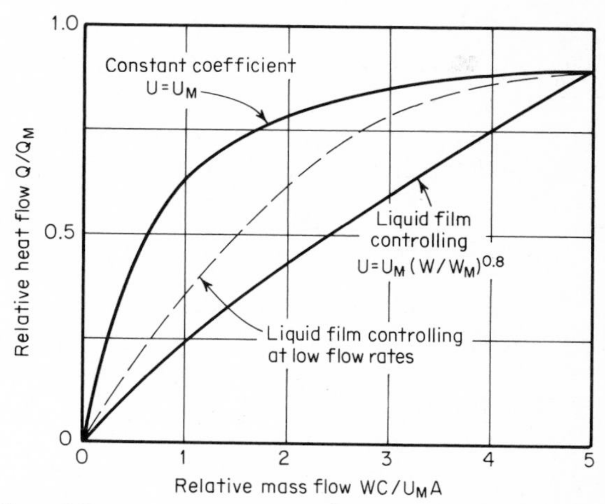

figure 6.13 *The rate of heat transfer may be sharply nonlinear with the flow of liquid heating medium.*

liquid-side film were completely controlling, then, such that U varied with $W^{0.8}$, Eq. (6.44) may be restated as

$$\frac{Q}{Q_M} = \frac{WC}{U_M A}\left[1 - e^{-(U_M A/WC)^{0.2}(U_M A/W_M C)^{0.8}}\right] \tag{6.45}$$

where $U_M = U$ at W_M, the maximum flow.

For the case where $W_M C/U_M A = 5$, the solution to Eq. (6.45) is plotted in Fig. 6.13. Observe how the variation of film resistance with flow moderates the nonlinearity. The actual relationship for any given installation will lie somewhere between the two curves, since U is never constant nor is the liquid film controlling throughout. Typically, liquid-film resistance would be controlling at low rates but not at high rates; thus the relationship between Q and W would be relatively linear at low flow rates but would approach the limiting upper curve at high rates.

Direct Combustion of Fuel Heating by direct combustion of fuel is common only where high temperatures are needed, as in crude-oil heating, and is practiced principally in refineries, where fuel is available from various sources. Most of these units are gas-fired, using mixtures of hydrogen and hydrocarbons vented from columns and other vessels, supplemented by natural gas and sometimes propane. One of the problems accompanying the use of mixed fuels is the varying heating value and density of the mixture.

The heat obtainable by combustion of a volumetric unit F of fuel gas is

$$Q = H_c F \tag{6.46}$$

where H_c is the heating value per standard volume of gas. Orifice meters are customarily used for gaseous fuels, and their differential pressure h relates to volumetric flow as

$$F = k\sqrt{\frac{hp}{\rho T}} \tag{6.47}$$

where k = meter coefficient
h = orifice differential pressure
p = absolute static pressure
T = absolute temperature
ρ = specific gravity of the gas

The specific gravity is simply its molecular weight divided by 29, the molecular weight of air. Equation (6.47) assumes that the fuel gases commonly used follow the ideal-gas law under metering conditions. Combining (6.46) and (6.47) yields the heat flow as a function of meter differential and specific gravity:

$$Q = k\frac{H_c}{\sqrt{\rho}}\sqrt{\frac{hT}{p}} \tag{6.48}$$

Pressure and temperature compensation may be applied where necessary. Correction for composition is the factor $H_c/\sqrt{\rho}$, known as the "Wobbe index."

For hydrocarbon gases, the Wobbe index varies uniformly with gas specific gravity as plotted in Fig. 6.14. Therefore, a specific-gravity measurement alone may be used to apply heat-flow correction. Mixtures containing hydrogen deviate substantially from the nearly linear relationship. But as Fig. 6.14 indicates, variation of the hydrogen content of a given mixture does not really need correction.

The relationship between gravity and Wobbe index breaks down completely when components such as nitrogen or carbon monoxide or dioxide enter the fuel system. Although their gravities are high, their heating values are low or nil, and their presence in the fuel cannot then be corrected by gravity alone.

Thermal calorimeters are available for measuring the heating value of a fuel gas, and some even generate the Wobbe index. But in general they respond no faster to a fuel-composition change than the heater does, and their correction is too late to be of benefit in controlling temperature. They are, in fact, a pilot model of a heater and are therefore faced with the same type of thermal lags that characterize a fired heater.

The airflow required for complete combustion varies directly with the heat flow. Therefore, if a heat-flow signal is available, it can be used to set airflow. If airflow is set in ratio to an uncompensated orifice differential, however, increasing fuel gravity will also increase the fuel-air ratio. This could mean smoky or even hazardous operation with high fuel gravity or too much excess air and inefficient combustion with low-gravity fuels.

Fuel-air ratio is most satisfactorily controlled by using an oxygen analyzer sampling the flue gas. The system shown in Fig. 6.15 provides this control. Linearizing the flow signal by square-root extraction is absolutely essential to stable control of product temperature. If the temperature controller sets differential pressure rather than flow, the gain of the temperature control loop will vary with flow because

$$\frac{df}{dh} = \frac{1}{2f} \tag{6.49}$$

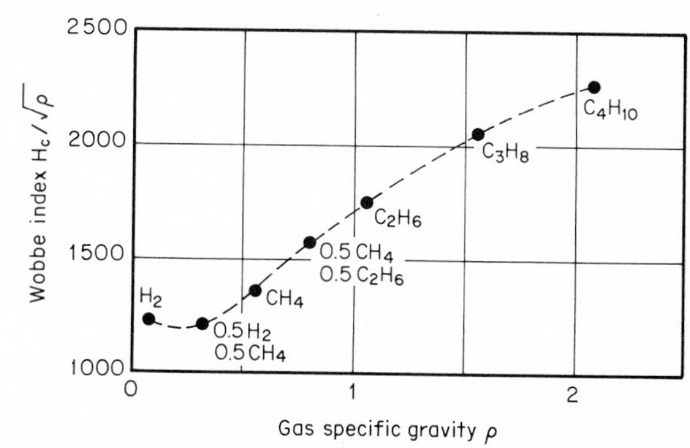

figure 6.14 *The Wobbe index varies uniformly with specific gravity for mixtures of light hydrocarbons.*

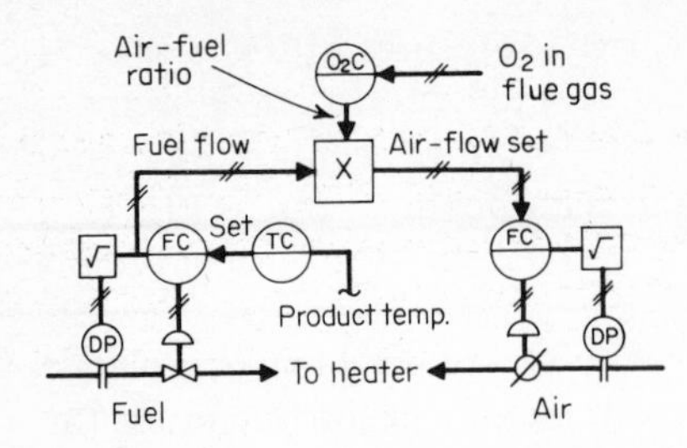

figure 6.15 *Air-fuel ratio is adjusted to hold a desired O_2 level in the flue gas.*

where h is the fractional differential pressure and f is the fractional flow. If the temperature controller sets h but f affects temperature, the control-loop gain increases as flow decreases. On the other hand, if the primary control loop is not temperature but column differential pressure, square-root extraction is not required because both column and flowmeter have the same nonlinear characteristic.

Burner backpressure has been used successfully as a flow signal, with the burner acting as a fixed orifice. But if the number of burners is variable or their apertures are adjustable, calibration will shift. A square-root extractor should be applied to the pressure signal when used for temperature control. Stack temperature has also been used as a secondary controlled variable responding to variations in fuel quality and load. Furthermore, it is almost linear with firing rate.

Precise control of excess air is essential to conserve fuel. Combustion of methane proceeds accordingly:

$$CH_4 + 2(1+X)O_2 + 7.52(1+X)N_2 \rightarrow CO_2 + 2H_2O + 2XO_2 + 7.52(1+X)N_2 \tag{6.50}$$

where X represents the fractional excess air and 7.52 is the number of moles of nitrogen accompanying 2 mol of oxygen in air. The products of combustion number $10.52 + 9.52X$ mol; consequently, 1 percent excess air increases the molal flow by about 1 percent and reduces internal temperatures by the same amount. For a given temperature, heat losses increase by that same amount. An oxygen analyzer on the flue gas measures the $2X$ mol of oxygen in the presence of the total moles of flue gas. The mole fraction of oxygen in the flue gas is related to excess air as

$$O_2 = \frac{2X}{10.52 + 9.52X} \tag{6.51}$$

Energy loss due to excess air increases as higher flue-gas temperatures are reached. Therefore, fired heaters with high-temperature flues can more readily benefit from flue-gas analysis.

The problems of maintaining control and safe operation while fuel or air is constrained are described in Chap. 9, "Controlling within Constraints."

CONTROLLING VAPOR FLOW

For steam-heated reboilers, controlling steam flow virtually ensures a proportional vapor flow through the column. The only vapor-flow disparities likely to be encountered in this case would be the result of vapor sidestream withdrawal, vaporized feed, or internal condensation caused by subcooled feed or reflux. The vapor-sidestream problem can be corrected by the controls shown in Fig. 3.6.

Direct firing is almost as dependable as steam in providing a proportional vapor flow, as long as fuel composition does not change drastically. But liquid heating media give a very nonlinear response of heat flow to liquid flow, and the relationship is quite sensitive to temperature difference.

If the heat input is only called upon to *close* the heat balance by holding column pressure or a liquid level constant, there is no need to measure heat flow. Then the column heat loading is set by reflux or condensing rate. But in most cases heat outflow is not as measurable as heat inflow, necessitating some means of measuring and controlling vapor flow.

Column Differential-pressure Control Overhead vapor flow has been measured with an orifice in some columns, and this gives generally satisfactory results, although it is an expensive installation for large vapor lines. An alternative is to measure the differential pressure across the column, using the trays as an orifice. Boilup measurement in this manner is subject to inaccuracies due to the head produced by variations in liquid downflow. Nonetheless, column differential has been used successfully to control boilup and is a sensitive indication of downcomer flooding. The relationship between differential pressure and flow is square root, as an orifice. In most installations, liquid and vapor flow go up and down together, so the liquid-flow effect is not often a problem.

The most serious obstacle in the way of obtaining a reliable measurement is in the installation of the transmitter. The two pressure taps are often located over 100 ft apart, and the lines typically contain vapor which will condense at ambient temperatures. If the transmitter is located below a pressure tap, liquid will accumulate in the line and give a false reading. Normal boilup may generate only 100 in. of water differential pressure across a 100-ft-high column. Thus a few feet of liquid head, or even its variation with temperature, would cause serious error.

The line leading from the bottom of the column has problems, too. Even if it is straight and vertical, heat losses can develop a reflux within, which also causes an error. Therefore, vapor lines leading to the transmitter should be insulated. Purging them with noncondensable gas is not recommended since the gas will accumulate to the point where condensing is interfered with and pressure control becomes affected.

For most columns, the overhead condenser is not located above the column but instead is at grade or first level. The vapor line typically runs alongside the column most of the way down. If the top pressure tap is made where the

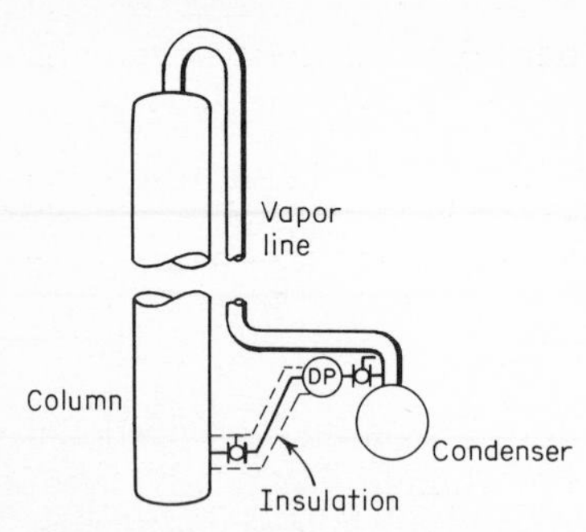

figure 6.16 *Connecting the low-pressure tap to the condenser inlet minimizes the difference in elevation between the taps.*

vapor line enters the condenser, much of the elevation difference between the taps will be eliminated, as shown in Fig. 6.16. The transmitter will be as accessible as the condenser, and only the bottom line need be insulated. For double-column installations, the transmitter for the bottom section may be mounted at grade level between the sections. Plug-style valves should be used, and horizontal runs avoided wherever possible.

Column differential pressure responds rapidly to heat input, controlling with a period of 20 s or thereabouts—essentially limited only by the dynamics of the reboiler. It is therefore quite capable of correcting for temperature or flow variations in the heating fluid as well as ambient disturbances.

Differential-pressure control is recommended for the top section of a two-section column in order to overcome variations in heat losses in the long vapor and liquid lines connecting the sections. As mentioned earlier, excessive vapor and liquid flow rates can raise the differential pressure across a tray so high that the liquid cannot flow through the downcomer. This is called downcomer flooding and is readily detectable by a sharp rise in column differential. Failure of liquid to flow downward fills the trays with froth, increasing their resistance to vapor flow. Entrainment, however, does not notably affect column differential; yet the vapor velocity at which entrainment becomes serious can be measured and controlled through differential pressure.

Calculating Heat Transfer Heat transfer can be calculated directly by using temperature and flow measurements to solve Eq. (6.19). The calculation may be implemented either with analog or digital components as described in Fig. 6.17. The dynamic response of this control loop is somewhat peculiar. A step increase in liquid flow will cause a direct and linear increase in calculated heat flow. However, the increased flow will soon cause outlet temperature T_2 to rise in accordance with Eq. (6.42). As a result, the indicated heat flow will lose much of the increase it gained immediately following the change. Thus the indicated heat flow always leads the true heat flow by the residence time of the liquid in the reboiler.

The dynamic response of outlet temperature to flow also creates a serious control problem. In an earlier work [19], the author described an attempt to control the outlet temperature of a liquid leaving a heat exchanger by manipulating its own flow. Because the dead time in the response of temperature to flow varies with flow, it is impossible to find controller settings which provide uniform damping. The control loop tends to cycle continuously with a period and amplitude that vary inversely with flow. Because this temperature loop is one constituent of the heat-flow loop, undamped oscillations may be expected. This analysis is confirmed by others who have encountered considerable difficulty arriving at stable settings for the heat-flow controller, even at a single flow condition. This loop is a candidate for adaptive control wherein the mode settings are adjusted proportional to flow.

Controlling Preheat As mentioned before, energy put into preheating the feed is less useful than if used for reboiling because the resulting vapor passes only through part of the trays. Nonetheless, preheaters are valuable for recovering lower-level heat than the reboiler can accept and for balancing the vapor loading in the column. A subcooled feed means less vapor and less liquid above the feed tray than below, which has the potential for flooding the stripping section while failing to utilize the full capacity of the enriching section. Complete vaporization of the feed is desirable in some applications, notably extractive distillation.

Variations in feed enthalpy affect the material flows within the column, which could be reflected in product quality variations if not corrected. Material-balance control systems are naturally resistant to these enthalpy-related influences. Yet Lupfer [11] cites instances of cyclic behavior caused by feedback of bottom-product enthalpy to column feed via bottom-feed exchangers. He reports that the response of a feed-temperature controller manipulating steam flow into a trim preheater was insufficient to quench the instability. Instead, he bypassed a controlled flow of feed around all the exchangers, blending with preheated feed for improved response. The amount of blending was minimized by a second controller which maintained the average position of the bypass valve at a specified point by manipulating steam to the trim heater. This is the familiar valve-position controller described in Fig. 6.6. Any bypassing is undesirable since the bypass stream fails to recover heat from the bottom products, which must be made up by the trim heater.

Preheating with steam follows essentially the same principles as reboiling with steam. And preheating with a liquid features the same nonlinearity as reboiling with hot oil, for example. When controlled preheating is to be achieved using a process fluid whose flow cannot be manipulated, one of the fluids must bypass the exchanger as shown in Fig. 6.18. A three-way valve is recommended since it minimizes pressure loss while allowing flow to be completely diverted in either direction. A single two-way valve bypassing the exchanger has limited rangeability in that some flow will always pass through

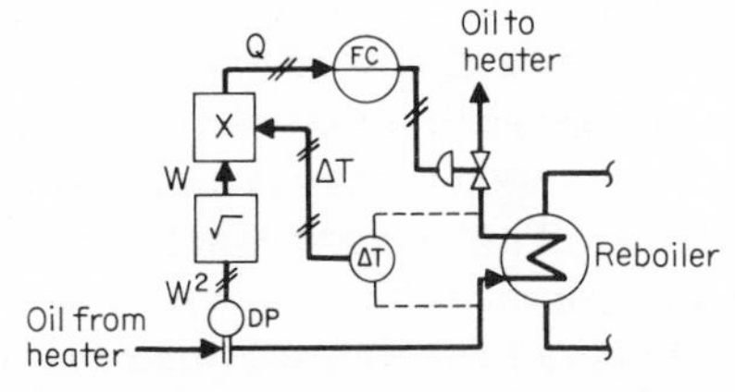

figure 6.17 *Heat input may be calculated from flow and temperature difference and controlled directly.*

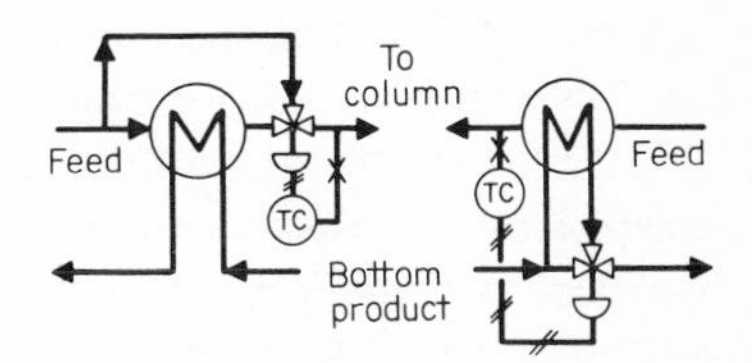

figure 6.18 *Preheat may be controlled by bypassing either fluid around the exchanger, but the response differs in each case.*

the exchanger. Two two-way valves may be used instead of a three-way valve, but the cost and the pressure loss both tend to be higher.

Bypassing of the cold feed provides fast response of temperature to valve position. Reducing the flow through the exchanger will cause its exit temperature to rise after the fluid residence time has elapsed, however, so the net change in heat recovery may be small. As Fig. 6.13 indicates, at relatively high flow rates Q may change scarcely at all with flow. The dynamic response of this loop is very much like the heat-flow control loop where heat flow is calculated. The initial response in temperature is moderated by a later reaction in the other direction. In this case, however, the controlled variable is the true feed temperature whereas the heat-flow calculation led the true heat-flow rate.

Bypassing the heating medium does not offer the same speed of response as bypassing the feed. Its linearity is limited by the same mechanism described by Fig. 6.13. Three-way valves with equal-percentage characteristics are therefore preferred for both schemes shown in Fig. 6.18.

If the feed is simply to be heated and not vaporized, temperature control ought to suffice. From an economic standpoint, total heat recovery would be desirable, which would not require control at all. However, destabilizing reactions could occur as described earlier, particularly where the heating medium forms a feedback loop with the column feed stream.

In other cases, partial vaporization is desirable. Then temperature control may not be satisfactory since the degree of vaporization would change with feed composition. A lower-boiling mixture would require more vaporization to reach the same temperature as a higher-boiling feed. This would seem to be desirable in directing more vapor upward and less liquid downward when the feed contains more volatile components.

In cases where column pressure varies, the feed temperature measurement must be pressure-compensated if it is to be controlled. Where steam is used for preheating, simply setting steam flow in ratio to feed flow provides a constant heat gain. If the enthalpy of the feed is reasonably constant, the ratio can be set to give a certain percent vaporization. A percent-vaporization measurement may also be obtained from a ratio of the differential pressure across orifices on the inlet and outlet of the heater.

Rather than control at some arbitrary temperature or percent vaporization, a preheater can be used to balance column loading. This is most readily achieved by controlling the differential pressure across both the stripping and the enriching sections of the column. Reboiler heat input can be manipulated to control stripping section differential, and feed preheat can be adjusted to control that across the enriching section. This control system has the advantage of correcting promptly for disturbances in feed composition and enthalpy and for variations in the enthalpy of the heating medium as well. It can be applied even when the feed is not vaporized at all, since any change in feed enthalpy will affect the relative loading both above and below the feed tray.

REFERENCES

1. Shinskey, F. G.: "Process-Control Systems," McGraw-Hill Book Company, New York, 1967.
2. Keenan, J. H., et al.: The Fuel Shortage and Thermodynamics, MIT Industrial Liaison Program, Paper 12–13–73.
3. Denbigh, K.: "The Principles of Chemical Equilibrium," p. 63, Cambridge University Press, London, 1961.
4. *Ibid.*, p. 116.
5. O'Connor, J., and H. Illing: The Turbine Control Valve . . . a New Approach to High-loss Applications, *Instrum. Technol.*, December 1973.
6. Denbigh, K.: *op. cit.*, p. 70.
7. Shinskey, F. G.: Temperature Control for Gas-phase Reactors, *Chem. Eng. (N.Y.)*, Oct. 5, 1959.
8. Shinskey, F. G.: "Process-Control Systems," pp. 264–268.
9. Adiutori, E. F.: "The New Heat Transfer," chap. 7, Ventuno Press, Cincinnati, 1974.
10. Frank, O., and R. D. Prickett: Designing Vertical Thermosyphon Reboilers, *Chem. Eng. (N.Y.)*, Sept. 3, 1973.
11. Lupfer, D. E.: Distillation Column Control for Utility Economy, presented at 53rd Annual GPA Convention, Denver, Mar. 25–27, 1974.
12. Shinskey, F. G.: "Process-Control Systems," pp. 71–73.
13. Buckley, P. S.: Material Balance Control in Distillation Columns, presented at AIChE Workshop on Industrial Process Control, Tampa, Nov. 11–13, 1974.
14. Shinskey, F. G.: "Process-Control Systems," p. 102.
15. Buckley, P. S., R. K. Cox, and D. L. Rollins: Inverse Response in a Distillation Column, *Chem. Eng. Prog.*, June 1975.
16. Van Winkle, M.: "Distillation," pp. 515–516, McGraw-Hill Book Company, New York, 1967.
17. Shinskey, F. G.: "Process-Control Systems," p. 86.
18. Shinskey, F. G.: When to Use Valve Positioners, *Instrum. Control Syst.*, September 1971.
19. Shinskey, F. G.: Controlling Unstable Processes, pt. II: A Heat Exchanger, *Instrum. Control Syst.*, January 1975.

CHAPTER SEVEN

Control of Energy Outflow

Most of the previous chapter was devoted to the characteristics of reboilers and the heating media supplied to them. Most of this chapter is similarly devoted to condensers and their cooling media. While air- and water-cooled condensers provide the heat sinks for most columns, alternative methods of cooling are being explored in an effort to conserve energy in separations where they apply. Because the less familiar heat pumps may reach prominence in the near future, considerable attention is given to their operation and control. Refrigeration, because of its high cost, is also discussed, with the objectives of maximizing thermodynamic efficiency through proper selection of controls. But all these systems ultimately reject energy to a condenser. Therefore, this is the logical subject to begin a discussion of energy outflow.

CONDENSERS

Control over heat removal from a column is just as critical as control over heat input and is subject to more upsets. Most cooling systems reject heat to the environment, which can be quite variable, depending on the medium used. Dry air is a more variable heat sink than wet air, which in turn is more variable than surface water, groundwater, etc. Furthermore, the cooling medium is, in most cases, not readily manipulable, nor is the resulting heat flow easily measurable. As a result, heat *input* to a column is usually more controllable than heat *removal*, such that the latter is more often used to close the heat balance. In this role, it is a slave to the heat-input system as well as to the environment.

Air-cooled Condensers The condenser that has increased most in popularity in recent years is the dry air-cooled type. Air is directed upward through a horizontal bundle of finned tubes by a fan. The fan may be located either below or above the tubes. Buckley [1] reports that condensers with fans mounted above the tubes are less sensitive to rainfall.

The factors governing the rate of condensation are essentially the same as those described for the liquid-heated reboiler. The temperature of the condensing vapor tends to be constant if the tubes are not flooded with liquid. Therefore, Eq. (6.43) applies as revised below:

$$Q = WC(T_c - T_a)(1 - e^{-U_M A/W^{0.2}C}) \tag{7.1}$$

where T_c = condensing temperature
T_a = air inlet temperature
W = mass flow of air
U_M = a constant
C = specific heat of air

The rate of heat transfer is definitely limited by the air film, and hence the 0.2 power in the exponent is justifiable. As a consequence, heat flow should be reasonably linear with airflow as in the lowest curve of Fig. 6.13.

However, controlling the airflow is quite another matter. Variable-speed fans are rarely used because of the high cost of the drive units. Most air coolers use multiple fans, which can be energized in stages, but this gives only incremental control. Some fans are equipped with variable-pitch blades. Although this might seem an attractive means of manipulating airflow, operators report that the pitch-control mechanism tends to respond erratically. While it may control reflux *temperature* satisfactorily, it is not recommended for *level* or *pressure* control. Adjustable louvers may give satisfactory results, but they are not in widespread use, perhaps because of the large areas of tubes which they must cover.

Dry air-cooled condensers are designed for about a 50°F approach to dry-bulb temperature. Without controls, condensate temperature T_c will vary with the heat load Q and ambient temperature T_a as Eq. (7.1) indicates. Naturally column pressure will vary with T_c. Even worse, rain tends to convert the dry condenser into a wet condenser, reducing its resistance to heat flow and lowering T_a toward the wet-bulb temperature. The onset of rain can be a severe and sudden upset to a column, and some means must be provided to counter its effect. Since manipulation of airflow is usually not satisfactory—except for stopping fans—manipulation of the process is required.

There are four basic mechanisms for adjusting (reducing) cooling on the process side of a condenser:

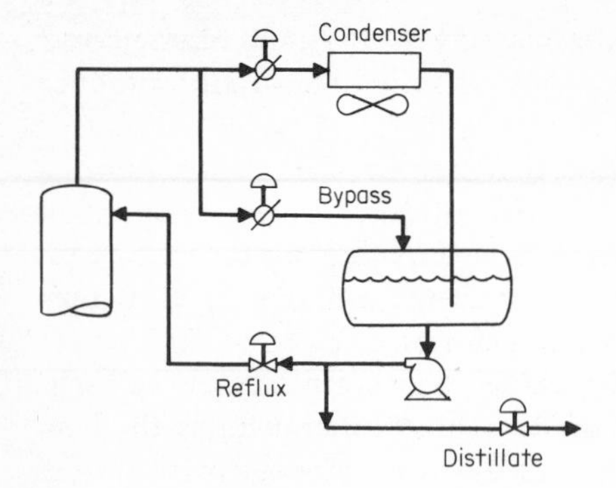

figure 7.1 *A hot-gas bypass is the most common means of controlling air-cooled condensers.*

1. Flooding with condensate
2. Bypassing
3. Reducing condenser pressure
4. Adding noncondensable gas

Most dry air-cooled condensers are horizontal, with only two to four rows of tubes oriented vertically, so that flooding does not give smooth control. Mounting in a sloping plane would help. But the tubes also tend to be of large diameter and can therefore contain a substantial amount of liquid when flooded, which limits the rate of response to flow changes.

Bypassing hot vapors around the condenser as shown in Fig. 7.1 is common practice. Note that the condensate and bypass lines are *not* joined before entering the accumulator. If they were, the direct contact of hot and cold streams within the confines of the pipe would cause water hammer, just like mixing water and steam.

Note also that *two* valves are required. Installations without a valve in the line to the condenser have experienced very limited rangeability. Even with the bypass fully open the condenser carries flow, and it may be enough to cool the reflux too low to hold pressure. Vapor flowing through the bypass must also condense against the walls of the accumulator and the surface of the condensate. Extremely cold condensate may increase this rate of condensation beyond the capacity of the bypass valve so that control is lost. But throttling the vapor to the condenser can avoid these limitations.

Column pressure can be controlled by the bypass valve, with accumulator pressure controlled by the condenser-inlet valve, or vice versa. Or the differential pressure between column and accumulator can be controlled by one of the valves. All these schemes are workable, although there will always be some interaction between the pressure controllers. A simpler arrangement has the column-pressure controller manipulating both valves, one closing as the other opens.

Column pressure may be controlled quite satisfactorily with the condenser inlet valve alone. The disadvantage of this system is that it lowers accumulator pressure and forces the reflux pump to work against an increased head (the pressure drop across the control valve).

Injecting a noncondensable gas into the condenser or accumulator is not recommended. Although it increases the vapor pressure of the overhead mixture, it contaminates the product; and when the heat load is increased, it must be vented. Upon venting, it carries product away as described in Example 3.1.

Evaporative Condensers Evaporative condensers are air-cooled condensers with tubes contained in a wet cooling tower. The wet surface improves heat transfer, and the evaporation taking place provides an environment which is more constant in temperature. Design conditions are typically a 20°F approach to wet-bulb temperature, which is always lower than dry-bulb temperature. Consequently, condensate temperatures are much lower than with air cooling, and the heat capacity of the circulated water reduces its sensitivity to atmospheric conditions. The disadvantages include water pumping, treatment, blowdown, protection from freezing, and difficulties in detecting and repairing leaks. But the evaporative condenser may be used more economically for condensing propane, propylene, and other components of similar volatility where dry air-cooled devices may be unsatisfactory. Evaporative condensers provide lower condensing temperatures, lower vapor pressures, and enhanced relative volatilities.

Figure 7.2 describes a typical evaporative condenser [3]. Water is sprayed downward concurrently with the induced air. The process flow is multipass upward, so that the outflow is in contact with the coldest water and air. Upflow condensation causes some problems in that the condensed liquid prefers to gravitate downward. Consequently, the condenser tends to discharge in slugs rather than smoothly.

Heat flow is proportional to the difference between water and condensate temperatures. When water flow is constant, the exponential term in Eq. (7.1)

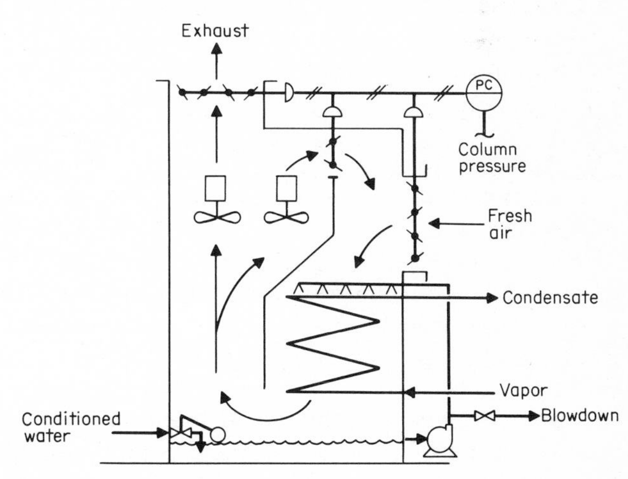

figure 7.2 *An evaporative condenser may be controlled by recirculation of humid air.*

will be constant such that

$$Q = U_c' A_c (T_c - T_w) \tag{7.2}$$

Here U_c' is not simply a heat-transfer coefficient but includes the exponential term as well; A_c is the surface area of the condenser and T_w is the temperature of the water approaching the tubes.

In the steady state, the same flow of heat must be transferred to the air by evaporation. The driving force for this transfer is the difference between water temperature and the wet-bulb temperature of the air:

$$Q = U_w' A_w (T_w - T_{wb}) \tag{7.3}$$

Here again, U_w' includes an exponential term relative to airflow, which should be constant if air and water flow rates are constant; A_w is the surface area of water exposed to evaporation and T_{wb} is the wet-bulb air temperature. Equations (7.2) and (7.3) may be combined by eliminating water temperature:

$$Q = \frac{T_c - T_{wb}}{1/U_c' A_c + 1/U_w' A_w} \tag{7.4}$$

For the most part, air and water flow rates tend to be constant, in which case $U_c' A_c$ and $U_w' A_w$ are constant. However, wet-bulb temperature changes seasonally, diurnally, and with the weather.

Wet-bulb temperature increases with both dry-bulb temperature and humidity. It varies from a minimum of 32°F to as high as 80°F in temperate zones. It is not subject to as rapid nor as wide fluctuations as dry-bulb temperature. Rainfall which cools the air may scarcely affect wet-bulb temperature at all, since it increases humidity at the same time.

By humidifying the air, much more heat may be removed per unit airflow than with a dry condenser. Heat flow may be expressed as the mass flow W of dry air multiplied by its change in enthalpy or humid heat (dry basis):

$$Q = W(H_2 - H_1) \tag{7.5}$$

From a psychrometric chart it may be seen that enthalpy is essentially a function of wet-bulb temperature alone. Then Eq. (7.5) may be restated by substituting wet-bulb temperatures for enthalpies:

$$Q = WC_w (T_{wb2} - T_{wb1}) \tag{7.6}$$

where C_w is defined as the wet-bulb heat capacity of air:

$$C_w \equiv \frac{dH}{dT_{wb}} \tag{7.7}$$

The wet-bulb heat capacity varies from 0.57 Btu/lb-°F at 50°F wet bulb to 1.19 Btu/lb-°F at 80°F. By contrast, the dry-bulb heat capacity varies from 0.21 to 0.25 Btu/lb-°F at 60 to 120°F. The wet-bulb heat capacity is therefore two to five times as great for equivalent conditions. The combination of higher heat capacity and lower resistance to heat transfer gives evaporative cooling several advantages over dry-air cooling:

1. Airflow may be reduced, and with it fan horsepower.
2. Multiple passes may be employed on the process side, and total heat-transfer area may be reduced.
3. Much lower condensing temperatures are achievable.
4. The reduced airflow and size of the unit allow it to be controlled with dampers.

This last factor gives the evaporative condenser a very wide turndown range. As indicated in Fig. 7.2, the fresh-air, exhaust, and recirculation dampers are all operable from the same control signal to adjust fresh-air flow from zero to full capacity. With internal air and water flows maintained constant, the internal wet-bulb temperature at the top of the condensing tubes—T_{wb} in Eqs. (7.3) and (7.4)—varies with fresh-air flow. This system with controlled recirculation may be analyzed in exactly the same manner as was done for the oil-heated reboiler in Chap. 6. Accordingly,

$$Q = W_M C_w (T_c - T_{wb})(1 - e^{-UA/W_M C_w}) \tag{7.8}$$

where W_M is the fan capacity and UA is the product of overall heat-transfer coefficient and area between wet air and the condenser. Mixing of fresh air W at T_{wb1} with recirculated air $W_M - W$ at T_{wb2} determines T_{wb}:

$$T_{wb} = T_{wb1} + (T_{wb2} - T_{wb1}) \frac{W_M - W}{W_M} \tag{7.9}$$

Substituting (7.6) into (7.9) eliminates T_{wb2}. Then substituting the result into (7.8) gives a solution of Q in terms of W and W_M:

$$Q = \frac{W_M C_w (T_c - T_{wb1})(1 - e^{-UA/W_M C_w})}{1 + [(W_M - W)/W](1 - e^{-UA/W_M C_w})} \tag{7.10}$$

For high internal-circulation rates, $W_M C_w / UA$ will typically exceed 4 as in Fig. 6.13 so that the exponential terms in (7.10) become insignificant. Then (7.10) reduces to

$$Q \approx W C_w (T_w - T_{wb1}) \tag{7.11}$$

Therefore, in the recirculating system heat flow is effectively linear with fresh-air flow.

The condenser requires an auxiliary source of heat for protection from freezing, but it is only required when the process is not operating during freezing conditions.

Water-cooled Condensers Water-cooled condensers were most common in the past but are used less in new installations where water conservation is being pursued. Water may be available from river or well, or it may be recirculated through a cooling tower. Well water is most uniform in temperature but is usually reserved for those units that especially need its 55°F cooling. River water typically contains suspended solids which can deposit on heat-transfer surfaces at velocities below 3 ft/s. Both river and well water contain varying quantities of calcium bicarbonate, which tends to form calcium carbonate scale on heating above 140°F or so. Because cooling-tower water is recirculated, it can be treated to minimize deposits and corrosion and is less restrictive than once-through cooling water. However, its temperature is more variable, riding on atmospheric wet-bulb temperature.

Water-cooled condensers may be mounted directly in the head of the column or externally in either a horizontal or a vertical alignment. If the condensing vapor is clean and noncorrosive, as are most petroleum distillates, horizontal condensers with water in the tubes are customary. Corrosive vapors are usually condensed in vertical downflow tubes. Where noncondensable gases must be vented on a regular basis, horizontal shell-side condensing is preferable.

Internal condensers present the problem of measuring and controlling reflux and distillate. Ideally, all the condensate should be trapped and withdrawn for metering and control. But some columns are fitted with an internal weir, allowing only distillate to be withdrawn. Although this eliminates a reflux pump and flowmeter, it reduces the options available for control. A rectangular weir tends to make reflux flow too sensitive to distillate withdrawal—a vee-notch weir is preferable.

Coolant-side manipulations are rarely used to adjust the heat-transfer rates in water-cooled condensers. Response of condensing rate to water-flow variation is both nonlinear (as described in Fig. 6.13) and slow. Furthermore, the speed of response varies with coolant flow, causing control-loop stability to deteriorate as flow decreases. Finally, limits placed on low velocity and high temperature restrict the range over which water-flow manipulation is possible.

However, coolant-temperature manipulation is both linear and efficient if a circulating pump is used to maintain a constant flow through the condenser as shown in Fig. 7.3. Heat removal may be changed by manipulating the cold-water valve directly or by adjusting the set point of a temperature controller sensing condenser outlet conditions. The relationship between heat flow and cold-water flow is essentially linear, as described by (7.11) for the evaporative condenser. Adjusting the temperature set point in cascade will

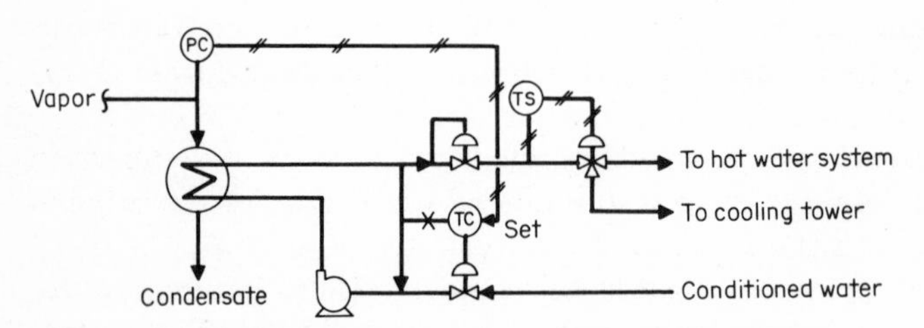

figure 7.3 *Adjustment of cooling-water temperature provides responsive and linear control over condensing rate.*

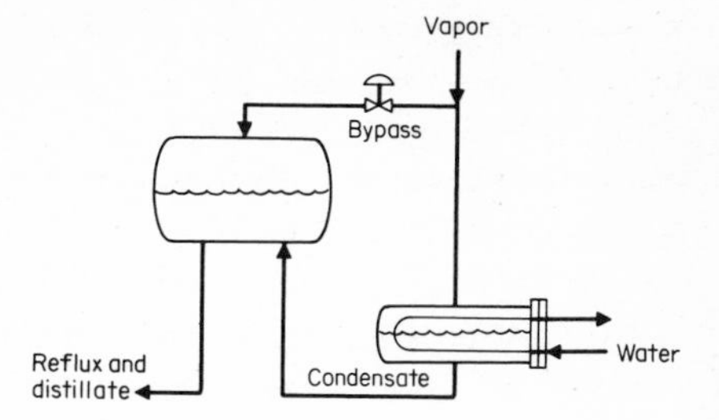

figure 7.4 *Heat-transfer rate in a submerged condenser is controlled by flooding.*

improve the speed of response of the primary loop; proportional action alone is satisfactory for the temperature controller.

Pressure control by manipulating coolant temperature may not be as responsive as using a hot-vapor bypass, but rangeability, reliability, and linearity are superior. Yet its most outstanding feature is the capability of recovering energy at the highest attainable level. Whereas all other heat-removal control schemes attempt to reduce the efficiency of the condenser—term $U_c'A_c$ in Eq. (7.2)—this system does not. Instead, it maximizes the temperature of the discharged water, T_w, by maintaining constant maximum $U_c'A_c$. During conditions of high heat load, the discharged water may not be sufficiently warm to merit heat recovery and so would be diverted to the cold-water return line by the temperature switch (TS). As the heat load is reduced, the temperature of the discharged water increases until the point is reached where it is sent to the plant hot-water system. Water as low as 120°F may be used to heat buildings, and 180°F water may be used to reboil certain low-temperature columns, like those separating propane and propylene. This is a far better approach for cooling atmospheric columns than letting the column float on a vent header or blanketing it with a noncondensable gas.

The hot-vapor bypass arrangement of Fig. 7.1, using two valves, is also commonly applied to water-cooled condensers mounted *above* the reflux accumulator. An alternative but less satisfactory arrangement has the condenser mounted *below* the accumulator as shown in Fig. 7.4. The condenser is then said to be "submerged," and its heat-transfer rate is manipulated by flooding its shell with condensate. Column pressure is controlled by manipulating the vapor-bypass valve, which adjusts the relative levels of condensate between the condenser and the accumulator. Closing the valve forces liquid out of the condenser; opening it allows liquid to flow back into the condenser. The principal advantage of the submerged condenser is the requirement for one small valve compared to two large valves for the elevated condenser. In addition, the condenser may be mounted at grade level to facilitate maintenance.

The pressure drop across the bypass valve is the difference in static head between accumulator and condenser levels—typically 10 ft or so. The valve must be large enough to pass all the vapor that will condense within the accumulator under the minimum head that will completely flood the condenser. If the valve is too small to control pressure under the extreme conditions of low heat load and winter weather, it may be steam-traced.

The submerged condenser can give reasonably responsive pressure control, since flooding does not depend only on accumulation of condensate within the condenser but some may be borrowed from the reflux drum as well. However, the liquid forms a U tube which will resonate with a period of 4 s or thereabouts. Furthermore, the shifting of condensate between the accumulator and the relatively large capacity of the condenser shell promotes a strong interaction between the accumulator-level and column-pressure control loops. Additionally, the condensate is always subcooled, which limits the rangeability of the system. As the heat load is reduced, the condenser flooding is increased and so is the subcooling. This increased subcooling causes condensation in the accumulator to increase, which will continue to lower column pressure.

The *elevated* condenser may be flooded by throttling condensate either into or out of the accumulator. In the former case, there is a liquid level in the reflux drum; in the latter case, the drum is flooded along with the condenser. Both arrangements are shown in Fig. 7.5. Flooding the drum eliminates its level controller and one valve. In fact, some columns with a flooded condenser have been operated quite successfully without a reflux drum at all, actually exhibiting improved quality-control response owing to the absence of this time lag.

The response of the pressure-control loop is limited by the volume of the condenser shell and the capacity of the valve. If distillate flow is manipulated for pressure control, three problems arise:

1. Distillate is usually a smaller flow than reflux and therefore has less influence over column pressure.
2. Heat-load variations upset the column's external material balance and thereby affect product quality.
3. Heat-load variations upset the column being fed by the distillate.

For these reasons, reflux or total condensate flow ought to be selected for pressure control, except when distillate flow is relatively high *and* is sent directly to storage.

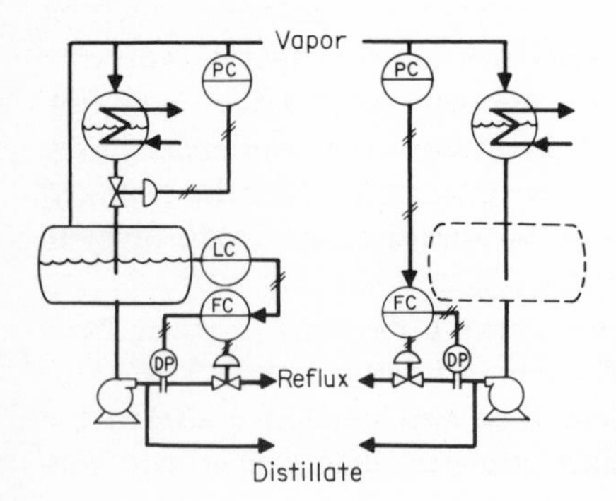

figure 7.5 *Flooding an elevated condenser can eliminate a level controller, a valve, and in some cases even the reflux drum.*

Discounting the possibility of manipulating distillate flow for pressure or level control, there are still other control-loop arrangements beyond what appears in Fig. 7.5 that will provide stable column performance. For example, *both* the reflux and distillate could be flow-controlled: reflux to set the heat load and distillate to control product quality. Then accumulator level or column pressure may be controlled by manipulating the heat input to the reboiler.

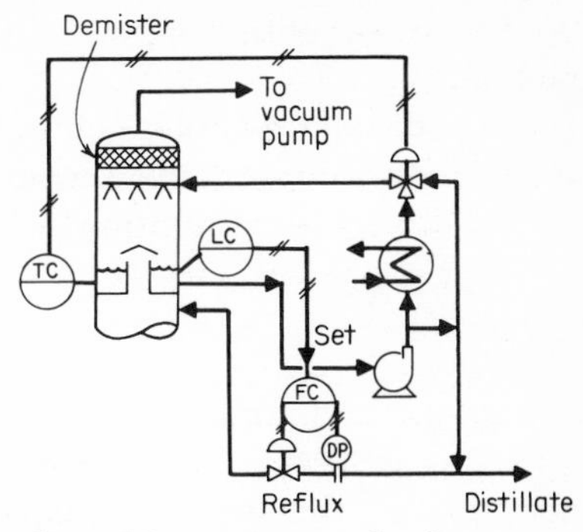

figure 7.6 *Condensing by direct contact with subcooled spray is sometimes used in a vacuum column.*

Direct-contact Condensers Where water is the overhead product, a direct-contact barometric condenser may be used, with the overhead product leaving in the company of the cooling water. While this is common practice with evaporators, the opportunity seldom appears in distillation. However, condensation by direct contact is nonetheless resorted to in vacuum distillation where minimum pressure loss is mandatory. The objective is achieved by subcooling the distillate and recirculating through a spray condenser as shown in Fig. 7.6.

In a vacuum column, noncondensable gases must be continuously removed to maintain the desired absolute pressure. As in other partially condensing systems, another variable besides column pressure must be used to manipulate heat removal. Figure 7.6 shows a reflux-temperature controller making that adjustment. Again, in a partially condensing system temperature and pressure must *both* be controlled to maintain overhead vapor composition as described in Chap. 3.

COMPRESSION AND EXPANSION

Historically, the only distillation systems which were designed to conserve energy were air-separation plants. Their feedstocks are free; therefore their only operating cost is the power required to pump heat through the plant. All the designer's attention could then be focused on this single cost factor—the power required to generate a unit of salable product. Unfortunately, few processes outside the cryogenic area have benefited from this technology. But with the increasing price being paid for energy, the capital investment required for compressors and expanders may be returned by reduced operating costs in a few months. This technology is relatively new, having been applied only to limited segments of separation units where it could be of particular benefit. Some of the problems as well as benefits are presented below.

Heat Pumps The most obvious application of a heat pump to distillation is that of using the heat of condensation of overhead vapor for reboiling. The attractiveness of this option depends on the thermodynamic conversion of the heat pump.

In the discussion of loss of available work through heat transfer, Eq. (6.18) related heat and work in a Carnot cycle. This expression can be differentiated into a rate equation relating power $\dot{w}$ and heat flow Q:

$$\dot{w} = Q \frac{\Delta T}{T} \tag{7.12}$$

The thermodynamic conversion of the heat pump, $Q/\dot{w}$, is seen to vary inversely with the temperature difference ΔT between source and sink and directly with the absolute sink temperature T. The most significant factor is that the required power approaches zero with the temperature difference. Hence the heat pump will be most attractive when separating close-boiling mixtures where high heat flows are needed and the ΔT is low. Propane-propylene distillation is one of the more common separations using a heat pump outside the cryogenic field. A simplified schematic of a column with heat pump appears in Fig. 7.7.

The compressor draws suction on the overhead vapor stream, compressing it nearly isentropically and raising both pressure and temperature. If the vapor is saturated at suction conditions, it will become superheated. The condensing temperature of the overhead vapor must be above the boiling point of the bottom product by whatever temperature difference is needed to transfer heat through the reboiler.

Typical sets of temperatures and pressures for a heat pump on a propane-propylene splitter are given in Fig. 7.7. They were derived using the pressure-enthalpy chart for propylene shown in Fig. 7.8. The starting point for the propylene cycle is taken as saturated overhead vapor at 70°F and 152 psia. Allowing a pressure drop of 8 psi through the column and reboiler would have bottom-product propane boiling at 160 psia and 87°F. A temperature difference of 23°F across the reboiler heat-transfer surface is assumed, giving a condensing temperature of propylene at 110°F.

The propylene must be compressed to 260 psia to reach this saturation

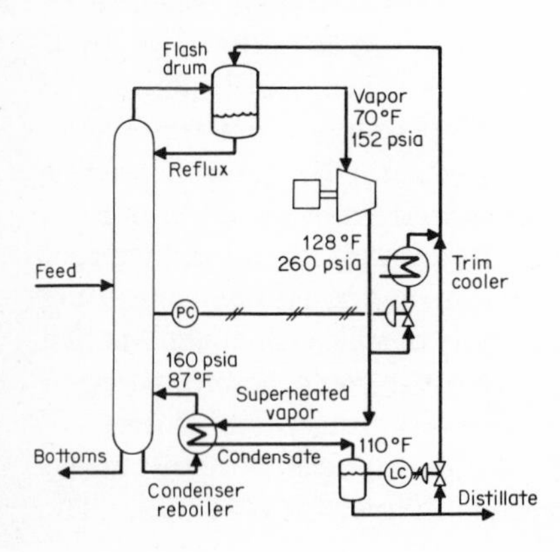

figure 7.7 *A heat pump is the most efficient mechanism to supply energy to columns separating close-boiling components.*

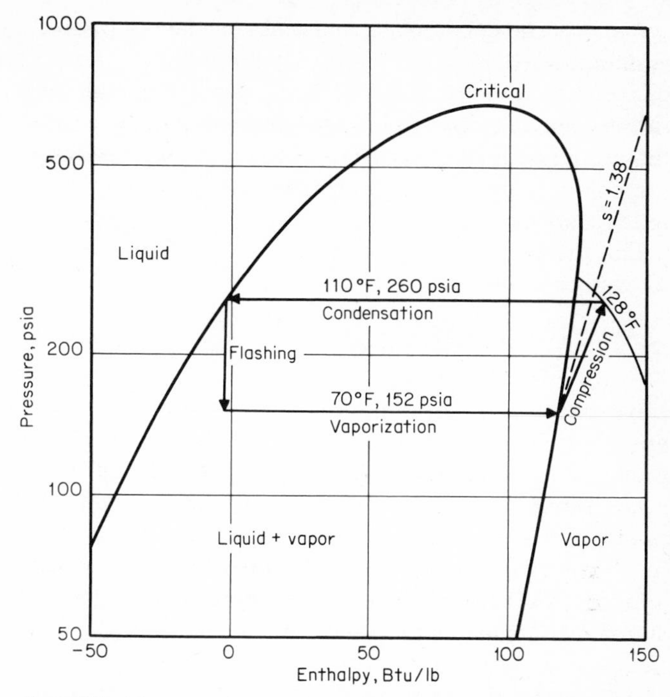

figure 7.8 *The propylene cycle is illustrated on the enthalpy-pressure diagram. (Enthalpy of gas is zero at O°R.)*

temperature. If the compression were to be performed isentropically along the 1.38 contour, the enthalpy of the propylene would increase 11 Btu/lb to 130. However, friction and internal leakage take their toll, reducing the adiabatic efficiency of the compressor to 0.76, typical for this service [3]. The actual enthalpy increase is then 11/0.76 or 14.5 Btu/lb, so that the gas is discharged at 133.5 Btu/lb and 128°F. In cooling to a saturated liquid at 260 psia, the enthalpy is reduced to −2 Btu/lb, a decrease of 135.5 Btu/lb. In boiling at 160 psia, the propane enthalpy increases from −10 to 131 Btu/lb, an increase of 141 Btu/lb. There is no requirement that propane and propylene flows balance.

The completion of the propylene cycle reveals a discontinuity in the energy balance. Most of the saturated propylene liquid at 260 psia and 110°F must be returned as reflux to the column at 152 psia and 70°F. Some of the pressure drop will be taken up in elevating the reflux to the top of the column—about 45 psi for a 200-ft head. The remaining 63-psi drop is developed across piping and a control valve. Since this pressure is dropped adiabatically, the enthalpy of −2 Btu/lb remains and the reflux will flash as it passes through the valve. The saturated liquid at 70°F and 152 psia contains −27 Btu/lb, and the saturated vapor 119 Btu/lb. From an enthalpy balance, it can be calculated that about 17 percent of the condensate will be flashed to vapor and returned to the compressor.

Because the condensate is already partially vaporized, the remaining reflux cannot absorb as much energy as the reboiler has generated. The increase in enthalpy in boiling the reflux is the difference between −2 Btu/lb and 119 Btu/lb—only 121 Btu/lb is absorbed, compared to 135.5 Btu/lb transferred to the reboiler. The 14.5 Btu/lb difference is the work put into the system by the compressor. It *must* be removed for the energy level of the system to reach a steady state.

The thermodynamic conversion of mechanical work w to heat transfer q for this example is

$$\frac{q}{w} = \frac{135.5\ \text{Btu/lb}}{14.5\ \text{Btu/lb}} = 9.34$$

If we assume a 30 percent efficiency for converting fuel energy into electrical energy at the power plant, an electrically driven heat pump is still able to provide a thermodynamic conversion of 2.8 from fuel energy to heat transfer.

Isentropic compression would have raised the enthalpy of the discharged propylene only to 130 Btu/lb such that it would give up 132 Btu/lb in condensing. The 121 Btu/lb absorbed in vaporizing would remain as before, leaving only 11 Btu/lb work done by the compressor. The thermodynamic conversion for isentropic compression is then

$$\left.\frac{q}{w}\right|_s = \frac{132\ \text{Btu/lb}}{11\ \text{Btu/lb}} = 12$$

This figure compares favorably with the ratio of absolute sink temperature to the temperature difference between source and sink:

$$\frac{T}{\Delta T} = \frac{530}{40} = 13.25$$

To take advantage of the efficiency of a heat pump, the ΔT should be minimized. Increasing the heat-transfer area by 50 percent could reduce its ΔT to about 15°F and the total ΔT to 32°F. This would raise the thermodynamic conversion to 11.7. The use of high heat-flux tubing [4] can further reduce the reboiler ΔT to 8°F and the overall ΔT to 25°F, increasing the thermodynamic conversion to 15. These improvements can be well worth the capital investment. However, there is a point of diminishing return since an infinite reboiler area would still require a 17°F system differential—the difference between boiling points of the two fluids.

The power introduced by the compressor must be removed, either by a trim cooler or as the difference in enthalpy between feed and products. The latter should not be relied upon because feed and product flow rates can be quite variable, especially during start-up and in emergencies. If the column is placed in a total-reflux mode of operation for even a short space of time, the accompanying loss in cooling of the feed-product energy balance could cause pressure to rise until safety valves lift. Unfortunately, one of the disadvantages of compressors is their very limited turndown. As a consequence, trim cooling must always be provided with a heat pump.

Because this is a closed system, the compressor speed has little ultimate effect on the pressure reached at steady state. An increase in speed can reduce suction pressure, increase discharge pressure, and increase flow. But the higher flow will increase the rate of heat transfer, thus increasing the temperature difference and hence pressure difference across the reboiler. The higher reboiler-condensate pressure will increase reflux flashing and ultimately raise column pressure. If the increased work introduced by the compressor is not balanced by cooling, pressure will continue to rise.

Consequently, compressor speed sets the heat load but cannot be used to control column pressure. Pressure depends ultimately on cooling. Therefore, the problem of the column-pressure controller manipulating cooling is the same with or without a heat pump. Column pressure will also be affected by coolant temperature with or without a heat pump.

Figure 7.7 shows the trim cooler in a bypass loop around the compressor. This allows the column pressure to be lower for a given coolant temperature than if the coolant were applied to the reflux.

Compressor Controls Centrifugal compressors are quite similar to centrifugal pumps in that their discharge head falls as flow is increased. But owing to the compressibility of gases as opposed to liquids, compressors exhibit a characteristic instability in the low-flow range which limits their turndown. This instability is known as "surge"—a dynamic reversal of the pressure-flow characteristic that causes oscillations to develop within the machine. Unchecked, surge may destroy a compressor in very little time.

Because normal operation is impossible within the surge region, a compressor must be protected from low-flow conditions by its control system, even during start-up. This is usually accomplished by bypassing some of the discharged product back to the suction after cooling. The cooling is essential to remove the heat of compression—otherwise the gas will continue to rise in termperature until damage is done or protective devices trip. For maximum efficiency, only enough gas should be bypassed as is absolutely necessary for surge protection. To minimize bypassing, the surge region must be carefully defined and the controls programmed as to its exact location.

Compressor characteristic curves are usually plotted in coordinates of adiabatic head versus volumetric flow at suction conditions. A typical plot appears in Fig. 7.9. The surge line describes what is essentially a parabola:

$$h_a = k_a F_s^2 \tag{7.13}$$

where h_a = adiabatic head, ft

F_s = volumetric suction flow, ft³/min

k_a = constant

Although this is a general model for centrifugal compressors, it is by no means all-inclusive. Some surge curves are nearly linear; others bend farther to the right at high heads. But there are further difficulties to be surmounted—neither adiabatic head nor suction flow is directly measurable. According to White [5], adiabatic head varies not only with compression ratio but also with

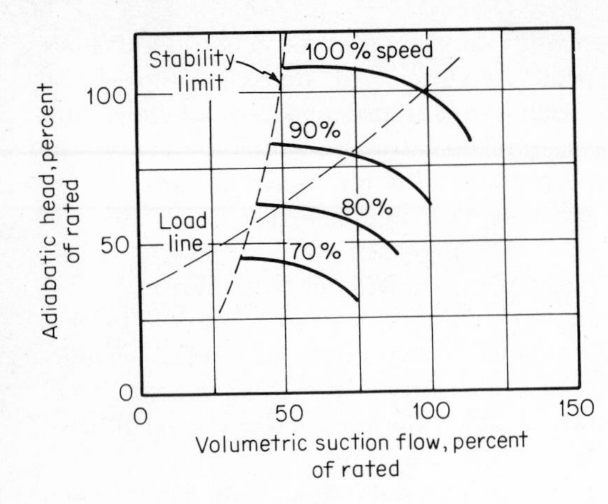

figure 7.9 *Characteristic curves for a typical variable-speed centrifugal compressor, outlining the surge region.*

molecular weight, temperature, supercompressibility, and specific-heat ratio. Compression ratio is nearly linear with adiabatic head for most machines, with the linearity deteriorating as compression ratio is increased. White then substitutes compression ratio for adiabatic head to develop his surge-curve model:

$$\frac{p_2}{p_1} - 1 = k_s F_s^{\,2} \tag{7.14}$$

where p_2 = absolute discharge pressure

p_1 = absolute suction pressure

k_s = constant

Molecular weight, temperature, and specific-heat ratio primarily affect the coefficient k_s of the relationship.

Suction flow is usually measured with an orifice or nozzle that produces a differential-pressure signal. The relationship between volumetric flow and measured differential pressure h_m for a gas is

$$F_s = k_m \sqrt{\frac{h_m T_1}{p_1}} \tag{7.15}$$

where T_1 is absolute suction temperature and k_m is the meter constant, which includes the specific gravity and supercompressibility of the gas. When Eqs. (7.14) and (7.15) are combined, we have

$$\frac{p_2 - p_1}{p_1} = \frac{k_s k_m^{\,2} h_m T_1}{p_1}$$

Reduced to its simplest form, the surge line becomes

$$\Delta p = k_c h_m \tag{7.16}$$

where Δp is $p_2 - p_1$ and k_c is a control constant which includes all other constants and their variability. Some of the variables like suction temperature and supercompressibility affect the head-compression ratio relationship and the volumetric-flow relationship in a compensating manner. However, some residual remains such that

$$k_c = f(T_1, M, \gamma) \tag{7.17}$$

where M is the molecular weight and γ is the specific-heat ratio. Both M and γ are functions of gas composition and tend to change together. The value of k_c increases about 10 percent with a change in M from 16 to 20 for light hydrocarbon gases and about the same amount for a 120°F reduction in T_1.

The first step in designing an antisurge control system is to convert data from the manufacturer's surge curves into measurable Δp and h_m. Then plot Δp versus h_m as in Fig. 7.10 and construct the best straight line on the safe side of the data points. The slope k_2 and possibly the intercept k_1 should be adjustable in the field to allow a better fit to installed compressor characteristics and to correct for changing operating conditions. A control system which implements this function appears in Fig. 7.11.

The FFC in Fig. 7.11 is known as a ratio controller; it is intended to hold one input (Δp) in ratio (k_2) to the other ($h_m - k_1$). If k_2 is set too high, a condition might arise which would place the compressor into surge; if set too low, gas may be bypassed needlessly, wasting energy. Therefore, accurate definition of the control line is important. Compensation may be applied for varying suction temperature if necessary. Constant-speed compressors controlled by adjustable inlet guide vanes have surge curves which vary with vane position, requiring compensation for their effect. With some compressors, locating the flow measurement on the suction line is not practicable so that a discharge measurement is needed. Then the control line must be compensated for variations in compression ratio, in effect referring volumetric discharge flow to suction pressure.

In addition to these steady-state requirements, the dynamic response of the surge-control system must be considered as well. The bypass valve should fail in the open position so that the compressor cannot be damaged by even a

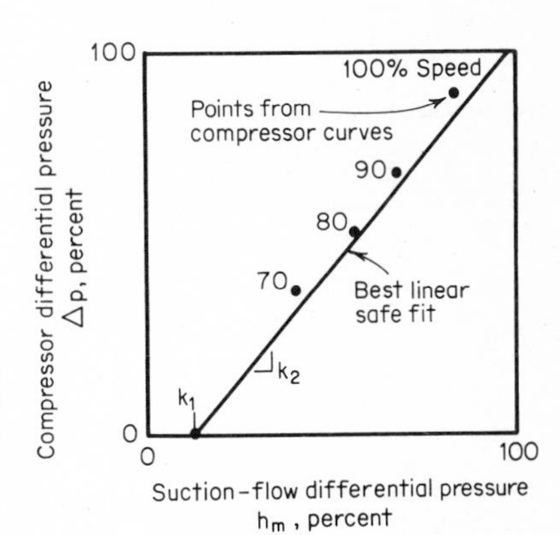

figure 7.10 *Data from the compressor's characteristic curves should be plotted as* Δp *versus* h_m.

momentary air-supply outage. The valve should have a linear characteristic—an equal-percentage valve restricts flow too much at low stroke when flow is most needed in this application. Furthermore, the valve should be equipped with a volume booster to increase its stroking speed.

The controller itself must be especially equipped for this service. Its normal condition is saturated, since the bypass valve is normally closed. A conventional controller in this service would "wind up" with its reset function keeping the valve closed until the control line is crossed and an unsafe condition exists [6]. An antiwindup device is available which can hold the controller output at the just-closed position of the valve by modulating the controller's reset action. Then a change in the variables in the *direction* of surge will cause the valve to open by proportional action *before* the control line is crossed. Any unnecessary bypassing caused by this action is temporary, lasting at most for a few seconds while the controller resets to the new load condition. Control action is then always on the safe side of the control line.

Start-up of a compressor is a precarious event. The bypass valve should be open, particularly if a parallel compressor is running, in which case a differential pressure is already established. However, an electrically driven compressor has been known to overload its motor when starting with the bypass *open*. These problems can be circumvented when enough thought is given beforehand to designing appropriate control logic.

The vapor compressor for a distillation column ought to be driven by a variable-speed steam turbine. Then flow of heat can be set by compressor speed. Decreasing speed reduces flow and the ΔT across the reboiler-condenser, causing discharge pressure to fall also. A load line of head versus flow, as related through the reboiler ΔT, is superimposed on the compressor curves in Fig. 7.9. The turndown achievable is limited to about 20 percent in speed and 60 percent in flow, but it is greater than could be obtained with a constant discharge pressure. Variable inlet guide vanes allow a similar but less efficient turndown. Under no condition is it reasonable to intentionally

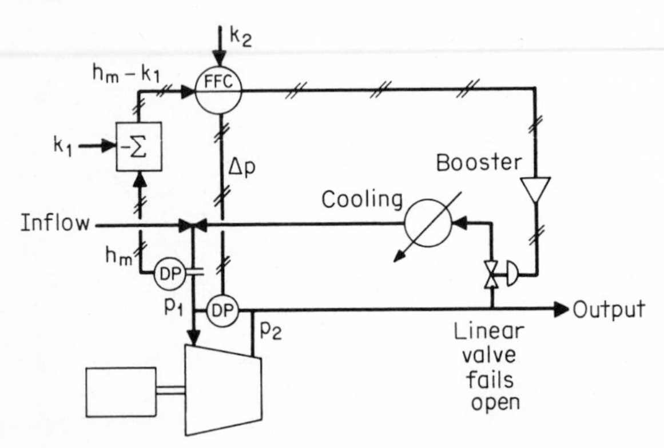

figure 7.11 *This system implements the antisurge control function described in Fig. 7.10.*

reduce speed or guide-vane position to enter the surge region, since the bypass valve would then remove excess energy to the cooling system rather than the process.

In the heat-pump cycle shown in Fig. 7.7, the trim cooler and its valve are already in the preferred position for antisurge control. This bypass loop could then provide both functions, with automatic selection of the controller requiring the higher flow. In the event that the antisurge controller takes command of the valve, pressure would fall below its set point since additional cooling would be applied. These selective systems are described in detail in Chap. 9, "Controlling within Constraints."

Expansion and Liquefaction Liquid was returned to the top of the column in Fig. 7.7 through a control valve. About 45 psi was lost to gravity head, while the remaining 63 psi was dropped across the valve. If a hydraulic turbine had absorbed that drop instead of a valve, some work could have been recovered to offset that used in compression. In addition, less flashing would have resulted, providing more reflux and requiring less trim cooling. Flashing service is difficult for turbines since it requires greater clearances and more costly fabrication, thereby reducing the attractiveness of this method of energy recovery. All-liquid or all-vapor operation is more satisfactory. However, steam turbines designed for condensing service can discharge a small percentage of water, and centrifugal compressors are capable of accepting some entrained liquid. This capability actually improves their thermodynamic conversion by reducing the enthalpy of the discharged vapor for a given pressure. The vapor is less superheated and, therefore, transfers heat more efficiently; less cooling and flashing are additional benefits.

In most *cyclic* processes, flows in and out are similar if not equal, permitting compressor and expander to be mounted on the same shaft. Although the use of expanders and hydraulic turbines for energy recovery is presently limited to large-scale installations, they will become very common in the future. As is shown below, very little in the way of additional control or management is required to recover much of the energy of pumping and compression. Franzke [7] indicates that at 100 percent load about 65 percent of the work put into a fluid by a pump can be recovered by letting down that fluid to its original pressure through an identical reverse-running pump on the same shaft. The percentage recovery falls off at lower loads because of throttling across control valves in both directions. He advocates a hydraulic turbine as superior to a reverse-running pump since it loses less power at lower loads through internal-gate control. Furthermore, a hydraulic turbine can accommodate substantial percentages of vapor and is therefore admirably suited to the heat-pump cycle.

Compressor-expander combinations are most common in cryogenic processing. A very important application is the separation of light hydrocarbons from natural gas as described in Fig. 7.12. The feed gas at a high pressure is chilled against overhead vapor from the stripping column. At this point, some of the heavier liquid components condense and are separated from the vapor and fed to the column. Flashing will occur as the liquid pressure is dropped through the control valve—a recovery turbine here would recover work and produce less flashing.

Gas leaving the separator is passed either through an expander or bypass valve to condense liquid. In the process of crossing through the critical region, some liquid will condense in the valve because of the Joule-Thompson effect. But the expander will condense more liquid and recover work as well. A look at the pressure-enthalpy diagram for methane in Fig. 7.13 will demonstrate this point. The gas expands adiabatically through the valve but nearly isentropically through the expander. Houghton and McLay [8] report expansion efficiencies of 78 to 85 percent across a flow range of 40 to 120 percent of design conditions.

The power recovered by the expander is delivered to a compressor mounted on the same shaft for recompression of the reheated overhead gas. The power recovered falls short of that needed to restore the gas to its original pressure by the combined efficiencies of compressor and expander. Supplementary compression is usually used to restore the gas to its original pressure for transmission. The discharged gas must then be cooled, and the heat of compression thereby recovered is more than enough to reboil the stripping column. Since this energy is returned to the overhead vapor, it does not leave the system; consequently, the work of compression must ultimately be removed by cooling against air or water. A glycol-water mixture is commonly used as the medium to transfer heat from the compressor to the reboiler because the boiling point of the bottom product is typically below the freezing point of water.

The fractionation achievable in the stripping column, as in any column, is a function of the boilup-feed ratio. But the boilup is wholly dependent on the liquid condensed in cooling and expanding the feed gas. Therefore, maxi-

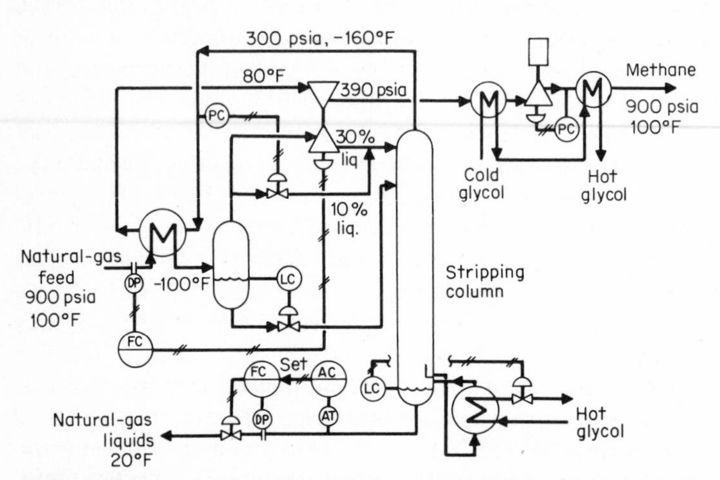

figure 7.12 *Light hydrocarbons may be recovered from natural gas by using expansion of the chilled feed to produce reflux.*

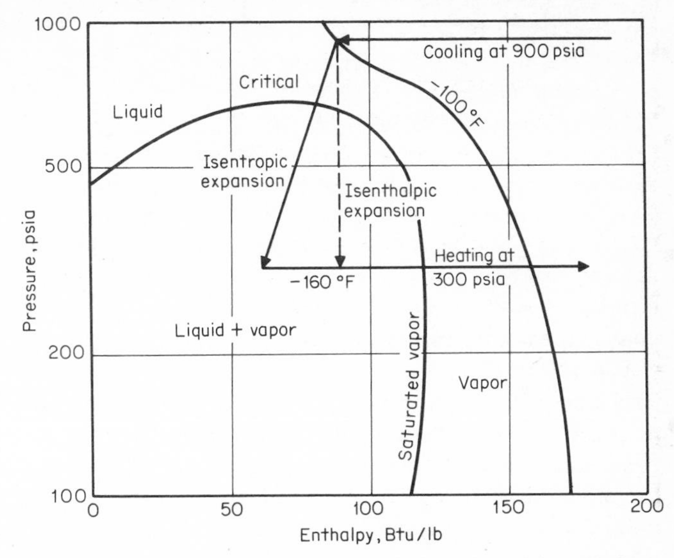

figure 7.13 *Isentropic expansion condenses more liquid than adiabatic expansion and recovers power as well. (Enthalpy of gas is zero at O°R.)*

mum liquefaction should be pursued to maximize recovery of the condensable gases in the feed. This goal is only achievable if no throttling through control valves is allowed. Feed flow should be controlled by adjusting the nozzles in the expander as described in Ref. 8. The bypass valve should remain closed. Column pressure can be controlled by manipulating the bypass valve since this valve takes power from the expander and therefore adds energy to the column. For a given pressure drop, the expander can condense typically 20 to 30 percent of the gas while the valve will condense less than 10 percent.

If the valve is left closed, column pressure will float on the gas-to-gas heat exchanger, feed-gas temperature, and the supplementary compressor. Increasing the feed rate tends to augment the temperature drop across the heat-transfer surface, raising the expander feed temperature and column pressure as well. However, with the bypass valve closed, the column will always be at its minimum attainable pressure, providing the maximum relative volatility, power recovery, and condensate recovery attainable at any given feed rate. A floating-pressure control system, as discussed in detail later in this chapter, should be used to actuate the bypass valve.

The objective of the column is to maximize recovery of bottom product with a specification on its methane concentration or methane-ethane ratio. Although only 30 percent of the feed gas may be liquefied, the quantity of recoverable hydrocarbons is much less. As a result, bottom-product flow is smaller than the boilup rate, which leads to the control-loop arrangement shown in Fig. 7.12. This particular arrangement minimizes the effects of the heat balance on product quality—an important consideration in this application, where liquefaction is variable, as is the temperature of the heating medium.

Vacuum Systems It is possible for a distillation column to operate in a vacuum with a total condenser and without any type of vacuum pump. However, start-up conditions must be considered and allowance made for leakage of air into the system during normal operation. Consequently, a controlled vacuum is usually maintained by a pumping system in conjunction with a condenser. If the overhead vapor is water, barometric direct-contact condensing may be used upstream of the vacuum pump and sometimes downstream as well. Any other overhead vapor requires a conventional surface condenser upstream and occasionally downstream.

The vacuum pump is typically designed for start-up and other maximum-flow conditions. Normal operation typically generates much less noncondensable gases so that the pumped flow must be reduced. The most economical means of reducing pumping rate is to reduce the speed of a mechanical blower or vane pump. Throttling its suction is also common, whereby pumping action is reduced by the lower pressure and density at the inlet of the pump. Another method is to bleed atmospheric air into the suction line, in which case the pump continues to carry its rated flow at the design pressure rise, with most of the flow being air. Both suction throttling and air bleeding are wasteful of the horsepower delivered to the pump.

Steam-jet ejectors are in very common use for exhausting distillation columns, but their thermal efficiency is well below that of mechanical pumps. Monroe [9] compared the 2 to 3 percent thermal efficiency of steam jets against the 30 to 50 percent efficiencies attainable with blowers and mechanical pumps when used in their best working range. Even when the cost of electricity to drive a mechanical pump versus steam for the jet is taken into account, the pump is four to five times less expensive to operate.

Steam jets are not normally controlled by throttling their steam supply or otherwise reducing its pressure. However, this is the only way to reduce steam usage. In throttling the steam, its flow is reduced as well as the pumping efficiency in pounds of air per pound of steam. Evans [10] gives steam-usage multipliers varying from 1.6 to 0.85 as steam pressure is adjusted from 50 to 300 psig. Since steam flow varies directly with absolute supply pressure, the available turndown in pumping rate for a constant suction pressure would seem to be $(315/0.85) \div (65/1.6)$ or 9:1.

Most commonly, however, steam is not throttled for control—instead, either the suction is throttled or air is bled into the suction to hold a constant vacuum. Neither of these methods changes steam flow. When the flow of noncondensable gases is very low, Buckley [1] recommends an air bleed over throttling the suction because the very small size required for the suction valve would promote plugging.

Condenser capacity changes sharply with absolute pressure in vacuum systems. Extremely high cooling-water flow rates are needed to attain the lowest absolute pressure, which may not be a cost-effective mode of operation.

The reasons for vacuum operation are principally to minimize thermal degradation of the product and to maximize reboiler capacity. Thus the critical zone in the column is the reboiler, where the boiling point and absolute pressure are greatest. However, absolute-pressure control is exerted by the condenser and vacuum pump where the boiling point and pressure are least. Even a perfect vacuum at the top of the column would leave a finite minimum absolute pressure and boiling point at the bottom. Consequently, bottom-product quality and reboiler capacity cannot improve as fast as condensing costs increase by lowering absolute pressure. There is then an optimum absolute pressure which balances product-value improvement and heating cost against condensing and pumping costs. An analysis of a similar situation in a vacuum refrigeration system is presented later in the chapter.

REFRIGERATION

Temperature levels well above ambient can be reached by using the heat from the products of combustion of a fuel. However, temperature levels below atmospheric reservoirs can only be reached with some sort of heat pump. In every case, the heat is pumped from the process to an atmospheric reservoir (air or cooling water). As a consequence, the temperature level attainable and the work required to pump a given flow of heat from the process are related to the temperature of the atmospheric reservoir. Distillation columns using refrigeration for cooling are therefore as sensitive to ambient conditions as those which reject heat directly to the atmosphere.

This interaction can be a stumbling block to quality control or a stepping stone to higher efficiency, depending on how it is used. Some of the same problems and limitations already encountered with condensers appear in refrigeration units. And methods for pressure control are similar as well.

Two of the types of refrigeration systems to be described are based on vapor compression, an already familiar technique. The most common system uses a single-component refrigerant. Deep cooling such as required to liquefy natural gas may use a multiple-component refrigerant with successive stages of evaporation. Then control of absorption and steam-jet refrigeration units are also presented, as they are in common use to provide chilled water.

Single-component Refrigerants The simplest application of refrigerated cooling is a single-stage system pumping heat from a column condenser. A refrigerant vapor is compressed and condensed against air or cooling water as shown in Fig. 7.14. The liquid refrigerant then flashes through the control valve into the column condenser. The lower suction pressure of the compressor reduces the boiling point of the refrigerant so that it can remove heat from the column overhead vapor. In condensing the vapor, the refrigerant evaporates and is recompressed.

It is possible to control the rate of cooling by manipulating the liquid-refrigerant valve directly. As the heat load is reduced from the maximum, the liquid level in the condenser would be reduced by closing the valve. This

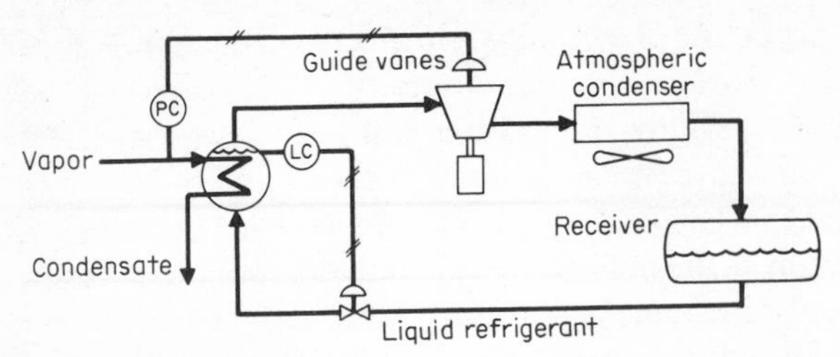

figure 7.14 *The column condensing tubes should be completely immersed in refrigerant to maximize heat transfer per unit horsepower.*

exposes condenser tubes, reducing the area available for heat transfer while maintaining a constant temperature difference. The system shown in Fig. 7.14, however, is devised to keep all tubes immersed, thus maintaining the maximum active heat-transfer surface. Then heat-transfer rate is reduced by raising the compressor suction pressure through guide-vane or speed manipulation. (In the case of constant-speed reciprocating compressors, cylinder intake valves are sequentially unloaded to reduce gas flow and shaft horsepower.) This arrangement of controls minimizes compression work by minimizing the temperature difference between the evaporating and condensing refrigerant.

By the same token, the refrigerant condenser should not be flooded or otherwise obstructed. Reduction in ambient temperature experienced at night, during rainy weather, and in winter should be allowed to impose its full effect on compressor power per Eq. (7.12). As the ambient temperature falls, evaporating temperature could be held constant (for a given heat load) by controlling column pressure constant. Then compressor horsepower would be conserved. Or column pressure could be allowed to fall by maintaining constant compressor horsepower. This would result in a reduction in heat load caused by an improvement in relative volatility. The choice between minimum pressure or minimum horsepower is clearly an optimization problem. In fact, for any combination of heat load and ambient temperature an optimum pressure exists that minimizes the total cost of operating the column.

If the compressor is a centrifugal type, its turndown is not wide due to the surge limit. A very low temperature difference will mean a low compression ratio, which will actually extend the range of flow turndown. So the surge limit may not become a constraint to the optimization problem described above. However, the guide-vane position or compressor speed should be limited so that the bypass valve does not have to open for surge protection. Although column-pressure control will be lost, the lower pressure is helpful in separation and compressor power is then not wasted.

Refrigeration units using more than a single stage of compression require interstage cooling. This is usually accomplished by allowing the condensed refrigerant to flash successively to each lower-pressure stage. The resulting vapors are returned to the compressor, providing interstage cooling and minimizing the flashing taking place in the column condenser. This practice also improves the efficiency of the cycle.

Where the temperature difference across the column is relatively low, refrigerant may be used for both heating and cooling. This is common in an ethylene plant where several columns require refrigeration. In principle, it operates like the vapor-compression cycle for a single column but instead serves several columns. And because it serves several columns, fewer compressors are required; but more heat exchangers are required than if every column had its own heat pump.

The compressor takes refrigerant evaporated from the column condensers, compresses it, and sends it to all the reboilers as shown in Fig. 7.15. As the reboilers demand more heat, the compressor speed or vane position must be increased to maintain suction pressure.

Condensed refrigerant leaving the reboilers is flashed to a liquid receiver. The column condensers draw on the receiver for liquid refrigerant, which flashes again when entering each condenser. The vapor lost through flashing cannot offer its latent heat for cooling. As a result, the system is capable of more heating than cooling, as was true with the single-column heat pump. The energy introduced by compression must be rejected to the atmosphere. This can be done by a trim compressor drawing suction on the vapors flashed in the receiver. Its discharge would then be condensed against cooling water and flashed back to the receiver.

All reboilers use the same vapor and all condensers empty into the same suction line. There is then no freedom for individual columns and reboilers

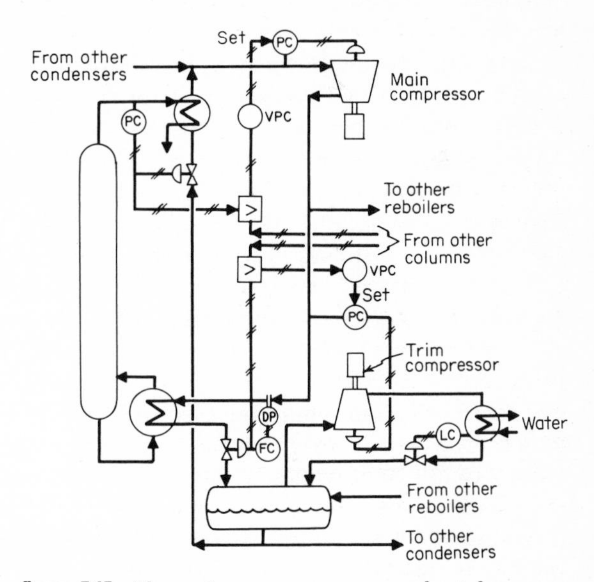

figure 7.15 *The main compressor pumps heat from condensers to reboilers; the trim compressor pumps heat to a water-cooled condenser.*

to operate at different refrigerant temperatures. Consequently, the individual heat exchangers must be partially flooded to vary their heat-transfer rates. Although this is less efficient than utilizing the entire heat-transfer surface, there is little alternative if each column is to operate under controlled pressure.

Nonetheless, the compressor differential pressure can be minimized for the system in the same manner that hot-oil temperature was minimized for the multiple-reboiler installation. The condenser-valve positions for all the columns in the system of Fig. 7.15 are compared in a high selector (or low selector if the valves open on decreasing signal). The most open valve position is controlled by the valve-position controller (VPC) manipulating compressor suction pressure. This can ensure that at least one of the condenser valves is nearly full open, and therefore the compressor differential is as low as possible without forsaking control of any column.

In the case where reboiler heat comes from the same source, the reboiler valve signals should be sent to another selector. A valve-position controller acting on the output of that selector would set main-compressor discharge pressure to hold one reboiler valve nearly full open. The trim compressor brings about a balance between heating and cooling requirements.

It is also possible to allow each column to float on its condenser by using maximum cooling at all times. In this example where refrigerant is used for both heating and cooling, reducing column pressure and improving relative volatility requires less work from the main compressor. The trim compressor would have to operate at maximum load, however, to deliver the lowest available refrigerant temperature relative to cooling-water temperature. But the trim compressor only needs to transfer the work introduced by the main compressor. If the thermodynamic conversion of power to heat flow in each compressor is 5, for example, the trim compressor would be only one-fifth the size of the main unit.

In cases of deep cooling, a single refrigerant may not be suitable from the coldest to the warmest points in the heat-flow path. The vapor pressures developed across the entire range of temperatures may exceed reasonable limits of compressor staging, suction piping size, etc. Then a cascade system is usually employed, using a lower-boiling refrigerant to pump heat from the process to a higher-boiling refrigerant. A second heat pump transfers the heat to the atmosphere. The two fluids are separated by a heat-transfer surface.

A cascade system could be used for the unit in Fig. 7.15. Vapors leaving the receiver would then be condensed in an exchanger rather than being compressed. Heat would be pumped from the exchanger to the water-cooled condenser by the trim compressor, using a higher-boiling refrigerant. Combinations of ethylene and propane, or two of the fluorocarbon refrigerants, are commonly used in cascade systems.

In cases where the boiling-point spread across a column is great, as in a demethanizer, the higher-boiling refrigerant could provide reboiler heat while the lower-boiling fluid removes it from the condenser.

Multiple-component Refrigerants Deep cooling is achieved in natural-gas liquefaction by using a mixed refrigerant rather than a cascade system. It is then possible to use a single compressor. However, the composition of the "cycle gas" refrigerant must be controlled. A simplified schematic of a natural-gas liquefaction unit is shown in Fig. 7.16.

A typical cycle gas consists of a mixture of paraffin hydrocarbons from methane through n-butane plus a small amount of nitrogen. Its composition is adjusted to obtain a certain percentage of condensation at each condenser in the system. The first stage of compression draws suction on the first condenser in the cold box. After compression, the mixture is partially condensed against atmospheric cooling. The liquid phase is further cooled in the first cold-box condenser and then flashed to provide cooling for that condenser along with vapors rising from lower condensers.

Vapors not condensed by atmospheric cooling are compressed further in the second stage and resubmitted to atmospheric cooling. The liquid condensed there is further cooled in the first and second cold-box condensers and is flashed to provide cooling to the second. The vapor from the second atmospheric cooler is partially condensed in the first cold-box condenser. Vapor and liquid are separated and conveyed individually through the next condenser where the pattern is repeated.

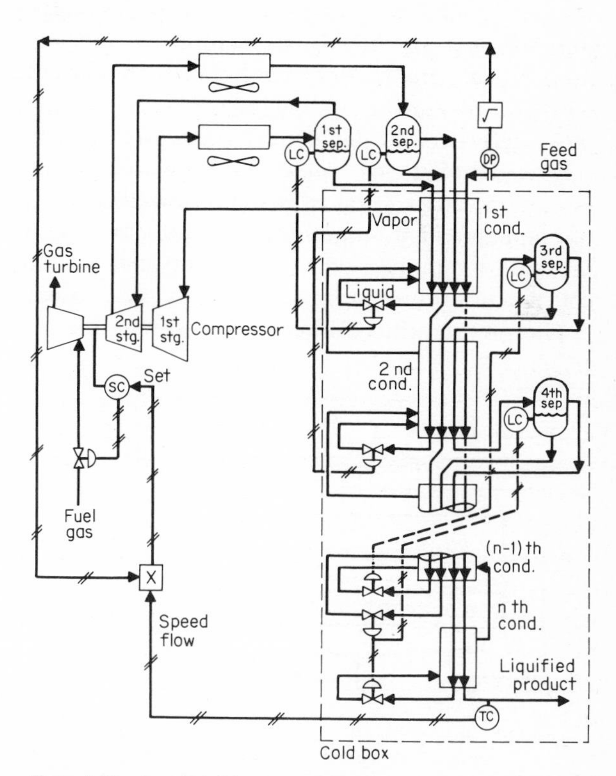

figure 7.16 *A multiple-component refrigerant is successively fractionated in a series of condensers.*

The refrigerant flowing downward through the condensers is essentially at compressor discharge pressure throughout. The successively falling temperature causes successive fractionation of the cycle gas. Each vapor-liquid separator therefore contains a different liquid mixture. For example, only 1 to 2 percent nitrogen may appear at the compressor suction whereas the vapor leaving the nth condenser may contain 30 to 40 percent.

Here again, the air coolers should not be restricted in any way since that would increase the work required by the compressor. Thus the pressures at the first-stage suction, discharge, and second-stage discharge will float depending on the heat-transfer rate and the ambient temperature. The discharge pressures will always be minimum for the current heat-transfer rate and therefore compressor power will be minimum.

An increase in compressor speed will increase the rate of condensation to each vapor-liquid separator. Then each liquid-level controller increases its liquid-outflow valve to increase evaporation to each condenser. Temperature distribution through the various condensers is related to refrigerant flow distribution, which is related to composition. Thus the composition of the refrigerant is adjusted to distribute the cooling uniformly across all the condensers.

Figure 7.16 shows the compressor speed set proportionally to feed flow with the proportionality adjusted by the product temperature controller. The feedforward flow signal adjusts compressor speed to match changes in liquefaction load; the feedback controller trims the calculation to maintain product temperature constant following variations in efficiency, ambient temperature, etc. Reference 11 develops the feedforward control concept in detail.

Absorption Systems A circulating-water stream may be chilled to 40 to 55°F by evaporating a spray of water at low absolute pressure. An absorption system capable of this method of chilling is shown in Fig. 7.17. The absolute pressure is maintained by the absorption of water vapor into a concentrated solution of lithium bromide. Heat of absorption is removed by cooling water.

The solution thus diluted by absorption must be reconcentrated by evapora-

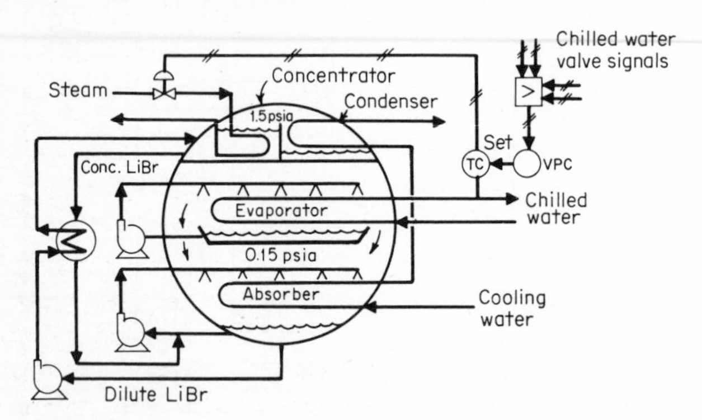

figure 7.17 *Absorption of water vapor into a solution of lithium bromide reduces its vapor pressure and thereby lowers its boiling point.*

tion against low-pressure steam. The resulting vapor is condensed against cooling water at the elevated (but subatmospheric) pressure of the concentrator and returned to the lower-pressure section to chill more water.

In the absence of air, the pressure within the chilling and absorbing section is limited by the concentration of the absorbent and the cooling-water temperature. The pressure in the concentrator is limited by cooling-water temperature alone. As with other condensers, cooling water should be allowed to flow at maximum rate through the unit. Then the steam flow determines the rate of evaporation in the concentrator and hence the concentration of the absorbent and its vapor pressure.

Absorption units using single-effect concentration and 85°F cooling water require about 19 lb/h of steam per ton of refrigeration at 55°F at full load and perhaps 16 lb/h at half load [12]. A ton of refrigeration is the heat flow developed by melting 2,000 lb ice at 32°F in 24 h. It is equivalent to 12,000 Btu/h. Using 15-psig steam having a latent heat of 945 Btu/lb, the heat pumped per Btu heat input is 0.67 at full load and 0.40 at half load. The cooling water must remove both the heat pumped and the latent heat of the steam. The heat pumped per Btu removed by the cooling water is then 0.40 at full load and 0.29 at half load. These efficiencies are obviously low, but they are offset by the capability of using low-energy heat.

Farwell [12] describes a two-stage absorption unit which has double-effect concentration of the absorbent. He reports a steam-flow reduction to 12 lb/h-ton at full load and 9 at half load. However, 125-psig steam is required. Considering a latent heat of perhaps 870 Btu/lb steam, the thermodynamic conversion of the two-stage unit is 1.15 at full load and 0.77 at half load.

Because chilled-water systems typically serve multiple users, the highest valve-position signal of the users should set the water temperature. This accomplishes two goals: it minimizes the hydraulic power loss through the chilled-water valves and minimizes the temperature differential across which heat must be pumped. This latter aspect is far more important with refrigeration systems than with heating systems. The reason is that the thermodynamic conversion of power into heat flow is directly proportional to the temperature difference between the two fluids—here chilled water and cooling water. The poor thermodynamic conversion cited above for half-load operation corresponds to a reduction in chilled-water flow at constant temperature difference. If the flow were maintained and the ΔT reduced instead, thermodynamic conversion would be *higher*, not lower.

Steam-jet Refrigeration There is occasionally a need for chilled water between 55 and 80°F—above the range of absorption systems and below that of cooling towers. This can be provided by vacuum refrigeration using the steam jet. Very simply, the steam jet draws a vacuum on a circulating loop of chilled water as shown in Fig. 7.18. Barometric condensers provide direct contact of steam and water and therefore eliminate the temperature drop of surface condensers. However, 35 ft of head must be provided to maintain vacuum at the inlet, with atmospheric discharge.

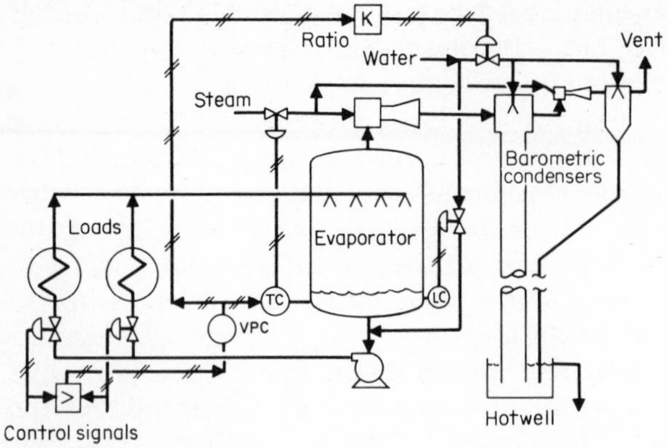

figure 7.18 *Vacuum maintained by a series of steam jets and condenser allows the chilled water to be cooled by evaporation to 50–55°F.*

Absolute pressure within the evaporator determines the chilled-water temperature. It may be controlled by adjusting either the cooling-water flow or the steam to the jet. Increasing the cooling-water flow reduces hot-well temperature and therefore lowers the vapor pressure. Increasing the steam supply pressure lowers the evaporator pressure but raises hot-well temperature. There is an optimum combination of cooling water and steam based on their relative values.

The flow rates of 85°F cooling water and 100-psig steam required to produce 1 ton of refrigeration at 40 to 60°F are compared in Table 7.1 from data appearing in Ref. 13. A cooling-water cost of $0.05/1,000 gal and a steam cost of $1/1,000 lb were used to generate the economic information. The variation of the optimum (underlined) cooling-water flow as a function of chilled-water temperature is apparent. If a single chilled-water temperature is always desired, then the optimum cooling-water flow should be in direct ratio to the flow of chilled water (as an index of heat load). If the temperature set point is changed significantly, the ratio of cooling water to chilled water should change according to Table 7.1. Variation in costs of cooling water and steam should also be taken into account when estimating the optimum ratio.

TABLE 7.1 Cost/Ton of Steam-jet Refrigeration for Selected Chilled-water Temperatures

Cooling water, gpm	Steam flow, lb/h			Cost of water and steam, $/ton		
	40°F	50°F	60°F	40°F	50°F	60°F
3			20.0			0.0290
4		26.0	16.5		0.0380	0.0285
6		19.3	13.8		0.0373	0.0318
8	25.3	17.6	12.8	0.0493	0.0416	0.0368
10	23.0	16.8		0.0430	0.0468	
12	22.0			0.0480		

Note that the steam flow required per ton of refrigeration at 55°F using 85°F cooling water is about the same as for a single-stage absorption unit. But the lower efficiency of the steam-jet unit appears in its use of 100-psig steam compared to 15 psig for the single-stage absorption unit.

The thermodynamic conversion of steam energy to refrigeration for the three near-optimum conditions underlined in Table 7.1 is summarized in Table 7.2. The heat content of the steam was taken to be 1155 Btu/lb (considering that boiler feedwater at perhaps 70°F must be heated to generate the steam). The improvement in conversion available at higher chilled-water temperatures is worthwhile and can be achieved by the familiar valve-position controller.

When chilled-water temperature is thus adjusted in response to cooling load, provision must be made to reoptimize cooling-water flow. Although the conditions underlined in Table 7.1 are not exact optima, a logarithmic relationship between cooling-water flow and temperature is suggested. This relationship is achievable by using an equal-percentage control valve for cooling water, one whose position is set proportional to the temperature set point. Because cooling-water valves are usually set to close on increasing signal, higher temperatures will call for less flow. Some calibration must be provided to position the valve appropriately across the operating range of temperatures. The proportionality should be adjustable to allow for changes in the costs of cooling water and steam.

The signal to the cooling-water valve is actually a feedforward control signal. An increasing load calling for a lower chilled-water temperature will open the valve immediately, helping to meet that demand. This optimizing control system would seem to apply equally well to single-stage absorption units whose comparable thermodynamic conversion and consequent thirst for cooling water make cost optimization more attractive for them.

CONTROLLING PRESSURE AND REFLUX

The examples of pressure control given to this point have generally assumed that pressure was to be maintained constant. The custom of maintaining a constant pressure on a column is little more than that—a custom. When a single temperature measurement was relied upon for control of product quality, it was important that column pressure not change. But in many separation units analyzers are now being used for quality control, eliminating

TABLE 7.2 Thermodynamic Conversion of Steam-jet Refrigeration at Optimum Operation

Chilled-water temp., °F	Conversion
40	0.45
50	0.54
60	0.63

the need for a constant pressure. Furthermore, pressure compensation can be readily provided for temperature measurements to allow quality control by temperature even in variable-pressure situations. Or the differential-vapor-pressure transmitter described in Chap. 5 may be used as a more accurate and sensitive indication of composition. In some cases [14] differential-temperature measurements, which are also insensitive to pressure changes, have been used.

Once having accepted the tenet that column pressure can be variable, the next question arises: What is its optimum value? Enough information has been presented in Chaps. 1, 2, and 7 to indicate that minimum-pressure operation is most efficient because of enhanced relative volatility. Having established that point, it is only necessary to describe how minimum-pressure operation is achieved.

Floating-pressure Control It would be possible to operate a total condensing distillation column with no pressure control whatsoever. The familiar control valves used to throttle, bypass, or flood the condenser would simply be eliminated. Although this would be ideal in the steady state, transient upsets to the heating or cooling system could cause undesirable reactions. Consider, for example, a column with an air-cooled condenser exposed to a sudden rainstorm. The wetted surface of the condenser becomes capable of transferring more heat, causing a sharp reduction in column pressure. The column itself is not cooled by the rain, but some of its sensible heat is quickly converted to latent heat by the fall in pressure. The vapor-rate increase caused by the transient reduction in sensible heat can be enough to flood the column. Its least effect would be to move substantial quantities of high-boiling components up the column and into the reflux drum. Protection must be provided against this type of upset.

While some sort of control needs to be provided to counter the short-term upsets, the pressure should be minimum in the long term. This can be achieved by judiciously adjusting the set point of the pressure controller. However, the adjustment must be slow and continuous rather than stepwise. Operators tend to adjust set points in steps, however small; and each step introduces the type of upset the pressure controller is intended to eliminate.

The solution is to use a valve-position controller (VPC) to adjust the pressure set point as shown in Fig. 7.19. It is intended to hold the condenser control valve in its fully open or fully closed position, depending on whether the valve bypasses, throttles, or floods the condenser. In any case, the condenser should be fully loaded in the long term.

The distinction between short-term and long-term control is made in the mode settings of the two controllers. The pressure controller has proportional and reset action adjusted to provide prompt recovery from upsets. The pressure set point will then be followed as closely as possible, as in any pressure-control loop. The valve-position controller, however, should have integral action alone. Then a sudden change in valve position will not elicit a proportional change in the pressure set point. Instead, the set point will move at

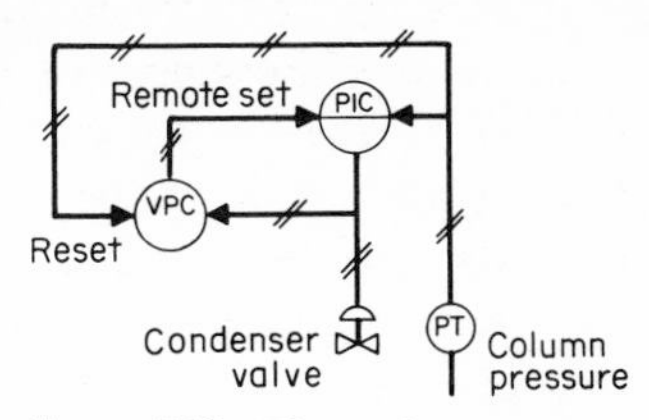

figure 7.19 *The valve-position controller slowly adjusts the pressure set point to keep the condenser fully loaded in the long term.*

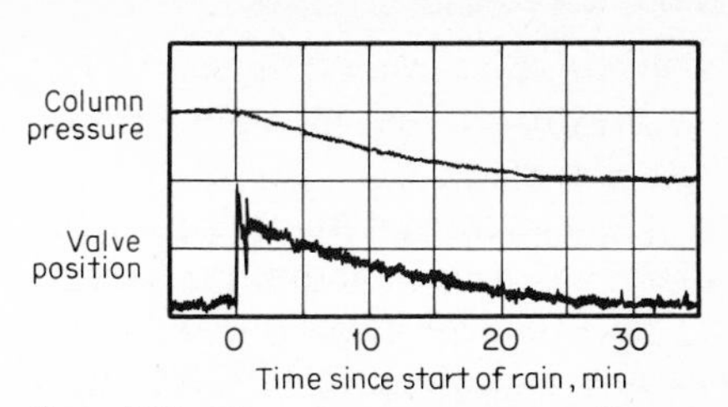

figure 7.20 *The column pressure will move exponentially to a new steady state as the bypass valve returns to a nearly closed position.*

a rate proportional to the deviation of valve position from its limit. The actual rate is adjustable through the reset time, which can be set up to 60 min in contrast to perhaps 2 min for the pressure controller. The effects of control action on valve position and column pressure are depicted in Fig. 7.20.

Since weather conditions usually deteriorate more rapidly than they improve, protection ought not be necessary for conditions causing a rise in pressure. Consequently, the valve-position set point could be safely held at 10 percent closed (or open as the case may be), which provides control over falling pressure only.

Variations in heat input will change column pressure along the condenser constraint as described in Fig. 1.9. Here again, a reduction in heat input would result in short-term pressure control while an increase would see no control at all. The accompanying increase in column pressure would cause some lag in response of vapor flow to heat input. It would be necessary to raise the sensible heat of the liquid on the trays as the increasing pressure raised its boiling point. Nonetheless, vapor flow, pressure, and temperature will all rise together along an exponential curve determined by the heat capacity of the column.

The scheme shown in Fig. 7.19 allows the operator to hold pressure constant at will by placing the pressure controller in its locally set mode. Either locally set pressure control or manual operation will take control away from the valve-position controller. Opening of the valve-position loop would then cause the VPC to reduce its output continuously in an effort to restore valve position to 10 percent. The operator might then find it difficult to return to the remotely set mode since the remote set point would be well downscale. This windup of the VPC can be prevented by using the pressure measurement as reset feedback [15]. As long as there is no deviation between the pressure remote set point and measurement, the VPC will integrate normally. But when a pressure deviation develops, integrating action will eventually come to rest with the VPC deviation equalling the pressure deviation. To place the system in valve-position control, the pressure controller must be transferred from the locally to the remotely set mode while in manual to avoid bumping the valve. This configuration presents the operator with only the conventional pressure controller at the panel. The VPC is blind (no display). It may be mounted away from the panel and requires no auto-manual transfer station.

The benefits of floating-pressure control are many. Reference 16 reports that the energy required to separate propane and isobutane decreases about 0.7 percent for every 1°F reduction in air-cooling temperature. Because the energy requirements fall with condensate temperature, the ΔT across the condenser surface falls. As a consequence, the drop in condensate temperature is always greater than the drop in coolant temperature. Although the latent heat of vaporization increases with falling temperature, this effect is more than offset by the reduction in V/F required to make the separation at the improved relative volatility.

The benefits of floating-pressure control are greater for air-cooled condensing than for water-cooled because air cooling requires a greater ΔT, and air temperature is more variable. Table 7.3 compares the energy required to split a 50-50 mixture of butanes into 97 percent pure products in a column of 40 theoretical trays as a function of air or water temperature. A 50°F approach to 90°F air and a 20°F approach to 80°F water were used as design bases. The data for V/F at various condenser temperatures were taken from Table 2.4.

The decrease in air temperature from 90 to 76°F drops the energy required per pound of feed by 12 percent, or 0.87 percent/°F. The decrease from 61 to 45°F drops Q/F by only 10.2 percent, or 0.64 percent/°F. By comparison, a reduction in water temperature from 80 to 62°F causes Q/F to change by the same 10.2 percent, or 0.57 percent/°F.

The data presented in Table 7.3 represent conditions of constant feed rate F with boilup V reduced as coolant temperatures fall. A similar table could be prepared for constant boilup where feed rate is increased as falling coolant temperatures allow the reduction in V/F. In that case, production would be increased by F/V at the cost of increased heat input by the amount that latent heat increases. However, flooding limits must be applied as presented in Fig. 1.8. For the butane splitter, allowable vapor flow per unit column area A decreases as temperature falls—this limit is taken into account in developing the data in Table 7.4.

TABLE 7.3 Energy Required to Separate Butanes at Constant Feed Rate

Condensate temperature, °F	V/F	Latent heat, Btu/lb	Q/F, Btu/lb	Air temperature, °F	Water temperature, °F
140	6.07	121	734	*90	
120	5.00	129	645	76	
100	4.21	135	568	61	*80
80	3.64	140	510	45	62
60	3.23	145	468	28	44

* Base case

TABLE 7.4 Energy Required to Separate Butanes at Maximum Vapor Rate

Condensate temperature, °F	V/A lb/h-ft²	F/A lb/h-ft²	Q/A Btu/h-ft²	Air temperature, °F	Water temperature, °F
140	1.04	0.171	126	*90	
120	1.02	0.204	132	68	
100	0.97	0.230	131	48	*80
80	0.91	0.250	127	29	61
60	0.83	0.257	120		42

* Base case

For the air-cooled condenser, production F can be increased by 19 percent as the air temperature falls from 90 to 68°F, an increase of 0.88 percent/°F, while at the same time the energy required to separate a unit of feed drops 12 percent, or 0.55 percent/°F. Again, the improvement achievable with water cooling is less dramatic, principally because the base conditions are already more favorable. The production increase is 0.46 percent/°F with a decrease in Q/F of 0.54 percent/°F. In many plants, the production increase may well be worth more than the energy savings, but both are obtained together.

Partial Condensers Partial condensers present some special control problems in being sensitive to both cooling and the accumulation of noncondensable gases. The pressure in most partially condensed systems is controlled at a constant value simply by venting noncondensables. Although this is a simple and obvious solution to the pressure-control problem, it is far from optimum. And unfortunately, the ease with which pressure is controlled by venting, compared with bypassing or flooding a total condenser, has led many engineers to force total condensers to be partial condensers. An inert gas is allowed to enter the condenser when the pressure is falling and vented when the pressure is rising. As described earlier, this mechanism wastes inert gas, overhead product, and available work while interfering with condensation.

Having established control methods for constant-pressure and floating-pressure operation of total condensers, the next step is to apply those concepts to partial condensers. Three types of partial condensers must be considered:

1. Condensing duty split between two media
2. Partial condensers with no liquid product
3. Partial condensers with liquid product

The control systems for the three differ substantially.

In the fractionation of natural-gas liquids, the first column in the train is usually a deethanizer. Its feed contains a mixture of paraffins from ethane through a gasoline fraction, with a sufficiently low methane content that its overhead product is fully condensable. The overhead product may contain a substantial quantity of propane with its much higher boiling point. As a result, the condensing duty is usually split between a water-cooled unit and a refrigerated condenser as shown in Fig. 7.21.

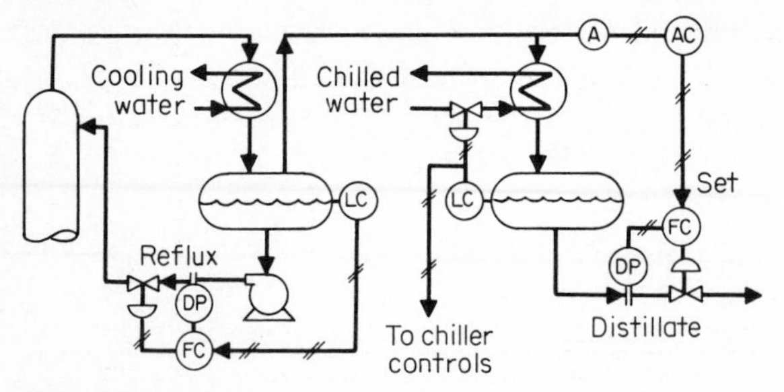

figure 7.21 *In a deethanizer, the condensing duty is usually shared between cooling water and refrigeration.*

Water cooling is usually applied only to the reflux, with refrigeration reserved for the distillate. This is a logical arrangement in that the reflux drum acts as an equilibrium stage—the dew points for reflux and distillate may differ by perhaps 30°F. To maximize thermodynamic efficiency of the condensing system, the water-cooled condenser should remove as much heat as possible, leaving as little as possible for the chiller. However, the allocation of reflux and distillate individually to the two condensers makes this impossible.

Floating-pressure control remains a possibility by allowing the water-cooled condenser to operate without restriction at all times. Whatever is condensed is totally refluxed by the level controller. Whatever is not condensed is passed on to the chiller. A composition controller will allow the discharge of only enough distillate to meet the specification on propane content. If more product is condensed in the chiller than is withdrawn, the balance will accumulate, raising the level in the distillate drum. A level controller throttles the chilled-water valve to balance the condenser against the load. The chilled-water valve position may then be sent to the chiller controls as shown in Figs. 7.17 and 7.18.

Alternatively, the level controller may be eliminated and the chilled-water condenser allowed to flood. Eventually enough surface will be covered to balance condensing rate against distillate withdrawal through self-regulation. This violates the concepts of thermodynamic efficiency and so is suboptimum. However, a level controller acting on condenser level can adjust chilled-water temperature in place of the valve-position controller, since in this case there is no chilled-water control valve. The results are essentially the same.

In neither case need there be a column-pressure controller. Should the rate of condensation not balance the flow of vapor into the chilled-water condenser for any reason, the level will change to force a balance. The pressure will always be the lowest that can be reached by the available cooling water. Furthermore, only as much chilling is used as is required to condense the distillate. The result is minimum column pressure, which minimizes the energy required to separate the feed and at the same time maximizes the capacity of the column.

This reasoning assumes that chilled-water capacity is always more than adequate to condense all the distillate that needs to be withdrawn. This is not always true. Should more distillate be withdrawn than can be condensed, the liquid level in the product accumulator will fall. Then an override is needed on feed flow to avoid loss of level.

A balance between cooling-water duty and chilled-water duty must also be struck to match the reflux-distillate ratio. If reflux condensation is excessive, there is no way to condense the required amount of distillate using the scheme shown in Fig. 7.21. In this case, a low level in the product drum should be corrected by reducing reflux cooling and thereby raising pressure to transfer more vapor to the product condenser. This must be done on the coolant side, however, because flooding or bypassing the reflux condenser will result in its loss as an equilibrium stage.

A partial condenser without a liquid product withdrawal is shown serving the demethanizer in Fig. 7.22. Typically, column pressure is controlled by manipulating the flow of vapor product. However, composition is affected by that flow as well as condenser temperature, reflux rate, etc. The *composition* should be controlled by manipulating product flow as shown, with pressure controlled by refrigeration. The level of refrigerant in the condenser may be controlled by refrigerant flow instead. Then column pressure will always be a minimum for the available refrigerant temperature because its full heat-transfer surface is being used for evaporation.

The scheme in Fig. 7.22 combines both level and pressure control to achieve floating-pressure operation. The level controller takes the role of a valve-position controller in adjusting the pressure set point to achieve complete flooding in the long term. However, the pressure controller is free to adjust the level in the short term to counter upsets in heat load. Using level rather than valve position for feedback keeps liquid refrigerant from leaving the condenser on a regular basis. The valve-position signal may then be used to minimize the pressure rise developed by the refrigeration compressor. The same cascading of level to pressure to coolant valve could be applied to the deethanizer of Fig. 7.21 by using distillate-drum or condenser level as appropriate. Whether the pressure controller would really be necessary depends on the sensitivity of the

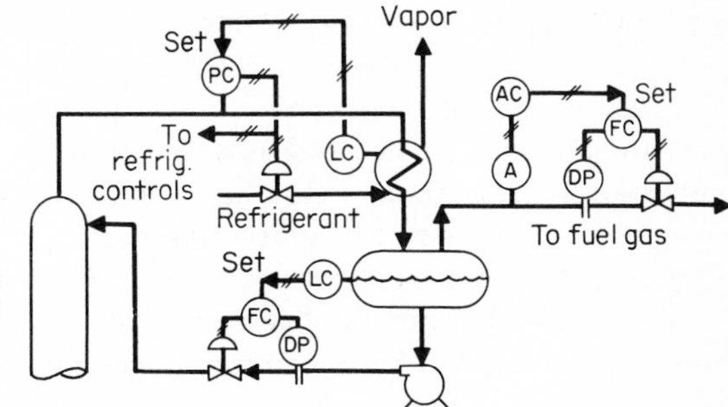

figure 7.22 *The overhead product from a demethanizer is usually withdrawn as a vapor.*

column to variations in ambient, heat load, etc., and the frequency and magnitude of those variations. The level controller must have proportional action as well as reset for stability.

When both liquid and vapor products are being withdrawn, an additional control loop is required. It is needed to control the composition of the vapor or the lighter-than-light key in the liquid. Figure 7.23 shows a composition control loop on the liquid and also vapor-pressure control applied to the vapor.

Column pressure is readily controlled by manipulating vapor flow, but its set point must be adjusted for variations in condensate temperature. The temperature measurement is characterized to an equivalent vapor pressure representing the desired composition. Then as coolant and ambient temperatures change, column pressure is adjusted accordingly. A differential-vapor-pressure cell would provide the same function even though it is not normally applicable to ternary systems such as this. A problem may arise in finding the correct filling solution.

The relationship between pressure and temperature is nearly linear over the range typically encountered with condensers. Furthermore, the slopes of the vapor-pressure curves for various compositions tend to be quite similar, although they are displaced from one another. Consequently, a bias adjustment should be available to recalibrate the functional relationship for desired changes in composition.

The instrument configuration shown in Fig. 7.23 provides true floating-pressure control. There is no restriction on the condenser, so the condensate is always as cool as possible. Hence the pressure in the column will always be as low as possible, maximizing relative volatilities and column capacity and minimizing the energy required for separation. Pressure drop and therefore power loss across the vapor valve are also minimized. This can be an important consideration when pressure is developed by a feed-gas compressor—minimum column pressure allows reduced compressor power, increased flow, or a combination of both.

Internal-reflux Control Column operation is best stabilized by maintaining control over internal vapor and liquid flow rates. Vapor-flow control was accomplished in various ways as required by the heating medium, reboiler, etc. If a tight rein is placed on heat input, reflux can be simply used to close the heat balance through pressure control or, in the case of floating pressure, accumulator level control. There are situations, however, where the heat input cannot or should not be independently controlled but should be used to control pressure or base level. Then the separation for the column is set by the ratio of reflux to feed as a substitute for vapor to feed. In these cases, it is important that internal reflux be controlled although it cannot be measured. This means that external-reflux flow must be compensated by mass or heat-balance calculations to infer internal reflux.

In the most common application of internal-reflux control, the external flow measurement is compensated for subcooling [17]. The relationship compen-

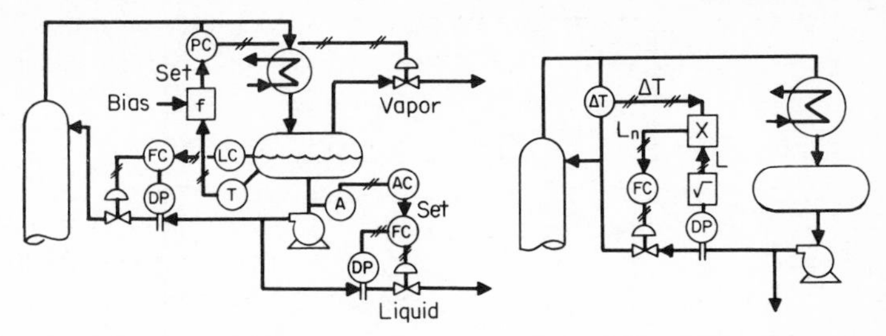

figure 7.23 *Overhead products in both phases require vapor-pressure control.*

figure 7.24 *This system compensates for subcooling but not for changes in composition.*

sating for subcooling is given in Eq. (7.18). It is dependent on the temperature-difference measurement between vapor and reflux:

$$L_n = L\left[1 + (T_V - T_L)\frac{C_p}{H_V}\right] \tag{7.18}$$

where L_n and L are internal- and external-reflux flows and C_p/H_V is the specific-to-latent heat ratio for the reflux. Table 2.6 lists C_p/H_V ratios for some selected hydrocarbons.

Although the system shown in Fig. 7.24 compensates adequately for subcooling, it reacts wrongly to variations in overhead composition. Should reflux temperature fall because of more cooling, $T_V - T_L$ increases and causes external reflux to be reduced. But should vapor temperature rise because of increasing high-boiling components, $T_V - T_L$ will rise also, causing external reflux to be reduced. This action is counter to control and actually augments the disturbance through positive feedback. Negative feedback from a composition or temperature controller is always required to ensure stability with this system.

Another type of internal-reflux control based on a material balance is used for columns with liquid sidestreams. Consider the unit shown in Fig. 7.25, where the main liquid product is withdrawn several trays down from the top. The overhead and bottom products are typically too small to have much effect on liquid level in the accumulator or column base. Consequently they are used for quality control instead of level control. The reflux and heat input *both* must be placed on level control. But such an arrangement is without self-regulation. An increase in boilup raises accumulator level, thereby increasing reflux, which will raise base level, etc. To stabilize the system, internal reflux must be controlled below the side-draw tray.

If the side-draw tray traps all the liquid, then a controlled flow may be reintroduced under it. But the usual design has an internal overflow which is unmeasurable. The system shown in Fig. 7.25 controls liquid flow L_i below the side-draw tray as the difference between external reflux L and product P.

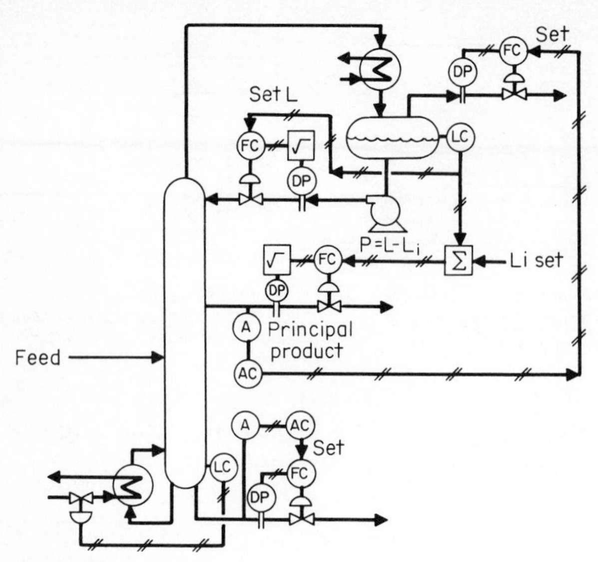

figure 7.25 *A column whose major product is withdrawn as a liquid sidestream requires internal-reflux control below that tray.*

But it does so by manipulating P in response to L:

$$P = L - L_i \tag{7.19}$$

Then an increase in accumulator level which raises reflux will automatically increase P. It is as if product flow were manipulated by the level controller, although product flow has no direct effect on level.

It should be noted that in all cases reflux *flow* is controlled, whether set by accumulator level, composition, or whatever. It is not sufficient simply to position the reflux *valve*—its hysteresis and nonlinear characteristics cause an undesirable degree of flow variation unless flow itself is controlled.

In most cases, internal-reflux calculations are unnecessary. Where a liquid level is maintained in the reflux drum, little subcooling normally develops. Furthermore, variations in overhead-vapor flow caused by heat losses or subcooling do not cause a problem because the liquid-level controller automatically adjusts reflux to reach a balance. Similarly, when a condenser is flooded by manipulating reflux for pressure control, reflux flow is automatically adjusted to reach a balance. And with floating-pressure control, subcooling is avoided altogether.

REFERENCES

1. Buckley, P. S.: Material Balance Control in Distillation Columns, presented at AIChE Workshop on Industrial Process Control, Tampa, Nov. 11–13, 1974.
2. Niagara Blower Company: Niagara Aero Heat Exchanger, Bulletin 169, Buffalo, N.Y.

3. Elliott Division, Carrier Corporation: Elliott Multistage Centrifugal Compressors, Bulletin P-11A, Jeannette, Pa.
4. Union Carbide Corporation: Conserve Energy by Using High-flux Tubing in Heat-pump Applications, Tonawanda, N.Y.
5. White, M. H.: Surge Control for Centrifugal Compressors, *Chem. Eng.* (N.Y.), Dec. 25, 1972.
6. Shinskey, F. G.: Effective Control for Automatic Startup and Plant Protection, *Can. Controls Instrum.*, April 1972.
7. Franzke, A.: Save Energy with Hydraulic Power Recovery Turbines, *Hydrocarbon Process*, March 1975.
8. Houghton, J., and J. D. McLay: Turboexpanders Aid Condensate Recovery, *Oil Gas J.*, Mar. 5, 1973.
9. Monroe, E. S.: Vacuum Pumps Can Conserve Energy, *Oil Gas J.*, Feb. 3, 1975.
10. Evans, F. L.: "Equipment Design Handbook for Refineries and Chemical Plants," vol. I, p. 89, Gulf Publishing Company, Houston, 1973.
11. Shinskey, F. G.: "Process-Control Systems," pp. 204–224, McGraw-Hill Book Company, New York, 1967.
12. Farwell, W.: More Cooling at Less Cost, *Plant Eng.*, August 1974.
13. Evans, F. L.: *op. cit.*, p. 170.
14. Boyd, D. M.: Fractionating Column Control, *Chem. Eng. Prog.*, June 1975.
15. Shinskey, F. G.: Avoiding Reset Windup in Cascade Systems, *Instrum. Control Syst.*, August 1971.
16. Fauth, C. J., and F. G. Shinskey: Advanced Control of Distillation Columns, *Chem. Eng. Prog.*, June 1975.
17. Lupfer, D. E., and M. L. Johnson: Automatic Control of Distillation Columns to Achieve Optimum Operation, *ISA Trans.*, April 1964.

CHAPTER EIGHT

Product-quality Control

This could well be the most important chapter in the book. The primary objective of any distillation system is to deliver products of consistent quality. As demonstrated in Chap. 1, if that quality just meets specifications and no better, operating cost will be low, product recovery will be high, and column capacity will also be high. Therefore, being able to control quality precisely brings with it many of the other objectives which combined mean profitable operation.

The foregoing is not intended to diminish the importance of controlling energy inflow and outflow or of applying constraints or optimizing. Yet it is possible to achieve acceptable quality control with little cooperation from the energy streams and without encountering constraints or developing optimizing schemes. The simple expedient of material-balance control implemented through a combination of feedforward and feedback functions is sufficiently powerful and responsive to overcome many of the vagaries of the energy sources and sinks. This is not true of the more conventional and less accurate methods of control, which allow material and energy streams to interact to the detriment of product quality.

While it is possible to control the quality of a single product using the methods described in this chapter alone, multiple specifications require multiple control loops and introduce the potential for interaction. Consequently, this chapter will be restricted to control of a single product quality through manipulation of the column's material balance. The problem of controlling more than one quality is best described and solved by using the special techniques given in Chap. 10.

MATERIAL-BALANCE CONTROL

Throughout this book, but principally in Part 2, the relationship between the composition of a product and its flow rate has been stressed. As product flow is increased, its quality must deteriorate because of the limited quantity of the desired component entering in the feed. Thus the premise was made that composition could be controlled by manipulating the relative quantity of that product extracted from the feed. In other words, composition is adjusted by the way the mass flows are distributed both within and without the column. The present discussion centers on how well control systems can dictate these material balances as determined by their structure.

Conventional Control Systems An example of the most common control system applied to distillation columns appears in Fig. 8.1. Its function is to control the quality of the bottom product by manipulating boilup appropriately. Observe that both products are under liquid-level control—their flow rates are then subject to influence from all disturbances that can affect the column.

Product quality (boiling point) is controlled by adjusting boilup relative to a fixed flow of reflux. As low-boiling components begin to accumulate in the column base, the falling temperature causes boilup to increase. Increasing boilup has two effects: it increases separation and shifts the material balance. The separation increase tends to make both products purer. But the shift in the material balance (as the difference between boilup and reflux) tends to move material from the bottom of the column to the top. Its effect is to

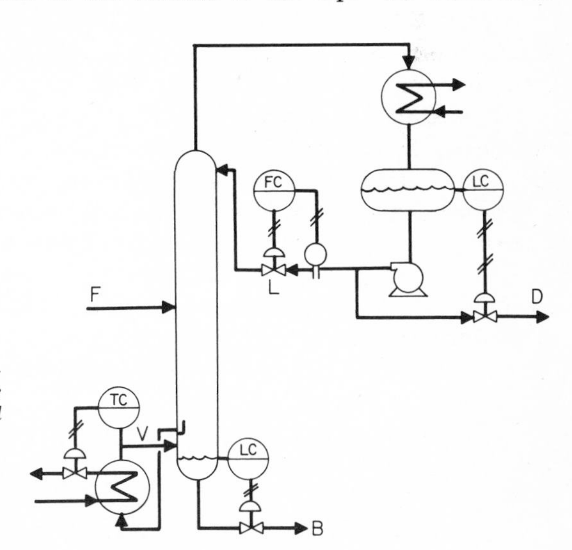

figure 8.1 *This very common control-loop arrangement allows the material balance to be affected by energy disturbances.*

improve the purity of the bottom product (by decreasing its flow) and to reduce the purity of the distillate (by raising its flow). Thus the simple action of adjusting the boilup rate alone produces two interacting effects.

To evaluate the relative magnitude of the two effects, consider an isomer splitter separating a 50-50 feed mixture into products each 95 percent pure using a V/F ratio of 3. There are at least three ways to adjust bottom composition—by manipulating the material balance with constant separation, by holding a fixed material balance and manipulating separation, and by a combination of both. Table 8.1 summarizes the manipulations needed to change bottom-product purity from 95 to 96 percent while holding boilup, bottom-product flow, and reflux constant.

Control with constant boilup represents the material-balance system—B/F or D/F is adjusted to affect product composition. Note that a change in B/F of 0.0124 in 0.50, or 2.5 percent, brings about the 1 percent change in composition. In the constant material-balance case, separation alone is manipulated for quality control—this arrangement is rarely, if ever, used because it lacks sensitivity. Boilup must change by 0.235 in 3.00, or 7.8 percent, to change composition by 1 percent.

The case of constant reflux describes the control system shown in Fig. 8.8. Here boilup must change only by 0.012 in 3.00, or 0.4 percent, to bring about the 1 percent change in composition. A comparison of the constant V and constant L data indicates that they are almost identical. The change in composition has been brought about principally by a shift in the material balance in both cases. The 0.4 percent increase in V in the last column is insignificant in comparison to the 7.8 percent increase required to produce the same effect when the material balance is fixed.

The conclusion can then be reached that Fig. 8.1 represents a material-balance control system since composition is controlled by adjustment of the material balance. However, this adjustment is indirect and is hypersensitive to its own and other influences. To demonstrate, consider that composition is six times as sensitive to variations in boilup with constant reflux as it is to bottom-product (or distillate) flow. This amplification factor is the V/B (or V/D) ratio. This means that composition is six times as sensitive to undesired variations (imprecision) in boilup as to bottom (or distillate) flow. Furthermore, boilup and reflux are subject to more variability than distillate or bottom-product flow because they are unmeasurable. The true values of

TABLE 8.1 Manipulations Required to Adjust Bottom Composition with Three Different Control Schemes

	Base case	Constant V, manipulated B or D	Constant B and D, manipulated V	Constant L, manipulated V
x	0.05	0.04	0.04	0.04
y	0.95	0.9377	0.96	0.9390
V/F	3.00	3.00	3.235	3.012
B/F	0.50	0.4876	0.50	0.4883
L/F	2.50	2.488	2.735	2.50

internal vapor and liquid flow rates must be inferred from external measurable rates, and their relationship depends on heat losses, subcooling, feed enthalpy, etc. This hypersensitivity and variability was pointed out as early as Chap. 2.

Figure 8.1 represents only one form of a non-material-balance control scheme. Any system which allows more than one product to float on level or pressure control is not a material-balance system.

Material-balance Controls Material-balance systems came into being when conventional systems were unable to provide sufficiently precise control over close separations [1–3]. The amplification factor cited above determines the relative sensitivity of composition to external influences without material-balance control. In a column where V/B is 1.0, for example, both these variables would have essentially the same effect on bottom composition. Material-balance control still has an advantage, however, in being insensitive to variations in enthalpy.

In a material-balance system, a product flow is always manipulated to control composition. Usually the controlled quality and manipulated flow belong to the same product, but not always. Extenuating circumstances described later in this chapter under "Sensitivity to Control Action" may force an alternative combination. In these cases, an additional loop may be required to provide satisfactory dynamic response, as also described later.

When distillate flow is assigned to control composition, accumulator level must be controlled by manipulating either reflux or boilup. If the condenser is flooded such that no level exists, then column pressure must be controlled by manipulating either reflux or boilup.

Making reflux a dependent variable in this way is still a controversial issue among plant operators and university professors alike. Distillation design procedures center around reflux and reflux ratio, and most columns are controlled with constant reflux. As mentioned, a conversion to material-balance control has been necessitated in the most difficult separations, although it is fundamentally superior for all separations.

There is no reason to suspect that making reflux dependent on level or pressure control will create instability, as long as reflux flow is controlled in cascade. However, level and pressure loops have exhibited limit cycling (oscillations of a fixed amplitude) when the valve has been manipulated directly without a cascaded flow controller. The combination of an integrating process (level or pressure), reset action, and valve hysteresis will always result in a limit cycle [4]. Closing a flow loop around the valve overcomes its hysteresis and eliminates the cycle.

Important as it is to have stable control of reflux, responsiveness is equally important. It was pointed out in Chap. 2 that manipulation of the external material balance will not, by itself, have any direct effect on composition—the internal balance must be adjusted accordingly. This is the role of the liquid-level controller manipulating reflux or boilup. Figure 8.2 illustrates a typical material-balance control system where distillate is manipulated for overhead

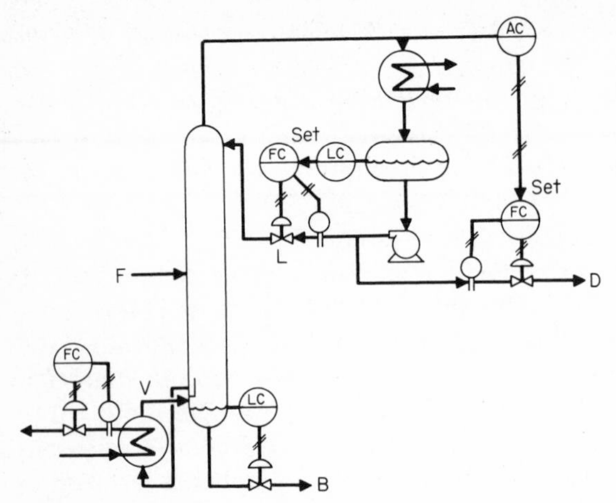

figure 8.2 *In a typical material-balance control system, one product flow is set to control quality.*

quality control. Following an adjustment to distillate flow, accumulator level will begin to change. Depending on the settings of the level controller, the reflux flow will be adjusted in response until a new steady state is reached where the net change in reflux equals the net change in distillate.

The level controller also serves to prevent disturbances in heat input, feed enthalpy, and reflux temperature from altering the material balance. An increase in heat input, for example, will simply raise the accumulator level and increase reflux by the same amount. Its net effect on composition will be much smaller than if reflux were flow-controlled and the increased boilup were passed along to the distillate. A glance at Table 8.1 will prove this point. With fixed reflux, a 0.4 percent change in boilup causes a 1 percent change in composition; but with fixed distillate and bottom flow (material-balance control), boilup has to change by 7.8 percent to produce a similar variation. Hence, in this example, material-balance control reduces the sensitivity of the column to disturbances in the energy balance by a factor of 20.

Figure 8.2 illustrates only one possible configuration of material-balance control. In some cases, reflux could be a more reliable measure of energy flow than the flow of heating medium, as might be the case with an oil-heated reboiler. Then reflux flow could be fixed, with accumulator level controlled by manipulating heat input. In the case of a flooded condenser, pressure could be controlled by reflux [1] or boilup.

The second basic material-balance scheme has bottom-product flow controlled to determine composition, with base level manipulating heat input, as in Fig. 8.3. This configuration is called for when the bottom-product flow is smaller than the distillate. The accuracy to which the entire material balance may be manipulated is that of the controlled product flow. Naturally the smaller flow can be controlled to the smaller absolute error if the relative (percentage) errors of both flows are similar. The level controller then forces the other product to close the balance, reducing its absolute steady-state error to that of the smaller flow.

A material-balance system for three products was shown in Fig 7.25. Overhead and bottom-product flows were small compared to the sidestream, and so they were manipulated for quality control. The sidestream was left to close the balance under (indirect) level control.

Sensitivity to Feed Composition An outstanding advantage of material-balance control is its natural ability to be adjusted by direct calculation to achieve a desired product quality. This property allows feedforward control to be applied with ease and simplicity. In effect, the correct setting of D or B may be predicted, with accuracy, to produce a desired composition. The same is not true when reflux and boilup are the manipulated variables.

The mathematical relationship used is simply the reverse of a blending calculation. An overall material balance is drawn, followed by a component balance:

$$F = D + B \tag{8.1}$$

$$Fz = Dy + Bx \tag{8.2}$$

Here z, y, and x are the fractions of any given component in the associated stream. When the two expressions are combined, the D/F or B/F ratio needed to satisfy product compositions may be quickly estimated:

$$\frac{D}{F} = \frac{z - x}{y - x} \tag{8.3}$$

$$\frac{B}{F} = \frac{y - z}{y - x} \tag{8.4}$$

As a first estimate, D/F can be set equal to z; z/y will give a more exact estimate if x is close to zero. Otherwise an estimate of x must be made to determine the D/F required to produce the desired y.

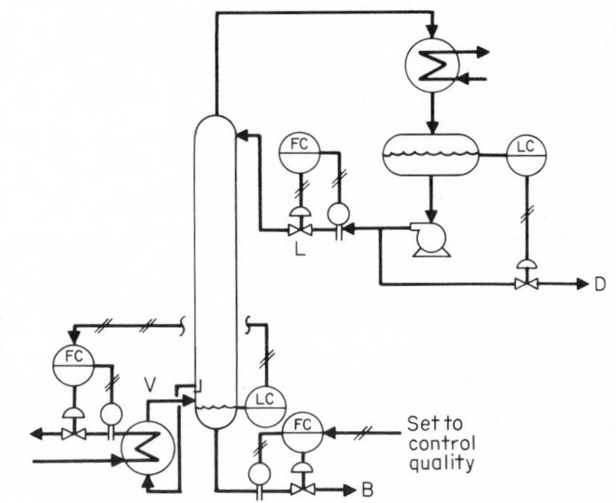

figure 8.3 *This configuration is recommended when the bottom-product flow is smaller than the distillate.*

The sensitivity of product compositions to variations in D/F or B/F is important in evaluating control effectiveness. Unfortunately, Eqs. (8.3) and (8.4) are insufficient to supply the information by themselves. Each is essentially a single equation with two unknowns. An adjustment to D/F or z will affect both x and y. The only sensitivity function that can be gleaned from the material balance alone is the adjustment in D/F or B/F needed to maintain control of x and y in the face of variations in z:

$$\frac{\partial D/F}{\partial z} = -\frac{\partial B/F}{\partial z} = \frac{1}{y - x} \tag{8.5}$$

Since y and x always lie between 0 and 1, D/F and B/F must always change more than the feed composition to maintain control. Thus a 1 percent variation in z will require something more than a 0.01 variation in D/F or B/F.

Equation (8.5) gives the *absolute* sensitivity of D/F to variations in z. The *relative* sensitivity is important for control purposes, i.e., the *relative* change in the material balance required for control:

$$\frac{\partial D/F}{(D/F)\, \partial z} = \frac{1}{z - x} \tag{8.6}$$

$$\frac{\partial B/F}{(B/F)\, \partial z} = -\frac{1}{y - z} \tag{8.7}$$

As $z - x$ approaches zero, D/F approaches zero, and a relatively large change is required to correct for a variation in feed composition.

The effect of variations in feed composition on product quality in the absence of control is of equal significance. By the same token, the effect of variations in the material balance on product quality must also be assessed. The material-balance equations may be readily rearranged to give product compositions y and x as a function of z and D/F or B/F:

$$y = x + \frac{z - x}{D/F} \tag{8.8}$$

$$x = y - \frac{y - z}{B/F} \tag{8.9}$$

However, partial derivatives of $\partial y/\partial z$ or $\partial x/\partial z$ are meaningless because variations in z, D/F, or B/F will affect *both* y and x at the same time.

A second relationship between y and x using the separation factor must be employed. For a binary system, separation is defined as

$$S = \frac{y(1 - x)}{x(1 - y)} \tag{8.10}$$

If the separation is held constant by maintaining a constant V/F and avoiding gross variations in z (leading to a loss in separation by a suboptimum feed-tray location), then enough information is available to define the system.

However, the nonlinear nature of the separation expression prohibits the use of simple derivatives. Instead, the combined effects of material balance and separation relationships are best discovered by combining partial derivatives to form a total derivative of z with respect to y or x:

$$\frac{dz}{dy} = \left.\frac{\partial z}{\partial y}\right|_x + \left.\frac{\partial z}{\partial x}\right|_y \left.\frac{\partial x}{\partial y}\right|_S \tag{8.11}$$

The last term in this expression is the derivative of Eq. (8.10):

$$\left.\frac{\partial x}{\partial y}\right|_S = \frac{x(1-x)}{y(1-y)} \tag{8.12}$$

To complete the derivation, the material balance is solved for z and differentiated:

$$z = x + \frac{D}{F}(y - x) \tag{8.13}$$

$$\left.\frac{\partial z}{\partial y}\right|_x = \frac{D}{F} \qquad \left.\frac{\partial z}{\partial x}\right|_y = 1 - \frac{D}{F}$$

Then (8.11) may be solved:

$$\frac{dz}{dy} = \frac{D}{F} + \left(1 - \frac{D}{F}\right)\frac{x(1-x)}{y(1-y)} \tag{8.14}$$

The sensitivity dy/dz is simply the inverse of the solution to (8.14). Having found dy/dz,

$$\frac{dx}{dz} = \frac{dy}{dz}\left.\frac{\partial x}{\partial y}\right|_S = \frac{dy}{dz}\,\frac{x(1-x)}{y(1-y)} \tag{8.15}$$

example 8.1

A column is separating a binary feed of $z = 0.7$ into a distillate of $y = 0.95$ and bottom product having $x = 0.03$. Determine the sensitivity of both products to a change in feed composition.

$$\frac{D}{F} = \frac{0.7 - 0.03}{0.95 - 0.03} = 0.728$$

$$\left.\frac{\partial x}{\partial y}\right|_S = \frac{0.03(0.97)}{0.05(0.95)} = 0.613$$

$$\frac{dz}{dy} = 0.728 + (1 - 0.728)0.613 = 0.894$$

$$\frac{dy}{dz} = \frac{1}{0.894} = 1.12$$

$$\frac{dx}{dz} = 1.12(0.613) = 0.69$$

Note that in this example the sensitivities are not greatly different from unity.

Sensitivity to Control Action The procedure for determining the sensitivity of product quality to adjustments in D/F follows the same pattern. The total derivative of D/F with respect to y is the sum of the partials:

$$\frac{d(D/F)}{dy} = \left.\frac{\partial(D/F)}{\partial y}\right|_x + \left.\frac{\partial(D/F)}{\partial x}\right|_y \left.\frac{\partial x}{\partial y}\right|_S \tag{8.16}$$

Partial differentiation of the material balance gives

$$\left.\frac{\partial(D/F)}{\partial y}\right|_x = \frac{-D/F}{y - x} \qquad \left.\frac{\partial(D/F)}{\partial x}\right|_y = \frac{D/F - 1}{y - x}$$

Then

$$\frac{d(D/F)}{dy} = \frac{-D/F}{y - x} - \frac{1 - D/F}{y - x}\,\frac{x(1 - x)}{y(1 - y)} \tag{8.17}$$

The sensitivity of y to D/F is simply the inverse of (8.17), and the sensitivity of x to D/F is

$$\frac{dx}{d(D/F)} = \frac{dy}{d(D/F)}\left.\frac{\partial x}{\partial y}\right|_S = \frac{dy}{d(D/F)}\,\frac{x(1 - x)}{y(1 - y)} \tag{8.18}$$

As before, these sensitivities are *absolute* values, although *relative* values are perhaps more significant in terms of the accuracy to which D/F may be manipulated. Accordingly, the absolute sensitivity given in Eq. (8.17) is converted to relative sensitivity:

$$\frac{d(D/F)}{(D/F)\,dy} = \frac{-1}{y - x} - \frac{\dfrac{1}{D/F} - 1}{y - x}\,\frac{x(1 - x)}{y(1 - y)} \tag{8.19}$$

Substituting for D/F and rearranging yields

$$\frac{d(D/F)}{(D/F)\,dy} = \frac{-1}{y - x}\left[1 - \frac{(y - z)x(1 - x)}{(z - x)y(1 - y)}\right] \tag{8.20}$$

example 8.2

Estimate the sensitivity of the column described in Example 8.1 to manipulation of D/F on both an absolute and a relative basis:

$$\frac{d(D/F)}{dy} = \frac{-0.728}{0.95 - 0.03} - \frac{1 - 0.728}{0.95 - 0.03}\,0.613 = -0.972$$

$$\frac{dy}{d(D/F)} = -1.028$$

$$\frac{dx}{d(D/F)} = -1.028(0.613) = -0.630$$

$$\frac{dy}{\dfrac{d(D/F)}{D/F}} = -1.028(0.728) = -0.748$$

$$\frac{dx}{\dfrac{d(D/F)}{D/F}} = -0.630(0.728) = -0.459$$

To this point, D/F was arbitrarily selected as the manipulated variable rather than B/F. On an absolute basis, no advantage would be gained in that $d(B/F)$ is simply $-d(D/F)$. But now the importance of relative sensitivity becomes apparent—in Example 8.2, B/F is much smaller than D/F, affording a reduction in relative sensitivity:

$$\frac{\dfrac{dy}{d(B/F)}}{B/F} = -\frac{dy}{d(D/F)}\frac{B}{F} = -\frac{dy}{d(D/F)}\left(1 - \frac{D}{F}\right) \tag{8.21}$$

example 8.3

Estimate the relative sensitivity achievable by manipulating B/F instead of D/F in Example 8.2:

$$\frac{\dfrac{dy}{d(B/F)}}{B/F} = +1.028(1 - 0.728) = +0.280$$

$$\frac{\dfrac{dx}{d(B/F)}}{B/F} = +0.630(1 - 0.728) = +0.171$$

It should be apparent from Example 8.3 that the sensitivity of product quality to disturbances in the material balance will be lower when the smaller product flow is selected for manipulation.

Sensitivities in Multicomponent Systems The material-balance equations for multicomponent and binary systems are identical. Consequently, the manipulation of product flow required to accommodate a feed-composition change is the same for both types of systems. But the sensitivity of y and x to variations in both feed composition and control action are related to the separation factor. Therefore, the relationship between y and x for the multicomponent separation factor must be used:

$$S = \frac{y_l/y_h}{x_l/x_h} \tag{8.22}$$

Here subscripts l and h represent light-key and heavy-key components.

Because the component concentrations are dependent on one another, a substitution must be made for y_l in terms of y_h and for x_h in terms of x_l. These particular terms have been selected for manipulation so that *impurities* y_h and x_l remain, since they are commonly the controlled variables. Any other substitution will achieve the same results, however.

The substitutions used appear as

$$y_l = 1 - y_h - y_{ll} \tag{8.23}$$

$$x_h = 1 - x_l - x_{hh} \tag{8.24}$$

where subscripts ll and hh refer to all components lighter than the light key and heavier than the heavy key, respectively. In making these substitutions, it is assumed that both y_{hh} and x_{ll} approach zero and there are no intermediate components between the keys.

If Eqs. (8.23) and (8.24) are inserted into (8.22), the resulting expression may be differentiated for x_h as a function of y_l at constant separation:

$$\left.\frac{\partial x_h}{\partial y_l}\right|_S = -\frac{(y_l + y_h)x_l x_h}{(x_l + x_h)y_l y_h} \tag{8.25}$$

The influence of the off-key components is apparent when (8.25) is compared to the similar binary differential given in (8.12). Note also the presence of the negative sign—in the binary differential x and y belong to the same component whereas in the multicomponent differential they do not. Since $\partial y_l = -\partial y_h$,

$$\left.\frac{\partial x_h}{\partial y_h}\right|_S = +\frac{(y_l + y_h)x_l x_h}{(x_l + x_h)y_l y_h} = \left.\frac{\partial x_l}{\partial y_l}\right|_S \tag{8.26}$$

Having the differential available in a single component allows solution of Eq. (8.11) for the sensitivity of y to z and solution of (8.16) for the sensitivity of y to D/F.

example 8.4

A deethanizer separates a feed containing 1 percent methane, 30 percent ethane, and 25 percent propane into a distillate containing 90 percent ethane and a bottom product containing 2 percent ethane. The remaining components in the feed are heavier than propane and will not appear in the distillate. Estimate the sensitivity of the ethane concentration in both products to that in the feed and to the D/F ratio.

$$\frac{D}{F} = \frac{0.30 - 0.02}{0.90 - 0.02} = 0.318$$

Let subscripts 1, 2, and 3 refer to methane, ethane, and propane, respectively, with subscript 4 referring to all higher-boiling components. Key components are ethane and propane.

From Eq. (3.20):

$$y_1 = \frac{0.01}{0.318} = 0.031$$

By difference,

$$y_3 = 1.00 - 0.031 - 0.90 = 0.069$$

and

$$z_4 = 1.00 - 0.01 - 0.30 - 0.25 = 0.44$$

From Eq. (3.21):

$$x_4 = \frac{0.44}{1 - 0.318} = 0.645$$

By difference,

$$x_3 = 1.00 - 0.02 - 0.645 = 0.335$$

Then

$$\left.\frac{\partial x_2}{\partial y_2}\right|_S = \frac{(0.90 + 0.069)(0.02)(0.335)}{(0.02 + 0.335)(0.90)(0.069)} = 0.294$$

From Eq. (8.14):

$$\frac{dz_2}{dy_2} = \frac{1}{0.318} + \left(1 - \frac{1}{0.318}\right)0.294 = 2.51$$

$$\frac{dy_2}{dz_2} = \frac{1}{2.51} = 0.398$$

$$\frac{dx_2}{dz_2} = 0.398(0.294) = 0.117$$

From Eq. (8.17):

$$\frac{d(D/F)}{dy_2} = \frac{-0.318}{0.90 - 0.02} - \frac{1 - 0.318}{0.90 - 0.02}\,0.294 = -0.589$$

$$\frac{dy_2}{d(D/F)} = \frac{1}{-0.589} = -1.697 \qquad \frac{dy_2}{\dfrac{d(D/F)}{D/F}} = -1.697(0.318) = -0.540$$

$$\frac{dx_2}{d(D/F)} = -1.697(0.294) = -0.499 \qquad \frac{dx_2}{\dfrac{d(D/F)}{D/F}} = -0.499(0.318) = -0.159$$

Dynamic Response It was mentioned in Chap. 2 that compositions in a column are affected by the relative flows of vapor and liquid within and not directly by D/F or B/F. It is only in reflecting the external balance on the column internals that compositions can be made to change. Although in the steady state the external balance will ultimately be imposed on the internal balance, there may be some delay in achieving this effect. This is the most obvious disadvantage of material-balance control, but it can be rectified by the addition of feedforward.

In Fig. 8.1, an adjustment to reflux flow will start overhead vapor composition changing promptly. An adjustment to distillate flow in Fig. 8.2 will *not* elicit the same response because reflux flow does not change at the same time. The change in distillate flow must first cause a deviation in accumulator level before reflux is adjusted. The dynamic response of the accumulator level loop then lies between distillate and reflux-flow manipulations.

To determine the dynamic properties of this loop, consider the change in the volume of condensate v in the accumulator:

$$\frac{dv}{dt} = V - L - D \tag{8.27}$$

A proportional controller acting on deviations in level will cause proportional changes to reflux. However, both measured volume and manipulated reflux appear to the controller as signals in fractions of maximum range:

$$m = \frac{100}{P}e + b$$

Here m is fractional controller output and e is fractional deviation; b is a bias and P is percent proportional band. Differentiation with respect to time yields

$$\frac{dm}{dt} = \frac{100}{P}\frac{de}{dt}$$

Next L/L_M may be substituted for m, and dv/v_M for de, where L_M and v_M are the maximum reflux flow and tank volume, respectively:

$$\frac{dL}{L_M\,dt} = \frac{100}{P}\frac{dv}{v_M\,dt}$$

When dv/dt from (8.27) is substituted into the above, a closed-loop equation results:

$$L + \frac{v_M}{L_M}\frac{P}{100}\frac{dL}{dt} = V - D \tag{8.28}$$

Following the procedure developed in Ref. 5, it can be shown that reflux responds to changes in distillate flow with a time constant:

$$\tau_L = \frac{v_M}{L_M}\frac{P}{100} \tag{8.29}$$

Tank capacity v_M divided by maximum reflux L_M gives the residence time of reflux in the vessel, which is also the time constant at 100 percent proportional band. Because liquid level is usually quite easy to control with liquid outflow, the proportional band of the level controller will typically be 10 percent or thereabouts. So a time constant of $0.1v_M/L_M$ could be expected.

While it might be desirable to set P as close to zero as possible to minimize this time constant, some allowance must be made for noise response. Liquid-level signals are seldom quiet, although base levels are considerably worse in this regard than accumulators. So if the proportional band is set too narrow, uncontrollable fluctuations in level can be amplified and passed on to the reflux, which could create composition or pressure disturbances. This problem could be especially severe when using evaporative condensers because of upflow condensation causing liquid to arrive at the accumulator in slugs.

Most accumulator level controllers have proportional-plus-reset action. This complicates the dynamic representation of the closed loop but does not significantly alter the time constant. Figure 8.4 compares the step response to that of a proportional control loop. Although reset action causes a greater change in reflux to restore level to its set point, the effect is not immediate. Reset action always lags proportional action—as a result, reset cannot contribute significantly to the reduction of the time constant of (8.29). Although most accumulator level controllers have reset action, the foregoing analysis indicates that it is not really needed.

In columns where heat input is unmeasurable, accumulator level could be controlled by the heat-input valve. Then reflux flow would set the heat loading. In this case, the response of distillate-product composition to changes in distillate flow is essentially the same as when reflux controls accumulator level. This arrangement is also very effective in controlling bottom-product composition. The boilup change is felt throughout the entire column in a matter of seconds, whereas reflux changes can take many minutes to reach the bottom of the column. Consequently, in cases where bottom-product composition needs to be controlled but the D/F ratio is smaller than B/F and therefore gives a lower sensitivity, heat input should be under accumulator level control. This arrangement is shown in Fig. 8.5.

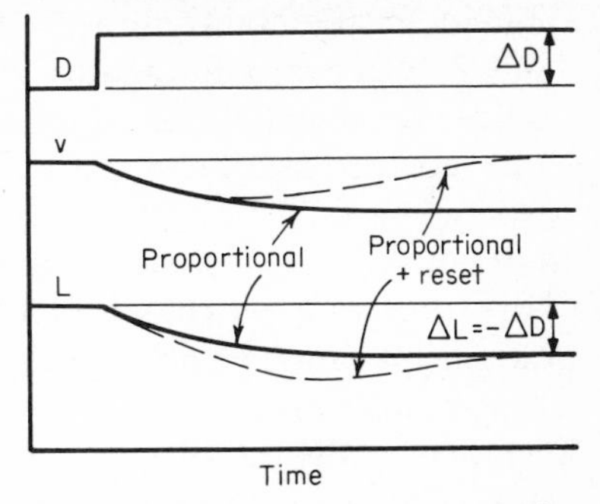

figure 8.4 *Reset action aguments the correction to reflux flow, but at a later time.*

When B/F is significantly smaller than D/F, bottom-product flow should be manipulated for quality control. This leaves base level to be controlled by heat input. Again, adjustments to bottom-product flow have no direct effect on product quality—the action of the level controller imposing equivalent adjustments to heat input is essential. The response of boilup to changes in bottom-product flow differs somewhat from the response of reflux to changes in distillate flow. The active volume of the column base is usually much less than that of the reflux accumulator, and therefore its residence time—expressed in Eq. (8.29) as v_M/L_M—is less. However, the response of level to a change in the flow of heating medium is affected by the thermal

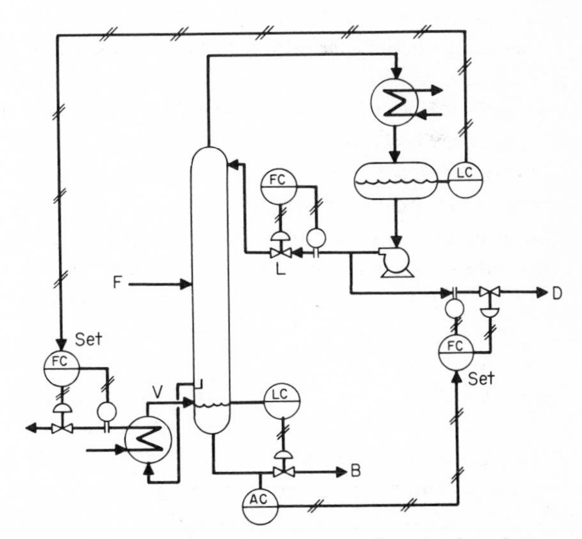

figure 8.5 *This arrangement is preferred when bottom composition must be controlled by manipulating distillate flow or when heat input is unmeasurable.*

capacity of the reboiler. As a result, base level may not change for 5 to 20 s following a change in heating-medium flow. This delay, combined with the reduced residence time, requires that the base-level controller have a much wider proportional band than did the accumulator level controller. A proportional band of 100 percent is typical for this loop, which gives a time constant equal to the residence time per Eq. (8.29). Because the residence time of the base is typically less than that of the accumulator, their time constants under level control would be similar.

Other problems may appear when using heat input to control base level. Turbulence due to boiling, hydraulic resonance, and inverse response may combine to make this loop unstable, as described in Chap. 6. Buckley et al. [6] describe an inverse-response characteristic so severe that reflux was more satisfactory than boilup for base-level control; B/F was very low in their column.

FEEDFORWARD SYSTEMS

Feedforward control has many common applications and comes under the heading of planning. Those who fail to use this control mode in their daily affairs are late for appointments, are in debt, fail examinations, etc. These are manifestations of feedback alone—taking action after the damage is done. Proper planning and preparedness avert calamity and make the most of opportunity.

It is difficult to explain why so little feedforward control is being applied to industrial processes—its merits should be obvious. Traditionally, it is only resorted to when conventional methods fail. Yet the technique is so powerful and so easy to apply that it can be justified for virtually any distillation column. As will be seen, process knowledge is required to implement feedforward, and this well may be an impediment to more universal application. But it need not be, as the following pages will demonstrate.

Feedforward Control The most complete treatment of feedforward control is contained in an earlier work [7] by the author. Since its publication in 1967, little development has taken place in the technology itself, although the application of feedforward control has broadened considerably. The concept has more recently been defined by the Instrument Society of America (ISA) [8] as

> *Control, Feedforward*—Control in which information concerning one or more conditions that can disturb the controlled variable is converted, outside of any feedback loop, into corrective action to minimize deviations of the controlled variable.

As shown in Fig. 8.6, disturbances which can upset product quality are measured and converted into equivalent changes in manipulated flow rates to offset that effect on quality. If properly designed and executed, feedforward is theoretically capable of perfect control; feedback cannot be since it requires a deviation in product quality to stimulate corrective action.

In practice, it is impossible to provide perfect corrective action for disturbances—some errors will remain. However, the deviation in product quality following a disturbance will be reduced by feedforward control in proportion to its accuracy. A feedforward calculation accurate to ±10 percent, for example, will reduce the sensitivity of product quality to those measured disturbances by a factor of 10—a substantial reward for modest accuracy.

The operation by which the manipulated variable is driven to a position relative to the imposed load is a direct calculation based on a mathematical model of the process. The model consists of two components: steady state and dynamic. Steady-state accuracy is more important because control improvement hinges almost entirely on it. The faithfulness of the dynamic model is less significant. In short, we must know exactly where we are going before we select the route.

A derivation in Ref. 7 describes how the integrated deviation in product quality is proportional to the magnitude of the load change when feedback alone is used. It is obtained by solving the three-mode control algorithm in reverse. In the steady state, the deviation e and its derivative are both zero. Then the difference in controller output m between two steady states relates only to the time integral of the deviation sustained:

$$\Delta m = \frac{100}{P}\left(\frac{1}{R}\int e\,dt\right) \tag{8.30}$$

Solving for the integrated error,

$$\int e\,dt = \frac{PR}{100}\Delta m \tag{8.31}$$

where P is the percent proportional band and R is the reset time in the same units as time t. Because product quality is both sensitive to manipulation and slow in response, the optimum values of P and R tend to be large. The only way to reduce $\int e\,dt$ significantly is to reduce Δm—that is, require less corrective action by the output of the feedback controller. Hence the argument that effective feedforward control minimizes the contribution Δm required by the feedback controller and thereby minimizes $\int e\,dt$. These rela-

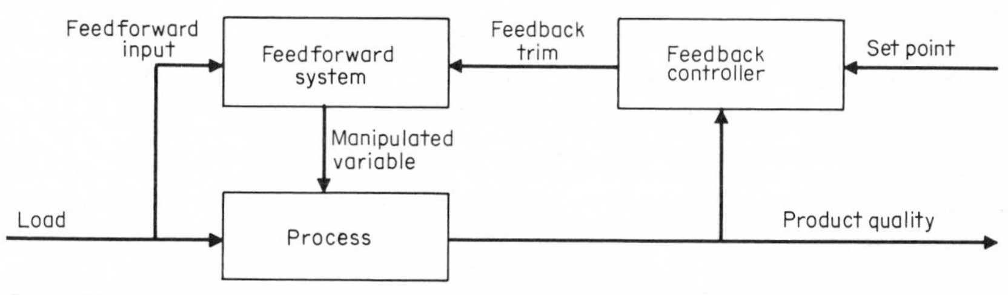

figure 8.6 *A feedforward system can provide most of the correction for measurable disturbances.*

tionships will become apparent when a feedforward system is developed for a distillation column in the next section.

Steady-state Models The principal feedforward calculation made for a distillation column determines product flow in terms of feed rate and feed composition: for example,

$$D = F\left(\frac{z - x}{y - x}\right) \tag{8.32}$$

However, this relationship contains both x and y, which are product compositions. Their actual measurements cannot be used in the feedforward calculation since this would constitute feedback: x and y are affected by manipulation of D. Instead, the values of x and y to be used in the calculations are set points, which will be distinguished by an asterisk. Then the feedback equation for constant separation becomes

$$D^* = F\left(\frac{z - x^*}{y^* - x^*}\right) \tag{8.33}$$

It is usually sufficient to simplify the feedforward calculation into a ratio relationship:

$$D^* = mFz \tag{8.34}$$

where m represents the recovery factor, D/Fz:

$$m = \frac{(z - x^*)/(y^* - x^*)}{z} \tag{8.35}$$

In this case, feedback control is applied for variations in feed rate and composition, and the ratio m may be adjusted either manually or by a product-composition controller.

The accuracy of approximation (8.34) can be evaluated by observing how m must change for variations in z:

$$\frac{\partial m}{\partial z} = \frac{x^*}{z^2(y^* - x^*)} \tag{8.36}$$

As x^* approaches zero, the accuracy of the approximation improves. Remember that the objective of feedforward is to minimize variations in m; therefore the above evaluation should be made in each application to determine whether the approximation is satisfactory.

For multicomponent feeds, z should include not only the light key but all lighter components as well.

A further simplification is the elimination of feed composition as a feedforward input:

$$D^* = mF \tag{8.37}$$

In this case,

$$m = \frac{z - x^*}{y^* - x^*} \tag{8.38}$$

Then m will be more variable with z:

$$\frac{\partial m}{\partial z} = \frac{1}{y^* - x^*} \tag{8.39}$$

In most cases, elimination of a feed-composition input is justified because z tends to vary less widely and less rapidly than F. For example, the feed to a given column might vary typically ±10 percent in composition but ±40 percent in flow. Furthermore, the rate of change of composition is limited by upstream capacity whereas flow rate can change as fast as a valve can move.

Although Eqs. (8.30) and (8.31) described the integrated error sustained by a feedback controller when moving from one steady state to another, there was no term relating to the rate of change between states. A relationship can be derived by differentiating (8.31):

$$e = \frac{PR}{100}\frac{dm}{dt} \tag{8.40}$$

The amplitude of the deviation from set point, e, is thereby related to the rate of change of output required to follow the changing load. The point of this discussion is that if z is changing slowly, m will change at a proportional rate per (8.39) and only a small deviation will be sustained.

Still another simplification is the use of orifice differential signals without square-root extraction. Then the flows appear in the squared form:

$$D^{2*} = mF^2 \tag{8.41}$$

The elimination of square-root extraction is justified in that the accuracy of the flowmeters above 50 percent flow is superior without linearization.

Forcing an Internal Balance Feedforward control has also been used by van Kampen [2] to impose the external material balance on the column internals. It involves a simple material balance on the reflux accumulator wherein reflux flow is made dependent on distillate flow:

$$L^* = m_L - kD \tag{8.42}$$

where m_L is the output of the accumulator level controller and k is an adjustable gain. The control-loop configuration is shown in Fig. 8.7. Linear flow signals must be used when adding or subtracting.

If k is set at unity, then changes in distillate flow will be imposed equally

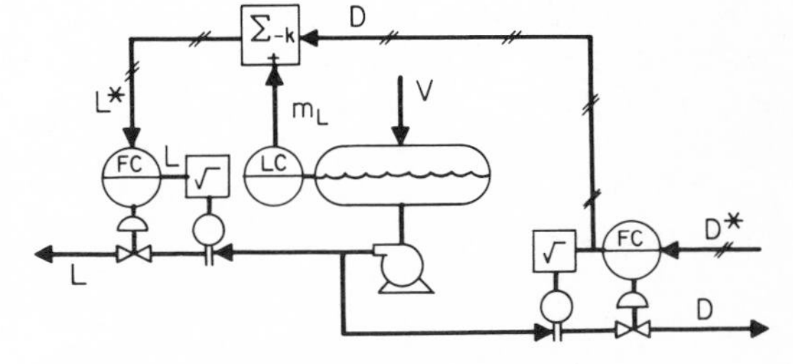

figure 8.7 *A feedforward loop from distillate to reflux can overcome the time lag of the level-control loop.*

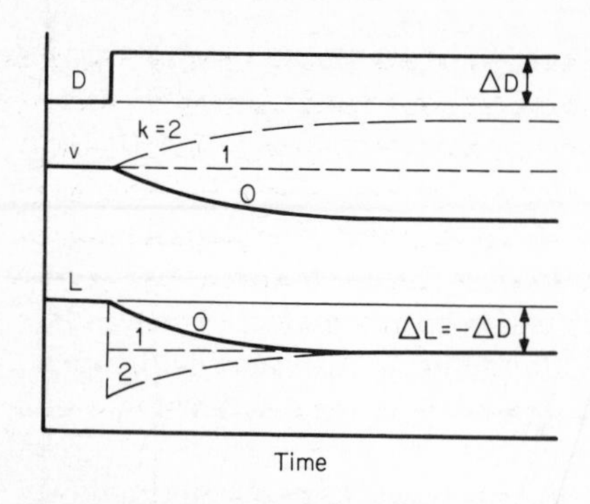

figure 8.8 *The accumulator can be converted from a lag to a lead by appropriate adjustment of* k.

on reflux, tending to keep liquid level constant. This action essentially eliminates the dynamics of the reflux accumulator from the quality-control loop. In actual practice, van Kampen has found it beneficial to increase k above unity so that variations in distillate are amplified. This causes an overcorrection which converts the accumulator from a lag to a lead. The response of reflux to distillate is summarized in Fig. 8.8 for values of k of 0, 1, and 2.

For the case of $k = 0$, the response of reflux to distillate under proportional level control was determined to be a first-order lag of time constant described by (8.29). Where $k = 2$, the response is that of a first-order lead-lag function, having a lead time constant twice that of the lag, with the lag still described by (8.29). The lead-lag ratio is, in fact, coefficient k. At $k = 1.0$, lead equals lag and the net effect is the elimination of any dynamic contribution. Values of k less than 1.0 give partial cancellation of the lag by a smaller lead. Using this feedforward loop, van Kampen [2] was able to reduce the period of oscillation of his distillate-quality loop from 5 h to 30 min.

Note that Eq. (8.42) differs from the actual accumulator balance (8.26) in replacing V with m_L for $dv/dt = 0$. For columns with constant boilup, this representation is quite adequate. But if boilup is variable, the desired response of reflux will not be achieved by (8.42), considering that V will usually be much larger than D. In this case, reboiler heat input Q may be incorporated into the reflux calculation:

$$L^* = k_Q Q - k_D D + m_L - b \tag{8.43}$$

Here each feedforward input has its own assigned gain; bias b must be included to give m_L a positive, nonzero value in the steady state. Note that if k_Q and k_D are set to achieve a constant level, m_L must equal b.

These equations are solved by analog or digital computers in the plant. Because the ranges of the various signals L^*, Q, and D differ, k_Q and k_D must include these ranges. Consider the example where the distillate flow range is 0 to 200 lb/h and that of reflux is 0 to 1,000 lb/h. To achieve a true k_D of 1.0, the coefficient k'_D set into the computer would be 0.2. Then a 1 percent (of scale) change in D would elicit only a 0.2 percent (of scale) change in L^*.

Dynamic Compensation It is not enough that distillate flow be driven with accuracy to the set point corresponding to present levels of feed rate and composition and desired product qualities. The correct dynamic path must be followed as well. Because the feed and distillate are physically located at different points on the column, their individual effects on product quality differ in their speed of response. When a difference exists between the responses of the load and the manipulated inputs, feedforward correction by a steady-state calculation alone may arrive at the wrong time. If the column responds faster to the manipulated flow, correction can be applied too early. Then product quality will tend to deviate in the direction forced by the manipulated flow until the effect of the feed-rate change is felt. In this case a dynamic lag should be placed on the feedforward input in an effort to match the response of the forward loop to that of the column.

The need for dynamic compensation is best illustrated by examining the response of a column to feed and boilup variations. Consider the scheme shown in Fig. 8.9, where bottom-product flow is set in ratio to a liquid feed through a forward loop. If no dynamic compensation is applied, a step increase in feed rate will elicit a concurrent step increase in bottom flow. Several minutes may elapse before the increase in liquid downflow reaches the base of the column. Yet during this time base level will decrease and the level controller will decrease boilup.

The dynamic overcorrection is apparent—boilup must ultimately increase by the portion of the feed increase not leaving as bottom product, but the dynamic imbalance has driven it in the opposite direction. This is an indication of a transient imbalance in the column's internal streams, which will be followed by a shift in its composition profile. In short, the temporary reduction in boilup is indicative of distillate product being allowed to move down the column.

The dynamic response of base level to feed rate is a series of interconnected lags reflecting the rise in liquid level on each tray. The change in boilup that would result from a step increase in feed rate is then a dead time followed by a lag as shown in Fig. 8.10. The response of boilup to bottom-product flow, however, has a dead time of only a few seconds. When the two effects are combined through a feedforward loop having no dynamic compensation, boilup displays a temporary reversal in direction as shown.

Compensation with a simple first-order lag can moderate the transient reversal but not eliminate it. Exact compensation requires a dead time equal to the difference in dead times between the feed and bottom responses, with

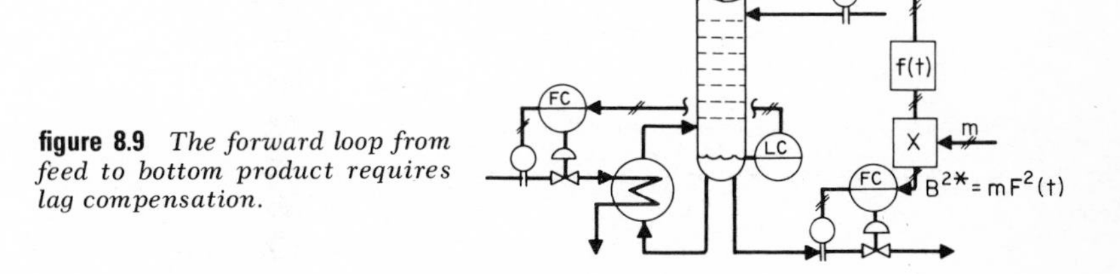

figure 8.9 *The forward loop from feed to bottom product requires lag compensation.*

a lag equal to the feed lag and a lead equal to the bottom lag [7]. Since most responses can be characterized by a dead time followed by a first-order lag, the universal dynamic compensator is dead time plus lead-lag.

In most cases the disturbance variable exhibits more dead time in affecting product quality (or in this case steam flow, which is an advance indication of a quality change) than does the manipulated variable. If the opposite were true, a time advance or negative dead time would be required for compensation, which is an unrealizable function. Although dead-time compensation is available in digital systems, it can only be approximated in analog systems by multiple stages of lags. To simplify operation and adjustment, compensation with analog systems is usually limited to second-order lag and first-order lead. Reference 7 describes in detail how the settings of the various terms affect the compensated response and gives guidelines for making the adjustments.

An important consideration is that the dynamic response of a column to a given input is determined to a great extent by the configuration of control loops other than the external material-balance loop. For example, the forward loop on the reflux accumulator shown in Fig. 8.7 is capable of converting that vessel from a lag to a lead. Thus the response of product quality to changes in distillate flow is completely altered by the selection of coefficient k in the reflux loop. When distillate flow is set in ratio to feed, dynamic compensation of that forward loop can be achieved by appropriate adjustment of k in the forward loop from distillate to reflux. Recalling that k was evaluated as the lead-lag ratio and that the proportional band of the level controller affected the lag time, adjustable lead-lag compensation is available without the use of a lead-lag instrument.

The material-balance equations solved in the feedforward loops presented to this point have assumed a constant separation. To achieve constant separation, boilup or reflux must be set in ratio to column feed. This constitutes a second forward loop which has a pronounced effect on the dynamic response of the column. If boilup is not changed when feed is increased,

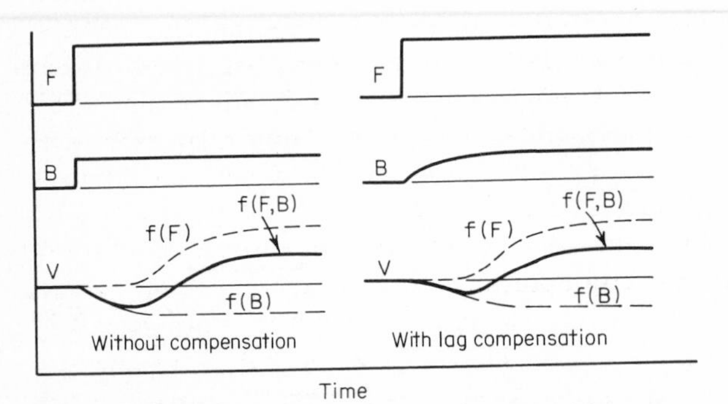

figure 8.10 *Lag composition can moderate the transient reversal of boilup responding to feedforward manipulation of bottom-product flow from feed flow.*

the manipulated increase in distillate flow must cause reflux to decrease. But when boilup is increased as well, reflux must rise instead of fall. Moreover, the internal balance is shifted more rapidly by boilup than by reflux, radically revising the requirements for dynamic compensation.

In fact, whenever heat input is set in ratio to a liquid feed, a dominant lag is required along with dead time (if available) for the same reasons given with regard to bottom-flow manipulation. But the penalty for inadequate dynamic compensation is more severe by the ratio of V to B. When manipulating bottom flow, the maximum transient error in boilup is $-\Delta B$ as shown in Fig. 8.10. This error amounts to $-\Delta F(B/F)$. But when manipulating boilup in ratio to feed, the maximum transient error $-\Delta V$ is equal to $-\Delta F(V/F)$, which is V/B times the error $-\Delta B$.

Requirements for compensation are also affected by the condition of the feed. A vapor feed augments the overhead vapor rather than the bottom liquid. Its effect on distillate composition will therefore be felt sooner and requires different compensation. When boilup is set in ratio to a liquid feed, distillate may be set to change at the same rate as boilup, anticipating the increase in overhead vapor. But a vapor feed has no direct influence on the bottom of the column. Distillate flow must then be adjusted in advance of an increase in boilup, which must wait for reflux to increase first.

Feedback Trim Reference 7 makes a case for applying feedback trim to the term in the feedforward equation representing the product-quality set point, that is, y^* or x^* in Eq. (8.33). However, successive simplifications replaced y^* and x^* with a simple ratio m. This, then, is the term adjusted by the feedback controller or by the operator to produce a change in product quality or offset a variation in feed composition.

When feed composition is not used in the feedforward loop, solving (8.37) or (8.41) will give $m = D^*/F$ or D^{2*}/F^2. A similar case may be made for $m = B^*/F$ or B^{2*}/F^2 as described in Fig. 8.9. This allows the previously derived sensitivities to control action to be applied to feedback trim of product quality. Whether the adjustment is made to D with constant F or to D/F under feedforward control, the result is the same.

These sensitivities are useful in predicting how effective feedback control will be on product quality. They are, in fact, one component of the steady-state gain of quality in response to control action. The gain of the product-quality controller and therefore its effectiveness in controlling quality is inversely proportional to the product of the process steady-state and dynamic gains. The dynamic process gain is related to the ratio of dead time to time constant in the product-quality loop. A discussion pertinent to this relationship is presented at the end of this chapter.

The relative sensitivities calculated in Examples 8.2 and 8.3 vary from 0.27 to 1.2, typical for many columns. Considering that product qualities are measured over relatively narrow ranges, from 0 to 10 percent down to parts per million, very high steady-state gains are common. Consider the combination of a relative sensitivity of 0.3 and an analyzer span of 0 to 5 percent.

Then a 1 percent change in the relative value of D/F would ultimately cause a 0.3 percent change in composition, which would be a 6 percent-of-scale change. The steady-state gain of this loop would then be 6.

Although the proportional band of the product-quality controller manipulating D/F may not have to be as wide as 600 percent due to the contribution of dynamic gain, it will not be far from that value. In some applications, a sufficiently wide proportional band to ensure stability may not be attainable. Fortunately, there are methods of scaling feedforward systems to reduce the gain of the feedback loop.

To allow for all possible feed compositions it may be desirable to allow m to vary over the complete D/F range of 0 to 1.0. But the gain of the feedback loop can be moderated by limiting that range. If, for example, D/F is never expected to exceed 0.3, then the range of m could be made 0 to 0.3. A lower limit may also be introduced, further restricting the range to, for example, 0.2 to 0.3. This represents a threefold reduction in gain from the 0 to 0.3 range and a tenfold reduction from the full range. This practice makes it easier for both the controller and the operator to find the exact D/F ratio to achieve a particular change of quality.

Note in Fig. 8.9 that ratio setting m is introduced downstream of the dynamic compensator. This allows the operator to see the direct result of his adjustment of m on product flow. Or in the case of automatic feedback trim, the dynamic compensator is kept out of the feedback loop. In the example shown in Fig. 8.7, dynamic compensation depends on the manipulated flow and therefore cannot be removed from the quality feedback loop. However, one of the functions of the reflux forward loop is to improve quality response; therefore, it belongs inside the quality feedback loop.

The sensitivities to control action vary with x, y, and z as shown in Eqs. (8.17) to (8.26). Therefore, gross variations in feed composition and changes to set points may require retuning of the feedback controller. However, effective feedforward demands less of the feedback controller, tending to minimize this problem in feedforward systems.

Adjusting the feedforward calculation manually for a feed-composition change is not recommended when feedback trim is applied. It is very difficult for an operator to time his adjustment properly to match the rate at which the change is being imposed on the process. This is particularly true if he must wait for a laboratory analysis. In the meantime, the feedback controller may already be making a correction for that upset. Then when he enters his new analysis into the feedforward calculation, he introduces a second upset which, if entered in a single step, may cause a greater disturbance than the composition change itself.

Scaling Although all the calculations described have been made using dimensional data, analog computers and some digital computers operate only on dimensionless signals. These signals all have the same range—for example, 0 to 10 V or 3 to 15 psig. The computer must therefore be provided with the correct scaling if the calculation performed is to be meaningful.

A very simple procedure has been developed to facilitate this scaling—but it must be followed rigorously or confusion will result. First, the process equation is written in dimensional terms such as

$$D^* = mFz \tag{8.44}$$

Then ranges must be determined for all inputs and the output. Full scale values will be designated by subscript M and zero scale values by subscript 0. Primed values will represent the signals over their standard range of 0 to 1.0 (or 0 to 100 percent of scale). Equate the dimensional variables to their signals and ranges:

$$D^* = D'D_M$$

$$m = m_0 + m'(m_M - m_0)$$

$$F = F'F_M$$

$$z = z'z_M$$

The only variable having a nonzero value at zero scale is m—flowmeters and analyzers usually read zero at zero scale.

Next, each term in the dimensional equation is replaced:

$$D'D_M = [m_0 + m'(m_M - m_0)]\, F'F_M z' z_M$$

Finally, the expression is solved for the output signal:

$$D' = [m_0 + m'(m_M - m_0)]\, F'z' \frac{F_M z_M}{D_M} \tag{8.45}$$

One of the important steps is to determine a reasonable range for m. When feed composition is constant or included in the calculation, a narrow range is possible.

example 8.5

A feedforward computer is to be scaled for a column having the following ranges of signals:

D: 0–3,000 lb/h

F: 0–10,000 lb/h

z: 0–50 percent

Scale the computer allowing a range of ±10 percent in m.

$$D^* = 3{,}000D^{*\prime}$$

$$F = 10{,}000F'$$

$$z = 0.5z'$$

$$m = 0.9 + 0.2m'$$

$$D^{*\prime} = (0.9 + 0.2m')F'z' \frac{(10{,}000)(0.5)}{3{,}000}$$

$$= (0.9 + 0.2m')1.67F'z'$$

This calculation may be solved in a single device or computation block, or it may require two, depending on the limitations of the devices used. If two devices are required, i.e., two successive multipliers, care should be taken to see that the output of the first does not saturate when receiving the highest combination of reasonable inputs. The scaling factor of 1.67 in this example could be factored into two terms to be applied to the two multipliers if necessary.

Many multipliers are capable of simultaneous division but are unable to multiply twice. The preceding relationship can be readily fit to that function simply by inverting the arbitrarily chosen parameter m:

$$D^* = \frac{Fz}{m} \tag{8.46}$$

Over the narrow range of ±10 percent variation in m given in Example 8.5, the gain of D^* with respect to m is virtually the same for either calculation, although the sign is different. A sign correction may be applied, however, so that little distinction between multiplying and dividing will appear:

$$D^{*\prime} = \frac{1.67F'z'}{1.1 - 0.2m'}$$

is an essentially equivalent expression to that found in Example 8.5.

THE EFFECT OF SEPARATION

The preceding presentation on feedforward control used mathematical models based on constant separation, although the control loops achieving constant separation were not shown. It is equally possible to operate a column under conditions of variable separation, which has a pronounced effect on the material balance. It is further possible to alternate between one of these operating modes and the other as constraints are encountered. That particular problem is deferred to the next chapter. The present consideration is to determine how separation affects the material balance and therefore the feedforward control model and to examine how separation is controlled and adjusted.

Maintaining Constant Separation If column pressure and feed composition are reasonably uniform, separation is principally determined by the energy or boilup applied per unit of feed as described in Chap. 2. And while pressure and feed composition have their effect, their contribution is relatively small compared to the V/F ratio, and compensation can be applied. This assumes, of course, that the column is operated in a range over which tray efficiency is also uniform. Excessive boilup can cause sufficient entrainment to decrease separation rather than increase it, and these inefficient regions of operation are to be avoided.

The most common feedforward system maintaining constant separation manipulates distillate and heat input proportional to feed rate as shown in

Fig. 8.11. A single compensating lag is sufficient for both loops due to the rapid response of the column to changes in boilup. As when bottom flow is manipulated in proportion to the feed, base level should be observed as a guide to adjusting the compensator. If the level falls following a feed increase, the lag needs to be increased. The addition of dead time to the compensator allows a better match to the response of the column, but at the same time it doubles the complexity of the compensator and the tuning effort required. If the feed is completely vaporized, separate compensation will be required for each forward loop, with a shorter lag in the distillate loop. In neither case is a reflux forward loop required for dynamic compensation.

The ratio of heat input to feed is coefficient K appearing in the ratio block. Because input and output signals are both in the squared form, the gain of the ratio device, Q^2/F^2, is actually K^2. A square-root scale is provided on the ratio station when differential-pressure signals are used so that the operator may enter the desired value of K directly. If he should require an increase in product purities or a reduction in losses, he may increase K. Its effect on the material-balance calculation is evaluated later in this chapter.

In cases where B/F is significantly lower than D/F, the forward material-balance loop should manipulate B. Then base level is left to manipulate boilup. Constant separation is achieved by setting reflux in ratio to the feed. If the condenser is flooded, internal-reflux control may be required to hold separation constant.

Other combinations are possible. For example, both distillate and reflux could be set in ratio to the feed, with heat input controlled by accumulator level. This arrangement would be preferred when heat input is not measurable or when x needs to be controlled by manipulating D/F. Dynamic compensation for these arrangements will differ, and each must be evaluated on its own merits. It depends not only on the choice of manipulated variables but also on the selection of the controlled variable.

When separation is maintained constant, in a binary system control of either x or y through manipulation of the material balance results in control of the other. This relationship allows one of the products to be controlled indirectly while the other is controlled directly. For example, an on-line measurement of x may not be available—but if y is controlled and separation is constant, x will be indirectly regulated. In a multicomponent system, however, constant separation does not provide this assurance. If, for example, y_h is con-

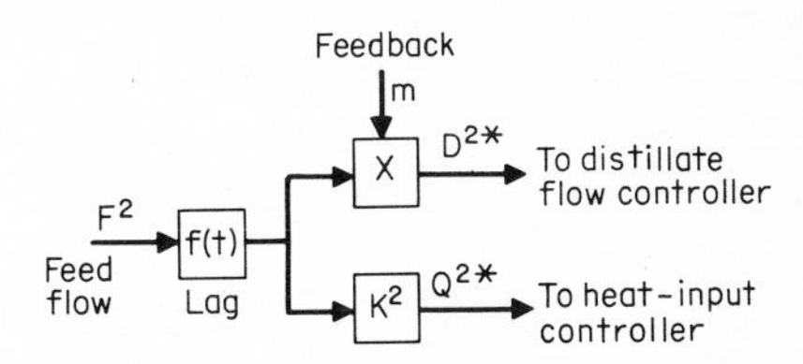

figure 8.11 *Distillate flow will be directly proportional to feed rate if constant separation is maintained.*

TABLE 8.2 Distillate Flow Required to Control y with Constant Boilup

V	F	y	x	z	D/F	D
5.2	1.0	0.98	0.103	0.50	0.452	0.452
	0.75		0.021		0.499	0.374
	0.5		0.003		0.509	0.255
	0.375		0.001		0.510	0.191

trolled by the material balance, variations in off-key components can change y_l, x_h, and x_l, although their ratio remains intact per Eq. (8.22).

Constant-boilup Operation In many chemical plants, certain products are so much more valuable than the energy required to purify them that maximum recovery is desirable. This goal demands maximum separation at all times, attained by holding boilup or reflux at the limits of column capacity. As feed rate varies, V/F then varies, with separation and recovery improving as F is reduced. Note that recovery is higher than achievable by constant separation for every feed rate up to the maximum. This is known as maximum-recovery or minimum-loss operation.

Although this mode of operation eliminates one feedforward loop, it complicates the other. Manipulation of heat input is not required, but variable separation means that x varies while y is controlled or vice versa. Then the linear models described by Eqs. (8.33), (8.34), (8.37), and (8.41) no longer apply. Because product recovery is enhanced at reduced flow rates, D/F must increase as F decreases when y is controlled or decrease when x is controlled. The exact relationship must be found by solving for the uncontrolled quality in terms of the controlled, at several values of V/F, and then calculating the corresponding value of D/F. A set of calculations is given in Table 8.2 for a propane-propylene separation in a column of 100 theoretical trays. Figure 2.3 was used to determine separation. The recovery for this column, D/Fz, is maximum at $x = 0$, being $1/y$ or 1.02; it is essentially achieved as F approaches zero.

A plot of D versus F from these data appears in Fig. 8.12. A representative feedforward model must be constructed to fit the curve over a reasonable operating range. The simplest model is a parabola of the form

$$D^* = mF - bF^2 \tag{8.47}$$

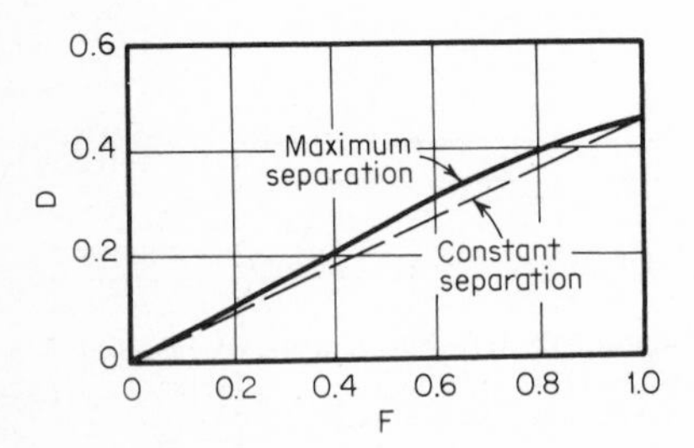

figure 8.12 *The improvement in recovery achieved by maximum boilup is the difference between the line and the curve.*

TABLE 8.3 Exactness of Parabolic Model with $m = 0.568, b = 0.116$

F	D	D^*	$\frac{D - D^*}{D}$, %
1.0	0.452	0.452	0.0
0.75	0.374	0.361	−3.5
0.50	0.255	0.255	0.0
0.375	0.191	0.197	+2.9

where m and b are individually adjustable coefficients. One feature of this model is that it uses available signals: F^2 is the flowmeter output prior to square-root extraction.

This second-order model with only two coefficients can be fit to the actual process curve at only two points. (Actually it is fit at a third point also; that is, $D = F = 0$.) The points should be selected near the upper and lower limit of the feed range. For the example described in Table 8.2, the model is matched to feed rates of 0.5 and 1.0 in Table 8.3. Errors are sustained at the other points. A better overall fit might be attained by selecting other operating points or by narrowing the range of operation. Alternatively a more complex model could be used, although this increases the cost of the system and requires more effort to adjust.

A second consideration is the effect of feed composition and quality set point on the model. It must be adjustable for variations in these conditions, but the adjustment should, for simplicity, be limited to coefficient m. Then the model will fit the true curve at only one point (other than zero). Table 8.4 compares the fit for three sets of conditions with m adjusted to produce zero error at $F = 0.75$ for each set and with b arbitrarily fixed at 0.15.

Observe that the selection of b gives a two-point match in the last column, but the fit in the center column is poor. The conclusion that might be drawn is that the nonlinearity b should be adjusted pertinent to a given product-quality set point but adjustment of m to compensate for feed composition is adequate. The values of m used to calculate D^* for the above three conditions are 0.611, 0.597, and 0.717 from left to right. Its value is obviously more dependent on feed than on product composition.

A control system embodying this model appears in Fig. 8.13. No dynamic compensation is shown—the forward loop manipulating reflux from distillate flow is recommended as described in Fig. 8.7. If x is to be controlled at maxi-

TABLE 8.4 Exactness of Model at Various Conditions

	$y = 0.98, z = 0.5$		$y = 0.99, z = 0.5$		$y = 0.98, z = 0.6$	
F	D	D^*	D	D^*	D	D^*
1.0	0.452	0.461	0.388	0.446	0.567	0.567
0.75	0.374	0.374	0.363	0.363	0.453	0.453
0.5	0.255	0.268	0.251	0.260	0.306	0.320

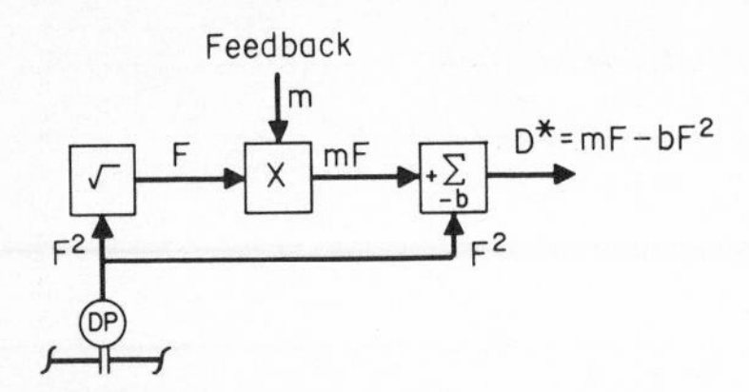

figure 8.13 *The parabolic model is easily implemented with standard analog instruments.*

mum separation by manipulating D, the sign of nonlinearity b must be positive, causing D/F to increase with F. The nonlinearity may be calculated as was done in the previous example or adjusted in the field to minimize the variation in m necessary to control quality as feed rate changes.

Adjustments to Separation The control scheme of Fig. 8.11 allows the operator to adjust separation through the heat-input–feed ratio K. Whether his intention is to improve recovery or to alter bottom-product composition, he nonetheless disturbs the column. Separation may also be changed when constraints are encountered, such as flooding, steam-flow limit, etc. To avoid upsetting product quality when separation is thereby adjusted, its change should be passed on to the material-balance loop. The mechanism for accomplishing this can be derived from the data on the propane-propylene column in Table 8.2. The material-balance equation in the form of $D^* = mFz$ may be made a function of V/F by plotting D/Fz against V/F. Figure 8.14 shows this relationship for the case where $y = 0.98$ and $z = 0.50$.

The relationship appears to be hyperbolic, reaching a maximum recovery of $1/y$ as V/F approaches infinity and x approaches zero. Notwithstanding its severe change in slope, the curve may be modeled reasonably well by the hyperbola:

$$\frac{D}{Fz} = \frac{1}{y} - \frac{a_1}{V/F - b_1} \tag{8.48}$$

For the curve in Fig. 8.14, $a_1 = 0.0849$ and $b_1 = 3.13$. Note that b_1 is essentially the minimum V/F for this system—as V/F approaches b_1, recovery falls sharply to zero.

Variable separation may be incorporated into the material-balance feedforward system simply by solving (8.48) for D^*. If there is no feed analyzer, m may replace z:

$$D^* = mF\left[\frac{1}{y^*} - \frac{a_1F}{V - b_1F}\right] \tag{8.49}$$

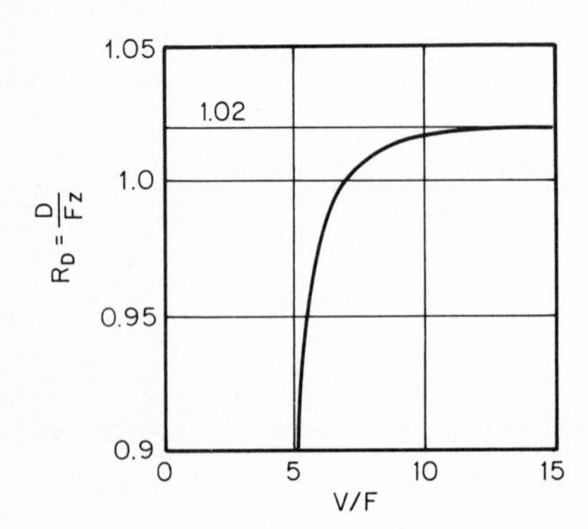

figure 8.14 *The relationship between the recovery factor and V/F appears to be hyperbolic.*

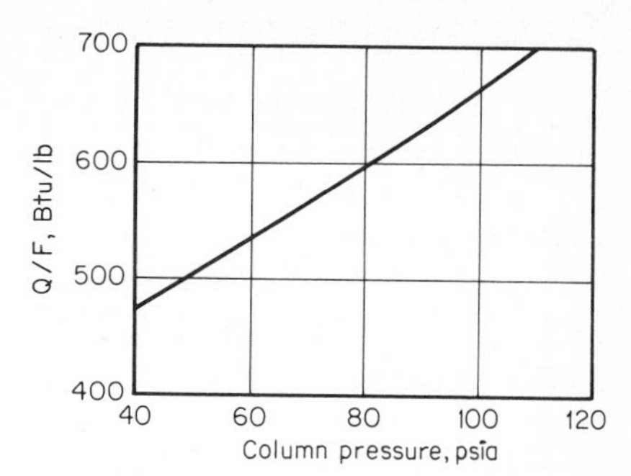

figure 8.15 *A plot of heat-input–feed ratio vs column pressure for the deisobutanizer suggests how pressure compensation might be applied.*

If a feed analysis is provided, then m could take the place of $1/y^*$ and provide correction for adjustments to y^*:

$$D^* = Fz\left[m - \frac{a_1F}{V - b_1F}\right] \tag{8.50}$$

In either case, coefficients a_1 and b_1 must be determined either from process modeling or from observation of actual recovery at two different values of V/F.

Compensating for Variable Pressure Separation and latent heat of vaporization both vary with pressure. It may be necessary to take these factors into account in a system where column pressure is quite variable. The columns subject to the widest variation in pressure in the shortest space of time are those with air-cooled condensers.

Constant separation is usually maintained by controlling heat input in ratio to feed rate. As demonstrated in Table 7.3, this ratio, designated Q/F, varies directly with temperature and therefore with pressure. A plot of Q/F as a function of pressure for the deisobutanizer of Table 7.3 appears in Fig. 8.15. Vapor-pressure data were taken from Table 2.2.

A linear approximation of the function appearing in Fig. 8.15 is expected to be satisfactory for most applications:

$$\frac{Q}{F} = a_2(b_2 + p) \tag{8.51}$$

The equation is given in this form to separate the ratio adjustment a_2 from the pressure intercept b_2. For all deisobutanizers, b_2 is the same number, 106 psia. But a_2 will vary with the number of theoretical trays in the column and the degree of separation desired. For the example given, a_2 is 3.19 Btu/lb-psi.

Whether pressure compensation is necessary depends on how a column is operated. Constant separation in a floating-pressure system can only be attained if compensation is applied or a product-quality controller is used to close the heat-input loop. Otherwise, intervals of low pressure will simply cause overpurification and little advantage will have been gained. Compensation is also required for optimizing calculations. Compensation for variable pressure must also be applied to temperature measurements that are used to infer product quality. That subject is covered in the next section.

FEEDBACK CONTROL OF QUALITY

In a more elementary treatise on distillation control, this heading might have introduced the main body of the work. Although quality control is the most important aspect of regulating a distillation process, success hinges on much more than the quality-control loop itself. Both the steady-state stability and the dynamic response of a column have been seen to depend on the arrangement of the lesser control loops—flows, levels, and pressures. Furthermore, feedforward control can remove much of the burden from the feedback controller. When all these loops are exercising their regulation over the process, the quality-feedback loop can be closed with confidence that it will make the corrections needed to maintain specification product.

Inferential Measurements The most common inferential measurement used for composition is the boiling point. In an ideal binary mixture, boiling point and composition are related by means of the vapor pressures of the components. Consider an ideal mixture of components A and B. By the law of partial pressures,

$$y_A = \frac{p_A}{p} \qquad y_B = \frac{p_B}{p} \tag{8.52}$$

where y = mole fraction of designated components in the vapor
p_A, p_B = their partial pressures
p = total pressure

Raoult's law states:

$$p_A = x_A p_A^\circ \qquad p_B = x_B p_B^\circ \tag{8.53}$$

where x is the mole fraction of the designated component in the liquid and p° is its vapor pressure. In a binary system, $y_A + y_B = 1$ and $x_A + x_B = 1$. When these relationships are combined with (8.52) and (8.53), a solution for x_A in terms of vapor pressures and total pressure is obtained:

$$x_A = \frac{p - p_B^\circ}{p_A^\circ - p_B^\circ} \tag{8.54}$$

Since the vapor pressures of the pure components are unique functions of temperature, a relationship exists between composition, temperature, and total pressure per Eq. (8.54). The curves calculated for the benzene-toluene system in Fig. 8.16 attest that the relationship is nearly linear.

For systems which depart significantly from ideality, and for ternary or more complex systems, the relationship among composition, temperature, and pressure can be determined using equilibrium constants. The procedure for calculating the vapor pressure of a given ternary mixture at a given temperature was demonstrated in Example 3.2. This procedure must be modified by selecting a pressure and estimating the temperature which develops that pressure for a given mixture. A solution can only be found by trial and error. An approximate temperature is selected, K values for the components are

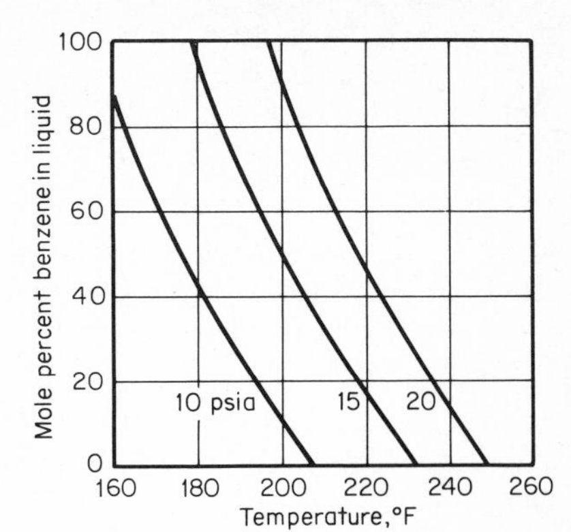

figure 8.16 *The relationship between boiling point and composition for the benzene-toluene system is nearly linear.*

determined for that temperature and pressure, and a corresponding vapor composition is calculated as $y_i = Kx_i$. The temperature is adjusted until $\Sigma y_i = 1.0$.

There are three basic limitations in using temperature to infer composition:

1. Variations in off-key components will cause errors.
2. Sensitivity of the measurement is too low for many applications.
3. Variation in pressure will cause an error.

The sensitivity of the measurement to off-key components creates problems in separating natural-gas liquids. Temperature is typically used to control the amount of propane leaving in the bottom product from a depropanizer in the company of butanes and gasoline components. Should a change in feed quality increase the average molecular weight and therefore the boiling point of the heavier components, the temperature controller will tend to let more propane pass to keep the boiling point constant.

In many applications, however, boiling point or vapor pressure is the desired controlled variable. This is true of many of the products of a petroleum refinery, which are complex mixtures of components. Some analyzers used, particularly in fractionating crude oil, are forms of boiling-point devices—they report the initial boiling point, end point, flashpoint, or vapor pressure.

Lack of sensitivity is a drawback in using temperature measurements as a guide to separating isomers. The difference in boiling points between isobutane and n-butane at 60 psig is only 23°F. A variation of 1 mole percent isobutane in the n-butane bottom product would change the indicated boiling point only 0.25°F. If the problem could be solved by devising a more sensitive thermometer, it might be a simple matter. But the same deviation of 0.25°F could also be caused by a pressure variation of only 0.25 psi or by a change in the concentration of off-key isopentane of only 0.6 mole-percent.

In an effort to improve the sensitivity of a temperature measurement to the concentration of a key component, the point of measurement is usually located several trays from the end of the column. This practice has the additional advantage of improving the speed of response to a feed-rate or composition upset and thereby accelerating control action. However, the measurement then becomes less representative of product quality and begins to be sensitive to feed composition. Wood [9] reports on the variation in product quality experienced with changes in feed rate and composition when selected tray temperatures are controlled.

Sensitivity to pressure variations can be minimized by controlling pressure at the point of temperature measurement. This presumes that pressure can be controlled tightly and in fact seems to be the principal reason for emphasis on tight pressure control in the past. And when pressure control is lost because of a condenser limitation, the temperature controller allows off-specification product to be discharged.

One solution to the problem is the use of a differential-vapor-pressure transmitter as described in Chap. 5. But for its signal to be meaningful, the mixture should be binary and the filling solution must be stable. A similar and more generally applicable technique is to compensate the temperature measurement mathematically for variations in pressure. The objective is to reference a temperature measurement made at a variable pressure to a base pressure. Then both the controller and the operator will recognize the boiling point at, for example, atmospheric pressure. The compensated temperature T_b is related to measured temperature T, pressure p, and base pressure p_b as

$$T_b = T - \frac{\partial T}{\partial p}(p - p_b) \tag{8.55}$$

The partial derivative is the inverse slope of the vapor-pressure curve at a normal product composition.

For the benzene-toluene mixtures appearing in Fig. 8.16, $\partial T/\partial p$ at 15 psia is 3.8°F/psi at 100 percent benzene and 4.2°F/psi at 100 percent toluene. The correction factor changes more with pressure than with composition, decreasing from 5°F/psi at 10 psia to 3.1°F/psi at 20 psia for pure benzene. If compensation is required over this broad a range, a second-order approximation should be used:

$$T_b = T - a_3(p - p_b) + b_3(p - p_b)^2 \tag{8.56}$$

Coefficients a_3 and b_3 are selected to fit the temperature-pressure curve across the operating range. For pure benzene, values of 4.1°F/psi for a_3 and 0.1°F/psi^2 for b_3 match the boiling-point curve at 10, 15, and 20 psia. Pressure compensation is essential when the boiling point is to be controlled under floating-pressure operation.

Luyben [10] and Boyd [11] describe the use of differential-temperature and double-differential-temperature measurements to obtain a more exact representation of product quality. Pressure variations have relatively little effect on these systems since all temperatures are influenced to essentially the same degree. In attempting to hold the column temperature profile constant, they provide better regulation over product quality than control of a single temperature point but must be matched to the characteristics of the column for maximum effectiveness.

Several other nonspecific measurements are used to infer product quality, principally liquid density and refractive index. Like temperature, their sensitivity is limited and they are affected by off-key components. Location at a point several trays from the end of the column provides a more favorable response, as with temperature measurements. Pressure sensitivity is not usually a factor.

On-stream Analyzers As analytical devices become more accurate and reliable, temperature control is being supplemented by composition control. The most common analyzer used for distillation control is the chromatograph, although occasionally an infrared or ultraviolet spectrometer or even a mass spectrometer is seen. Because the chromatograph operates on the principle of absorption and stripping, separations which are difficult in a distillation column also may be difficult in the analyzer. Certain azeotropes, for example, cannot be separated satisfactorily by a chromatograph without adding an extractant or applying some other specialized technique. High-boiling mixtures also present problems since the high temperatures required for separation cause fouling.

A distinct advantage of chromatography is its ability to report the concentration of more than one component. This makes it possible to control the ratio or the sum of two components. Controlling component ratio was discussed in Chap. 1 under the heading "Controlling Third Components."

Many specifications on product quality are given in terms of purity. An analysis of the major component in the product, typically exceeding 95 percent purity, cannot be made with absolute accuracy. Then the sum of the impurities is a more accurate representation of the purity. The sum of the heavy key and the lighter-than-light key in the distillate is usually sufficient to determine distillate purity.

The specificity which analyzers enjoy over inferential or nonspecific detectors is costly in terms of maintenance and dynamic responsiveness. A sample must be conveyed from the column to the analyzer and returned or suitably disposed of. It is most important to avoid a phase change in the sample line because fractionation will result and the analyzer will not see the entire sample. Vapor samples are preferred since their higher velocity means less dead time in the sample line. For distillation columns operated above atmospheric pressure, sampling the overhead vapor at the condenser inlet is recommended. Reducing the pressure at the entrance of the sample line helps to keep the stream vaporized. Heat tracing and insulation may still be required in winter.

Bottom products are much more difficult to sample—they must either be vaporized by applying heat and reducing the pressure or be accepted as a liquid sample. In the latter case, the dead time due to transportation is much longer than with a vapor sample. Additional dead time elapses as the components are transported from the sample valve to the detector. Finally, a chromatograph only performs an analysis at discrete intervals of time known as the "sample interval." Analyses are presented to the control system separated by these intervals, during which the control loop is open. If the controller is functioning continuously, the information it is operating on grows older as the sample interval passes. In effect, the average age of the analysis is one-half the sample interval. Therefore, the act of periodic sampling introduces an additional dead time in the loop that is equal to one-half the sample interval.

For an analyzer dedicated to a single stream, the dead time related to the sample interval is always less than the dead time involved in transporting the sample from the process to the detector. But in an effort to economize, engineers frequently share an analyzer among several streams. Then the effective dead time introduced by the sample interval can be much longer than that of transporting the sample to the detector and can severely degrade the response of the control loop.

Feedback-loop Dynamics The error magnitude and its time integral sustained by a feedback controller following an upset to the column are proportional to the product of the proportional band P and reset time R per Eqs. (8.31) and (8.40). Maximum performance of the feedback loop will then be attained when the PR product is a minimum. However, the minimum settings are related to the dead time τ_d and time constant τ_1 existing in the loop, which determine its period and dynamic gain per Ref. 12:

$$\frac{PR}{100} = K_p \frac{{\tau_d}^2}{\tau_1} \tag{8.57}$$

where K_p is a constant including the sensitivity of the process and the analyzer.

In Chap. 2, the time constant was determined to be the liquid holdup in the column and accumulator divided by its feed rate. It cannot be changed, except by eliminating the accumulator. Therefore, the control effectiveness can be maximized only by minimizing the dead time in the loop. The fact that τ_d is squared in Eq. (8.57) emphasizes the importance of reducing dead time as opposed to increasing the time constant.

The dead time appearing in a product-quality loop is related to a number of factors:

1. Selection of manipulated variable
2. Location of quality measurement
3. Configuration of other loops
4. Dead time in sampling
5. Sample interval

If boilup can be manipulated for quality control either directly, or indirectly by a level or pressure controller, dead time will be reduced because vapor rates propagate much faster than liquid flows.

The location of the measurement point is equally important. The author controlled a styrene-ethylbenzene column by manipulating distillate flow for control of distillate quality and reboiler steam for styrene quality. A liquid chromatograph was used to analyze for ethylbenzene in the styrene; the natural period of oscillation of this control loop was 2.5 h. A refractometer located eight trays from the top detected the styrene content at that point; the period of this control loop was only 20 min. The natural period of oscillation is usually equal to four dead times.

McNeil and Sacks [3] reported a period of 1 h using a 15-min analysis time on overhead vapor while manipulating distillate flow. In the presence of their 20-h time constant, the dynamic gain of the process was sufficiently low to use on-off control. Gallier and McCune [13] found that the response of temperature control in the top section of the column hinged heavily on the tightness of the accumulator level controller when manipulating distillate flow. And van Kampen [2] reduced the period of the overhead-vapor-quality loop from 5 h to 30 min by using feedforward control of reflux from distillate as in Fig. 8.6. His measured dead time was 7 min.

The author has also observed the operation of a very large alkylation deisobutanizer similar to that described in Fig. 3.8. Control of the isobutane content of the n-butane sidestream by manipulation of distillate flow exhibited a natural period of oscillation of 8 h. There was no feedforward control loop from distillate to reflux, and the point of composition measurement was located perhaps 60 trays from the top of the tower. The lag in the accumulator level loop and the dead time in transporting reflux down so many trays contributed much more dead time than did the analyzer. Without feedforward control from feed rate and composition, that column would not have been regulated acceptably. This type of sluggish response can be corrected either by using heat input to control accumulator level as a Fig. 8.5 or by setting heat input in ratio to manipulated distillate flow as shown in Fig. 8.17. This arrangement provides essentially constant separation along with responsive quality control. Coefficient K is the adjustable heat-input–distillate ratio.

Sampled-data Control The characteristic of sampling was described as creating an effective delay equivalent to half the sampling interval. But this is somewhat of an oversimplification since sampling is fundamentally different from dead time. In the process of opening the control loop, sampling masks much of the dynamic response of the process in proportion to its interval relative to the period of the process. Van Kampen observed that as the sample interval Δt was increased beyond $1.5\tau_d$, the period of oscillation of the

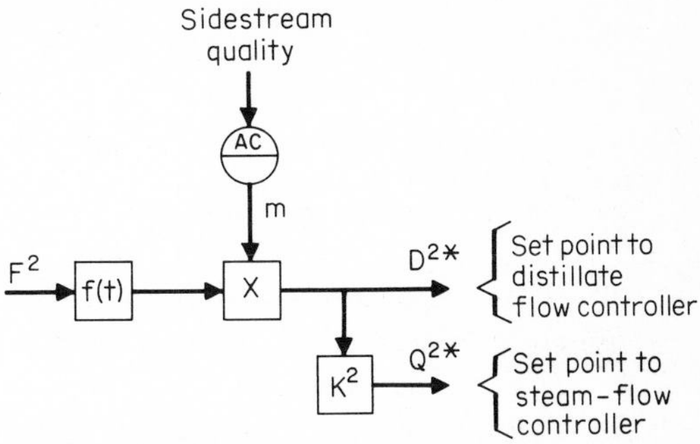

figure 8.17 *The dynamic response of sidestream or bottom-product quality to distillate flow can be enhanced by manipulating boil-up as well.*

quality-control loop changed from $4\tau_d + 2\Delta t$ to $2\Delta t$. Thus the act of sampling occludes the dead time when it substantially exceeds the dead time in magnitude.

Control over a sampled measurement may be improved significantly if the controller is not allowed to operate on aging information. When a new analysis is observed, the controller should apply immediate correction and then wait until the next analysis arrives. This action has the benefit of introducing reset or integrating action into the control loop without the penalty of an increased period of oscillation normally accompanying it. This "sampled-data control" is accomplished by placing the controller in automatic for only a short space of time following each analysis. In addition to providing improved control, the range of adjustment of reset time is extended in inverse proportion to the percentage of time spent in automatic. When periods of oscillation extend beyond an hour, this feature becomes quite helpful. Adjustment of the mode settings for this controller is described in Ref. 14.

A digital computer naturally controls in sampled-data fashion. It can control only one loop at a time and so cycles through its multiple-loop system in a periodic sequence. When its sampling interval for a particular loop is much less than the dead time naturally occurring in that loop, digital control resembles analog control. But when the dead time is less than the sample interval, sampling begins to dominate the response.

When controlling from chromatographic information, the computer may process information at its own sampling interval apart from that of the analyzer. But its performance will be improved if it is keyed to the interval of the analyzer. When a new analysis arrives, the computer should make but one calculation to reposition the manipulated variable and then wait for another analysis. This strategy has the additional advantage of protecting against a failure of the analyzer to generate a new signal—the manipulated variable is then held in its last position.

Compensating for Variable Gain The sensitivity analysis conducted earlier in the chapter reveals that the gains of the controlled compositions to the manipulated variables are not constant. Equation (8.16) can be solved for the multicomponent separation in terms of the heavy key, to reveal the gain dependence:

$$\frac{d(D/F)}{dy_h} = \frac{1}{(y_h - x_h)^2}\left[x_h - z_h + \frac{(z_h - y_h)(y_l + y_h)x_l x_h}{(x_l + x_h)y_l y_h}\right] \tag{8.58}$$

The compositions that exert the most pronounced influence over the sensitivity are x_l and y_h, as they approach zero. Then the process gain can be approximated from (8.58) as

$$\frac{dy_h}{d(D/F)} \approx k\frac{y_h}{x_l} \tag{8.59}$$

where k lumps all the remaining terms in (8.58). Assuming a constant bottom-product composition, the gain of distillate composition to D/F is seen to vary directly with its *impurity*, that is, y_h.

This gain variation appears in the closed-loop response as a nonsinusoidal oscillation in y_h. As the impurity approaches zero, the curve flattens, although larger values change more rapidly, as shown in Fig. 8.18. It is very difficult to adjust the composition controller in this situation. As y_h rises past its set point, the controller tends to overcorrect, eventually driving y_h toward zero. Then insufficient correction is applied, and recovery is slow.

Gain compensation having the form $1/y_h$ should be added to the controller at its input. This is ideally achieved using logarithms of the controlled composition and its set point, such that the controller acts on the difference between their logarithms:

$$e = \ln y_h - \ln y_h^* \tag{8.60}$$

The controller gain is then proportional to the derivative of (8.60):

$$\frac{de}{dy_h} = \frac{1}{y_h} \tag{8.61}$$

which cancels the gain variation of the process. This compensation would appear to be most necessary for products of high purity.

Extracting logarithms of measurement and set point may be inconvenient, however. A very effective approximation can be achieved by the system shown in Fig. 8.19, where measurement and set point are combined external to the controller. In this arrangement, the deviation from set point is calculated as

$$e = \frac{y_h}{y_h + y_h^*} - \frac{1}{2} \tag{8.62}$$

whose first derivative is

$$\frac{de}{dy_h} = \frac{y_h^*}{(y_h + y_h^*)^2} \tag{8.63}$$

As the measurement cycles geometrically about the set point, compensation is exact. For example, consider the case where $y_h^* = 1.0\%$; as y_h varies

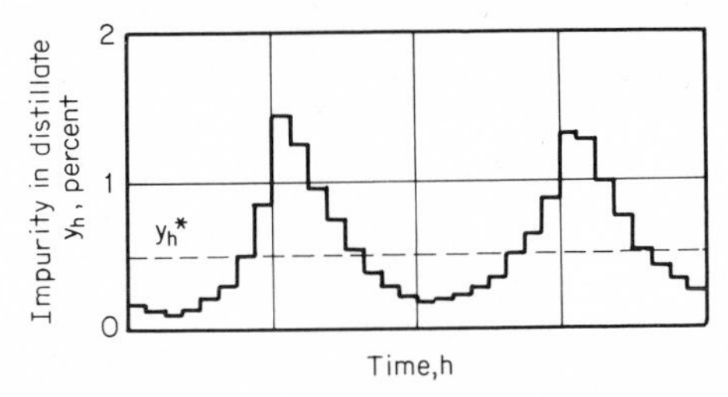

figure 8.18 *This chromatograph record of distillate impurity reveals the variable gain of the process in the closed loop.*

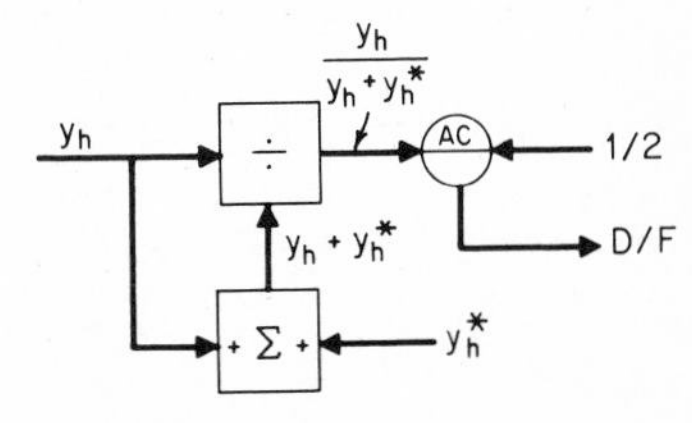

figure 8.19 *This system compensates reasonably well for variations in process gain with* y_h.

from 0.5 to 2.0%, the required gain variation per (8.61) is 2.0/0.5, or 4:1. The actual gain variation provided by (8.63) is $(2 + 1)^2/(0.5 + 1)^2$ which is also 4:1. Cycles that are not geometrically centered about the set point will not be compensated as well, but they would be the exception rather than the rule.

Control of temperature at a tray some distance removed from either the top or bottom of a column can encounter a gain reduction with increasing deviation in both directions. Typically, the measurement is made at the most sensitive point in the column temperature profile. Then a shift in that profile in either direction due to an upset will decrease the sensitivity of the measurement to the manipulated variable.

A symmetrically characterized nonlinear controller designed for pH control has been found useful in compensating for this nonlinearity. Reference 15 describes the controller and the procedure to follow in adjusting it. Lacking this nonlinear characterization, the period of oscillation tends to increase with increasing deviation, due to the falling sensitivity. If the reset mode of the controller is adjusted for maximum process sensitivity, it will tend to overcorrect during large excursions and precipitate a slow limit cycle. Removing the reset action will eliminate the problem, but this is only viable when temperature is being set in cascade by a composition controller.

REFERENCES

1. MacMullan, E. C.: Fractionator Control System Using an Analog Computer, U.S. Patent 3,282,799, Nov. 1, 1966.
2. Van Kampen, J. A.: Automatic Control by Chromatographs of the Product Quality of a Distillation Column, presented at Convention on Advances in Automatic Control, Nottingham, England, April 1965.
3. McNeill, G. A., and J. D. Sacks: High Performance Column Control, *Chem. Eng. Prog.*, March 1969.
4. Shinskey, F. G.: "Process-Control Systems," pp. 128–131, McGraw-Hill Book Company, New York, 1967.
5. *Ibid.*, p. 21.
6. Buckley, P. S., R. K. Cox, and D. L. Rollins: Inverse Response in a Distillation Column, *Chem. Eng. Prog.*, June 1975.
7. Shinskey, F. G.: *op. cit.*, chap. 8.
8. Instrument Society of America: "Process Instrumentation Terminology," ISA Standard S51.1, Pittsburgh, 1976.
9. Wood, C. E.: Method of Selecting Distillation Column Tray Location for Temperature Control, presented at Joint IMIQ-AIChE Meeting, Mexico City, Sept. 24–27, 1967.
10. Luyben, W. L.: Feedback Control of Distillation Columns by Double Differential Temperature Control, *Ind. Eng. Chem. Fundam.*, November 1969.
11. Boyd, D. M.: Fractionation Column Control, *Chem. Eng. Prog.*, June 1975.
12. Shinskey, F. G.: *op. cit.*, p. 102.
13. Gallier, P. W., and L. C. McCune: Simple Internal Reflux Control, *Chem. Eng. Prog.*, September 1974.
14. Shinskey, F. G.: *op. cit.*, pp. 114–116.
15. Shinskey, F. G.: "pH and pIon Control in Process and Waste Streams," Wiley-Interscience, New York, 1973, pp. 194–199.

CHAPTER NINE

Controlling within Constraints

Constrained operation is common to most separation units, and it should be. For if at least one variable in a column is not being held at its limit, then separation efficiency or productivity could still be improved. Maximum recovery, for example, can only be achieved at maximum heat input—whether the limit is imposed by reboiler, column, or condenser. Many plant managers are attempting to maximize production by whatever means is available. In this case, constraints are being encountered on all sides. The purpose of this chapter is to locate these constraints, provide protection against their violation, and permit operation to proceed in their presence.

The control systems described thus far for heat input, heat output, and product quality have assumed a freedom to manipulate flow rates as necessary to achieve their objectives. Yet in the real world limits exist on all flow rates. At the very extreme, control valves can only range between closed and fully open. In many cases, narrower limits must be imposed to avoid inefficiencies such as flooding and weeping and to protect equipment from overpressure, overtemperature, overflow, etc.

Whenever a limit of any kind is placed on a manipulated variable, the normal function of the control loop is inhibited and control is lost. If loss of quality control is intolerable, then the production rate must be decreased until control is regained. In other cases, it may be possible to restructure the control system by using a different variable to maintain control. In any case, the alternative modes of operation need to be investigated in order to determine the penalties for violating constraints and the remedial action required to keep the plant running.

Finally, the exact location of those constraints should be known at all times to warn the operator of impending crises. With sufficient information about the limits of his plant and its expected disturbances, he should be able to schedule production to make the best use of the resources available to him.

SELECTIVE CONTROL SYSTEMS

Selective control systems are used to impose limits on valve positions or manipulated flows. They can protect equipment or products from unsafe or abnormal situations and maintain control rather than shut down a unit. The principal element used is the signal selector—its function is to select the highest, the lowest, or in certain cases the median of a plurality of signals.

In some systems a controller output will be compared against a fixed signal in a selector—in which case that signal acts as a limit on the controller output. Sometimes multiple signals of the same sort are compared, as when controlling the highest level or lowest pressure in a unit. But most often, controller outputs are compared in an effort to keep all controlled variables on the safe side of certain constraints. Each of these systems is discussed below.

Self-imposed Limits A few limits to column operation are self-imposed and therefore do not require any overt implementation. The most common of these is the heat-input valve limit. Steam flow to the reboiler could be set to control product quality, set in ratio to the feed, or simply set at a fixed set point. Should the valve be unable to maintain the set point, whatever the reason, the objective of holding that particular flow will be sacrificed.

If quality control was the objective, it will be lost and the operator should be alarmed. He will have to take remedial action of some kind—either reduce the feed rate, supplement the steam supply, or prepare to rerun the product. In any case, some time may elapse before control is restored. During the time when control is lost, the quality controller—if left in automatic—will continue to index the set point of the steam-flow controller although the latter is unable to respond. When the overload is corrected, the quality controller, in failing to achieve response, may be requesting much more steam than is required to meet specifications, and a large overshoot is likely. This reset windup of the primary controller in a cascade system can be avoided with the scheme shown in Fig. 9.1.

The primary controller is fitted with what is known as "external reset"; i.e., its reset feedback signal is supplied from outside the controller. (Normally, this feedback loop is closed by internal feedback of the output of the controller.) When the deviation between measurement and set point in the flow controller is zero, then reset action takes place in the primary controller as usual. But if a flow deviation persists for any reason, primary reset action stops; the primary controller then has proportional action only, whose bias [see Eq. (6.34)] is the flow measurement Q and whose output is its set point Q^*:

$$Q^* = \frac{100}{P} e + Q \tag{9.1}$$

Then the composition deviation becomes proportional to the flow deviation:

$$e = \frac{P}{100}(Q^* - Q) \tag{9.2}$$

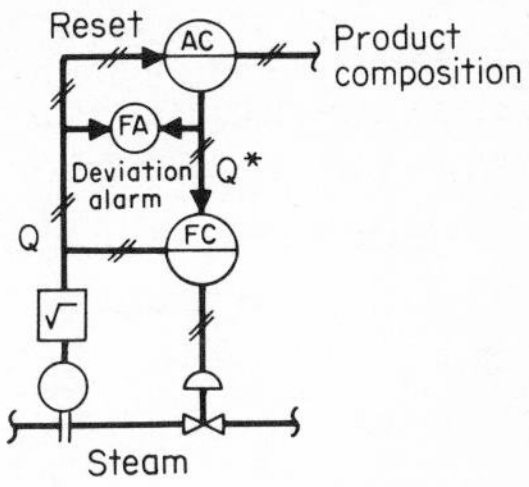

figure 9.1 *The secondary measurement can be used to reset the primary controller and thereby avoid windup when control is lost.*

If steam flow is set to control separation, failure to meet the set point could adversely affect the material-balance controls. Consideration of this effect is taken up later in the chapter.

Another possibility would be the manipulation of steam flow to control column pressure. If pressure could not be maintained, any temperature control loop on the column would be upset. Otherwise, a low pressure would not ordinarily cause a hazard.

If steam flow were used to control either base or accumulator level, failure to reach set point would result in a high level. Lacking self-regulation, the level would continue to rise, causing flooding of the column or condenser. Some additional protection may be required in the way of transferring level control to another variable in this eventuality.

Ordinarily, high and low limits should be placed on heat input as partial protection against flooding and weeping. The limits may be placed on valve position within the controller which manipulates heat input. They will narrow the allowable range of the valve. Loss of control then takes place beyond these limits, and any transfer of control to auxiliary devices must take place when the limits are reached.

When steam is the heating medium, limits on boilup may be more accurately placed on the steam-flow set point than on valve position. The reset windup problem of a composition controller setting steam flow will not exist then since the limits would be imposed on its output. But if the composition controller is used for feedback trim of a feedforward system, it is not directly connected to the steam-flow controller. Then to protect against windup the deviation between the steam-flow measurement and its calculated set point should be used for external reset feedback. If this deviation is zero, normal reset action takes place; if it is not zero, the composition controller is satisfied by an offset proportional to the steam-flow deviation. The system is arranged as shown in Fig. 9.2.

Again, the primary controller becomes a proportional controller whose bias is replaced by the external reset signal:

$$m = \frac{100}{P} e + m - Q^* + Q \tag{9.3}$$

Equation (9.3) then reduces to (9.2).

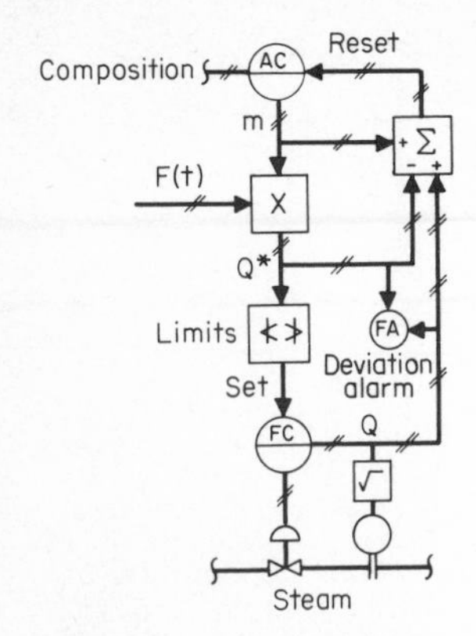

figure 9.2 *Windup of the primary controller in a feedforward system can be avoided by feeding back secondary deviation.*

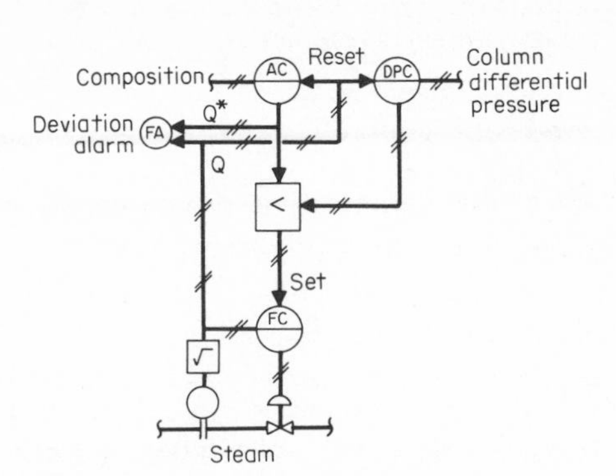

figure 9.3 *Override controls protect against violating constraints by automatic selection of the lower steam-flow set point.*

Override Controls Limits on steam flow or other variables do not have to be fixed—they may be automatically adjusted to satisfy measurable constraints. Automatic selector systems are then used to override normal control action to keep constraints from being violated. An example of column differential pressure being used to override composition control of steam flow appears in Fig. 9.3.

A low selector compares the steam flow required to control composition with that required to control column differential pressure. Normally, composition would be controlled because differential pressure would be less than its set point. The *dp* controller, in attempting to raise steam flow to satisfy its set point, would increase its output beyond that of the composition controller and would therefore be rejected. Should the differential pressure exceed its set point, however, the output of the *dp* controller would fall below that of the composition controller and would then be selected to set the steam flow.

Provision must be made to prevent the controller that is not selected from windup. This is most readily achieved by feedback of the selected output or steam flow as external reset for both controllers. The selected controller sees its own output and therefore has normal reset action. The other controller sees a foreign output and hence behaves as a proportional controller with steam flow as a bias per Eq. (9.1).

Detection of an override is necessary if loss of composition control will result. A deviation alarm on the composition controller could be used to alert the operator, but by the time it is activated the damage has already been done. Instead, a deviation alarm between the steam flow and the output of the composition controller will signal that an override has taken place—before it has a chance to affect composition.

A column-pressure controller may also be used to override heat input. However, it must be a different controller than the one used to manipulate the condenser. The dynamic responses of column pressure to reboiler and to condenser differ, and the controllers would also have different set points. The override controller must be set higher than the condenser controller so that it would function only when the other failed to hold pressure by manipulating the condenser.

When the control valve for a steam-heated reboiler is located in the condensate line, the possibility exists of blowing steam through the valve. Fouling of the heat-transfer surface could cause its maximum rate of condensation to fall below the capacity of the valve. In an effort to increase the steam flow, the valve would open further, but the additional steam might not be condensed. A small condensate tank fitted with a simple level controller as shown in Fig. 9.4 can protect against this condition. If the level controller has proportional action only, it needs no external feedback.

A similar override on heat input may be exercised by a column-base level controller. In the event the level falls to the point of exposing reboiler tubes or preventing circulation, the level controller could limit the heat input just as shown in Fig. 9.4. This action would prevent the boilup rate from exceeding the rate at which liquid is entering the column base. Under normal conditions, base level would be controlled at a higher set point by manipulating the bottom-product valve. Two level controllers would then be necessary having different set points and also different mode settings, as was necessary for pressure override.

The feed preheater is also a source of boilup, although its location limits its vapor contribution to the top section of the column. Any override of preheat to prevent flooding should thus be restricted to the trays over which it has direct influence. Therefore, if flooding above the feed tray is a problem, it can be avoided by overriding preheat with differential pressure between the feed and the top of the tower. On the other hand, if a subcooled feed is causing excessive liquid flow below the feed tray and a low vapor velocity above it, preheat could be controlled to balance the two differential pressures.

Overrides may also be exercised on reflux flow. If reflux is set to control composition, for example, an override from low accumulator level may be necessary to protect against inadequate boilup. This system is the mirror image of the base-level system just described. Reflux flow also affects column differential in raising the liquid height on the trays and in promoting a subsequent increase in boilup. So in columns where reflux is controlled at a fixed set point or used for composition control, override by a column differential-pressure controller is appropriate. If it is manipulated to control accumulator level, no such override would be necessary since reflux follows the heat load on the column rather than setting it.

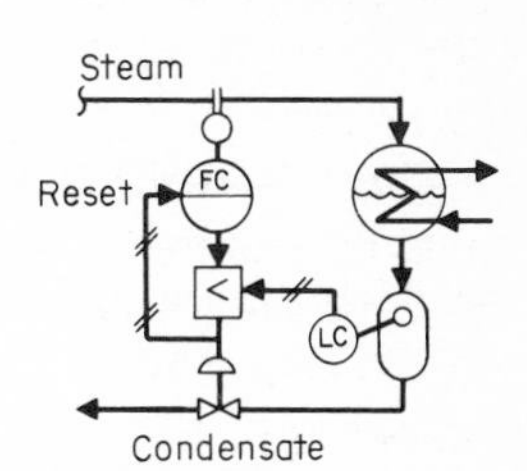

figure 9.4 *A simple level controller can keep the condensate valve from passing steam.*

Reflux, distillate, or heat input may be used to control column pressure with a flooded condenser.

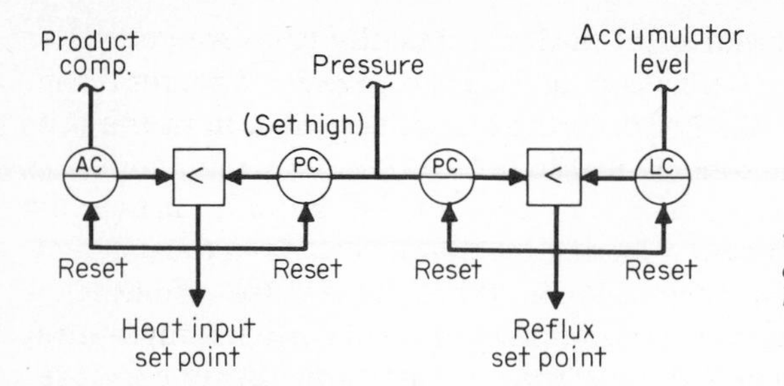

figure 9.5 *The level controller overrides when pressure control by flooding is no longer possible.*

When a combination of heat load, coolant temperature, and vapor composition reduces the maximum rate of condensation below the current rate of boilup, the condenser will empty. Boilup can continue to control pressure, but the liquid level in the overhead could fall to the point where the reflux pump cavitates. A level controller should then limit the flow of reflux or distillate.

If either reflux or distillate is used to control pressure, control will be lost when the condenser empties and the pressure will rise as necessary to reach a new equilibrium between heat input and removal. In the meantime, the pressure controller will continue to open its valve, draining the liquid out of the overhead system and eventually cavitating the reflux pump. Before this happens, a level controller on the accumulator should override the pressure controller as shown in Fig. 9.5. A second pressure controller with a higher set point would then be needed to override heat input.

Systems with Variable Structuring In certain columns, noncondensable gases can accumulate to the point where separation is limited and they must be vented. The addition of a condenser vent valve increases the opportunities for override controls. Where reflux was manipulated to control pressure, the accumulator level controller has been used to open the vent valve as shown in Fig. 9.6. Note that level is not affected by venting. However, venting lowers the column pressure, and the pressure controller counters by reducing reflux flow, which raises the level. These control loops are said to be "nested"; i.e., the pressure controller must function to provide level control. Neither loop responds particularly well in the constrained mode—reflux has no effect on pressure because the condenser is empty, and venting has no direct effect on level. In the constrained mode, the control-loop structure is incorrect.

There is a method for reconfiguring the control loops in the constrained mode by using selectors combined with other computing functions. In this case, pressure control would be transferred to the vent valve when liquid-level control of reflux becomes necessary. Figure 9.7 illustrates the technique.

As long as reflux is under pressure control, the two signals sent to the subtractor are identical and its output will be zero. When the level controller takes over reflux manipulation, however, a deviation between pressure-controller output and reflux set point develops, opening the vent valve. The level controller requires external reset feedback but the pressure controller does not—it is controlling in both operating modes. In fact, if the reflux set point *is* fed back to the pressure controller, reset action will *stop* in the constrained mode and a pressure deviation will develop.

In earlier discussions, when column pressure had to be controlled in both the constrained and the unconstrained modes, two controllers were used. Although this practice seems necessary, it has the disadvantages of adding a control station which is not normally used and of failing to provide any control between the two pressure set points. But the system in Fig. 9.7 uses a single pressure controller, thereby avoiding these disadvantages. Although the response of pressure to the vent valve will differ from that achieved when manipulating reflux, the difference may be compensated. The subtractor must have a variable gain G applied to both inputs:

$$m_v = G(m_p - L^*) \tag{9.4}$$

where m_v = vent-valve signal
m_p = pressure controller output
L^* = reflux set point

The gain of the pressure controller when acting on the vent valve will be G times its normal gain. Therefore, the controller should first be tuned while manipulating reflux; then G should be adjusted for stable control while manipulating the vent valve.

It is entirely possible that the reset time of the pressure controller should also change when manipulating the vent valve. Because it affects pressure directly as opposed to acting through condenser level, the vent valve could use a shorter reset time. Since the reset time set into the pressure controller for reflux manipulation is longer than necessary for venting, stability will not be lost while venting. If a longer reset time were required during venting, a lead-lag function could be applied to the vent-valve signal to provide the appropriate compensation. This technique is described in detail in Ref. 1.

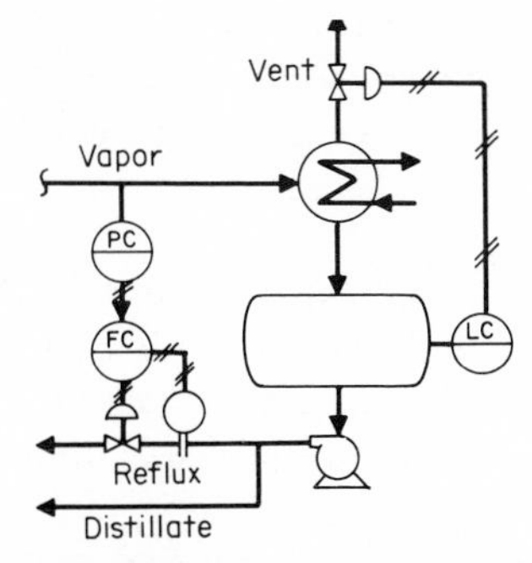

figure 9.6 *This configuration uses nested loops—the vent valve only affects level through the action of the pressure controller on reflux.*

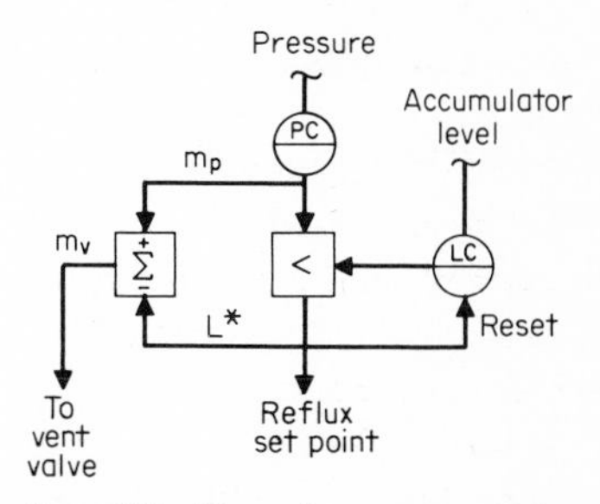

figure 9.7 *The subtractor transfers pressure control smoothly to the vent valve during a level override.*

Figure 9.5 described an override system requiring a second pressure controller for heat input. Transfer from reflux control of pressure to heat-input control can be improved and the second controller eliminated with the arrangement appearing in Fig. 9.8. The technique is the same as that used to manipulate the vent valve, but the starting point is different. Although the vent valve is normally closed, the heat input may be at any point on scale as needed to control composition. Therefore, the override on heat input must start at whatever value is current.

Figure 9.8 shows the deviation between pressure-controller output and reflux set point being subtracted from the output of the analyzer controller (AC). This corrected signal is then compared against the AC output in a low selector. Consequently, when the level controller (LC) overrides pressure control, the pressure controller (PC) automatically overrides composition. A developing deviation in composition caused by the override will attempt to raise heat input through the summing device; but the pressure controller, being more responsive, can readily overcome this induced upset.

A question frequently arises regarding the operability of a variably structured system. In many installations, the operator has direct access to the control valves through the cascade flow controllers, as would be the case for the system in Fig. 9.8. No such cascade controller exists for the vent valve in Fig. 9.7, however. If the operator must have direct access to it, a hand-control station would have to be inserted between the valve and the subtractor. Otherwise, he could only position the vent valve manually by adjusting both PC and LC outputs manually to establish a difference between them. Many packaged selector systems offered by instrument manufacturers have a single auto-manual transfer station at the output of the selector. While that configuration may be desirable for some applications, it has the disadvantage of removing all overrides when the station is in manual; furthermore, that arrangement cannot accommodate variable structuring.

Start-up, Shutdown, and Standby Operation Downtime is minimized in continuous distillation systems not only because in itself it is unproductive but because the start-up following is also unproductive. And when conducted perhaps only a few times per year, lack of operator familiarity with the necessary procedures prolongs the time required to reach a controlled state. Buckley

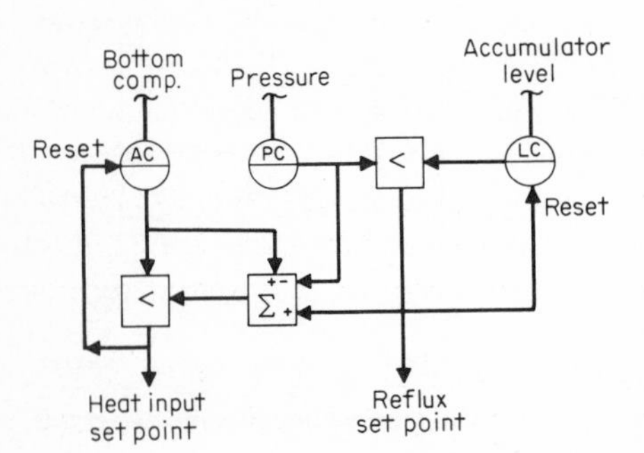

figure 9.8 *This system is an improvement over that shown in Fig. 9.5 since it requires a single pressure controller.*

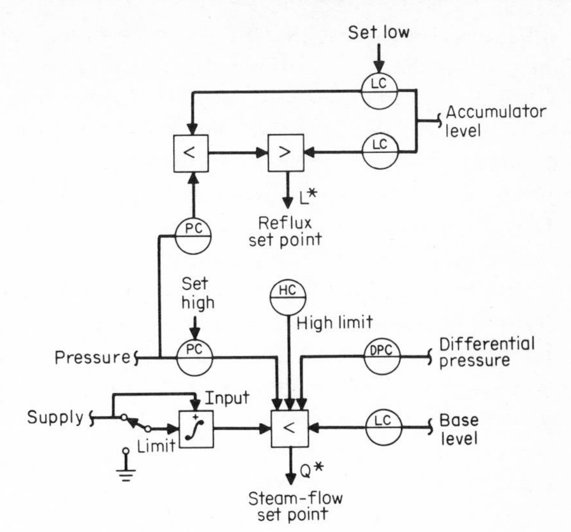

figure 9.9 *This system is designed to increase steam flow in a ramp, with protection of levels, pressure, and differential pressure.*

[2] has probably contributed more to automating the start-up and safe operation of distillation columns than any other engineer. His systems use a multiplicity of selectors to protect the column and its auxiliaries against hazards during abnormal conditions. Essentially they mimic an intelligent operator in bringing a column on line through successive stages: establishing liquid levels, pressurizing, raising temperatures, etc.

Start-up of a column requires a coordination between reflux and boilup different from that required for continuous operation. Once boiling begins, pressure will rise until reflux starts to flow. Rather than waiting for the level in the reflux accumulator to rise to its normal set point, reflux should be used to control *pressure* during start-up. The accumulator level should be prevented from falling too low since this could damage the reflux pump. Then when the level reaches the normal set point for continuous operation, the normal level controller would take control away from the pressure controller, as shown in Fig. 9.9.

Limits must be applied to heat input, however, to protect against loss in base level and excessive column differential pressure. Furthermore, the heat must be added gradually while the reboiler is still cold, or thermal shock may cause connections to leak. Buckley [2] uses a "slow-down" circuit, which raises the heat input gradually through a first-order lag. However, a first-order lag gives the familiar exponential response to a step input, its initial rate of rise being rapid. If an integrator is used as in Fig. 9.9, the rate of rise is constant throughout the ramp. The ramp is begun by raising the limit of the integrator. It will continue until one of the overrides takes precedence. In the configuration shown, reducing the integrator's limit to zero will immediately close the steam valve.

Once levels have been established, reflux flow will follow heat input. If feed and product flows are zero, the column is in the total-reflux mode of operation.

In this mode, absolute control of quality is impossible since it depends on manipulation of the material balance. Therefore, desired compositions for the two products cannot normally be achieved prior to feeding the column. Then the first product made will most likely not meet specifications and may require recycling to the feed tanks.

The actual rates of boilup and reflux have no bearing on compositions during total reflux as long as inefficient regions of tray performance are avoided. As feed is increased from zero, distillate or bottoms flow rate can be made to increase proportionately through a feedforward loop. The other product flow will follow as the material balance is closed. The initial ratio of product to feed cannot be set by a feedback controller as described because at total reflux the feedback loop is open. Consequently, the operator must make the initial adjustment to that ratio with the feedback controller in manual. The ratio he enters should be based on feed composition and the *desired* product composition rather than its actual value at total reflux. If the product-feed ratio is set accurately, the product composition will settle out near its desired value shortly after feeding is begun. Without the help of this feedforward loop, it may take hours of juggling boilup and reflux before the products become acceptable.

If heat input is to be held constant, the nonlinear feedforward system described in Fig. 8.13 will bring product flow along in response to feed rate with reasonable accuracy. But if heat input is also set proportional to feed, at low feed rates it will be riding on a low limit. Ideally, the feedforward system should function differently while heat input is constant. Although Fig. 8.12 indicates that the nonlinearity is not severe at low feed rates, the dynamic compensation will be incorrect. When heat input is manipulated in proportion to feed rate, a lag is used; but when heat input is constant, lead action is required. Since low-flow operation is definitely unprofitable, little time is spent there, so automatic modification of the feedforward controls for this condition is probably not warranted. When limits are applied to heat input at *high* flow rates, similar modifications are both necessary and worthwhile—these are described later in the chapter.

The feedforward system also automatically returns the column to total-reflux operation when feed is lost. Protection of the product-quality controller against windup during total reflux is possible by using the arrangement shown in Fig. 9.2. The limits would not ordinarily be necessary when manipulating product flow as they were for heat input or reflux. The one exception to this rule would be an override on distillate flow from low accumulator level for the column shown in Fig. 7.21, where the distillate had its own condenser. Otherwise, failure of the chilled water to condense the product at the rate being withdrawn by the composition controller would cause a loss of level. A deviation alarm across the selector, as provided in Fig. 9.3, should warn the operator of loss of quality control.

Shutdown is accomplished in the reverse of start-up. Feed is reduced until the column is on total reflux. If properly executed, both products will remain at specifications. Then heat input may be reduced with reflux following until weeping begins. At this point, base level will begin to rise as the column empties. Due to the accumulation of light components from higher in the column, the bottom flow must be diverted to feed storage. The product in the overhead accumulator should be unaffected and may be withdrawn to product storage. Again, the nature of feedforward material-balance controls facilitates shutdown as well as start-up.

MANAGING HEATING SYSTEMS

Although the management of heating systems was covered extensively in Chap. 6, its scope was limited to continuous control functions. At this point, protection against accident and the coordination of multiple sources is presented. An appreciation of selective controls is necessary for the systems described here to be meaningful to the reader.

Protection against Excess Fuel Figure 6.15 describes a parallel fuel-air metering system with ratio adjustment from a flue-gas analyzer. Fuel flow is set to maintain the desired outlet temperature, and airflow is expected to follow. In the event that airflow cannot follow because of a limitation in its supply capability, a failure of the damper or fan, or an obstruction of some kind, excess fuel could accumulate. Aside from being inefficient and causing smoke, excess fuel also represents an explosion hazard. To protect against such a possibility, Manter and Tressler [3] developed a two-way selector system for coal-fired boilers. Shown in Fig. 9.10, it is equally applicable to all fuels.

The master controller sets both fuel and air through a pair of selectors. If airflow fails to reach its set point, it will be preferentially selected by the low selector to set fuel flow. Then any of the failures described above will automatically reduce fuel flow to avoid an unsafe condition. It will then be impossible to maintain product temperature, so an alarm must be sounded; furthermore, the master controller must be protected against windup by external reset feedback from airflow.

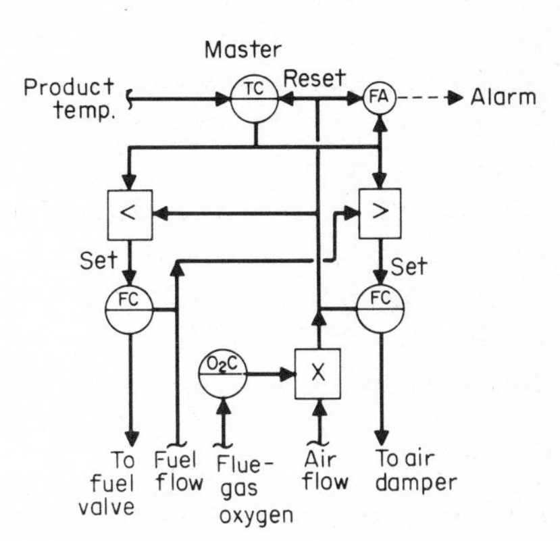

figure 9.10 *The selectors prevent airflow from falling below fuel flow.*

The high selector protects against an excess of fuel above that demanded by the master controller. An obstruction preventing the fuel valve from closing, or an operator error, might create such a situation. Airflow will be forced to follow fuel flow in this case. Again, an alarm is actuated and windup of the master controller is prevented by the system in Fig. 9.10.

The fuel-air ratio adjustment is made on the airflow measurement rather than on its set point as in Fig. 6.15. This is necessary because both controllers must have the same set point for the selectors to function. The results are essentially the same in either case.

Supplemental Firing In these days of fuel shortages, many heating systems make use of more than one fuel. For economic reasons, the cheaper fuel should be fired to the limit of its availability before it is supplemented with the costlier fuel. Thus the combustion controls must be capable of firing either fuel or both, with a smooth transfer from one to the other, accepting variations in the availability of one. Furthermore, their different compositions typically mean different heating values and combustion-air requirements. The system shown in Fig. 9.11 provides all these necessary features.

The availability of the waste fuel is limited by the backpressure controller acting through a low selector. Its flow is compensated for variable pressure and added to that of the supplemental fuel in the summing device. The scaling of that device must take into account the heating values of the two fuels as well as their flow ranges. Its output then represents total heat flow ΣQ into the system. This signal serves as the measurement to the heat-flow controller and as input to the airflow system (not shown).

When the backpressure controller limits the availability of waste fuel below what is needed to satisfy the heat flow controller, the supplemental-fuel valve must open. Its signal is derived from the difference between the flow-controller output and the waste-fuel valve position. But since the full range of the heat-flow controller may well exceed the capacity of the waste-

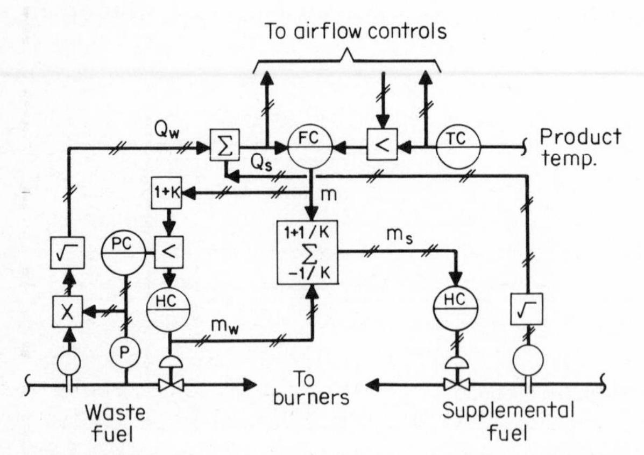

figure 9.11 *This system can control temperature effectively using a minimum amount of supplemental fuel.*

fuel valve, some additional scaling is necessary. Without this scaling, 100 percent output from the flow controller could not supply more fuel than full flow from the waste-fuel valve. In Fig. 9.11, K represents the ratio of maximum heat flow from the supplemental fuel to that of the waste fuel. In the case of no backpressure limiting, the supplemental-fuel valve signal m_s is related to the flow-controller output as

$$m_s = m\left(1 + \frac{1}{K}\right) - m(1 + K)\left(\frac{1}{K}\right) = 0$$

When the waste-fuel valve position is limited to m_w,

$$m_s = m\left(1 + \frac{1}{K}\right) - \frac{m_w}{K}$$

The heat-flow controller should not be capable of distinguishing which valve it is manipulating. Hand-control stations (HC) give the operator access to both valves.

Columns with Multiple Reboilers Whenever multiple reboilers provide heat to a single column, a decision must be made regarding the distribution of load. Multiple reboilers are generally used in severe fouling service, so that they may be individually cleaned without interrupting operation of the column. Although they may be identical in construction, their staggered cleaning schedule creates a difference in their degrees of fouling at any point in time. Consequently, they cannot all carry the same heat load. In fact, the heat-input valve on the most fouled reboiler could be fully open or limited to maintain condensate level by the system shown in Fig. 9.4.

In any case, there will be variations in duty among the reboilers and variations in the number of reboilers under control. The operator should be free to balance the load as he sees fit—without upsetting product quality. Similarly, quality control should not be affected by the number of reboilers in service, in automatic, or free of limits. Without some form of compensation, the control-loop gain will change each time a reboiler reaches or leaves a limit, is transferred between automatic and manual, or is removed from or returned to service.

The system shown in Fig. 9.12 is designed to prevent product-quality control from being affected by operator adjustments or imposed limits. Only two reboilers are indicated, but the concept is applicable to any number. The instruments labeled FFC are flow-ratio controllers which allow the operator to vary the load on any reboiler in proportion to the rest. Total steam flow is fed back to a blind (back of panel) controller which manipulates all ratio controllers to satisfy the demand for heat. Should an operator adjust a ratio setting, the steam flow to that reboiler will be affected directly. The resulting change in total flow will be corrected by the blind FC readjusting all individual set points. If the period of oscillation of the individual flow controllers is about 1 s, that of the total flow controller could be about 3 s. It is quite capable of

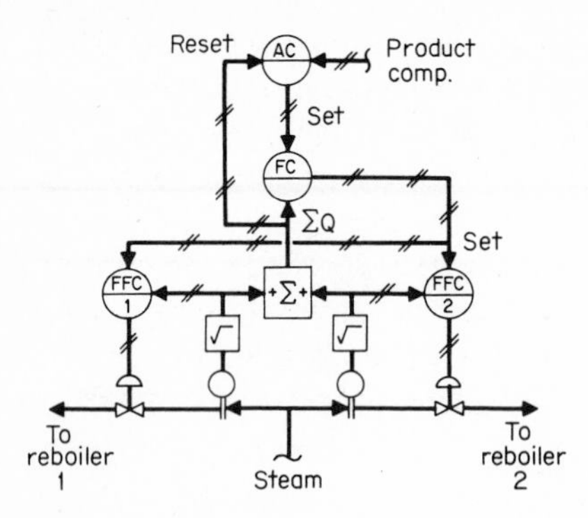

figure 9.12 *This system allows manipulation of individual reboilers without upsetting product-quality control.*

returning total flow to the demanded level before the disturbance can pass through the reboilers into the column.

When one of the valves reaches a limit or is placed in manual, that flow can no longer respond to the total-flow controller. The total-flow controller will simply continue to move all set points as necessary to meet the demand. Although the gain of the total-flow loop is changed by the unresponsive reboiler, the quality-control loop is unaffected. And because of the fast response of the total-flow loop, its gain change does not pose a problem to the quality loop.

There are other possible versions of this "multiple-output control system." For example, the individual ratio-flow controllers could be replaced with ratio stations or bias stations which modify the signals manipulating the valves directly. Then operator adjustments would be applied to valve positions rather than more accurate steam flows. However, removing the individual flow controllers can reduce the period of the total-flow loop to perhaps 1 s.

The reset feedback from total flow to the product-quality controller prevents windup in the event that all reboilers are in manual control, as during start-up.

MODIFYING QUALITY CONTROLS

Whenever constraints are encountered, decisions must be made with regard to which variables are to be sacrificed. Depending on priorities and the control system structure, modifications may have to be made to the structure. The technique has already been demonstrated for the inventory and heat-balance controls, but restructuring the quality controls is more complicated. Usually, the decision is simply to forego composition control as in Fig. 9.8. But occasionally this is inappropriate, and an alternative scheme retaining quality control must be found. Some of the implications of constraints on the quality-control system are examined next.

Meeting Multiple Specifications Products often have more than one specification, although only one is controllable at the point of discharge. For instance, a certain depropanizer distillate must contain at least 95 percent propane and no more than 2 percent isobutane. The unspecified component is the uncontrollable off-key ethane. If the ethane in the product exceeds 3 percent, the 2 percent isobutane specification cannot be violated and only the purity specification need be satisfied. But with less than 3 percent ethane, either of the two specifications could be limiting.

They may both be satisfied by appropriate reduction in isobutane content. Therefore, a single isobutane controller is needed to manipulate either the distillate or the reflux. But its set point will vary with the uncontrolled ethane content. To satisfy both specifications, the isobutane set point y_i^* must be

$$y_i^* \leq 2\%$$

$$y_i^* \leq 5\% - y_e$$

Figure 9.13 shows how y_i^* is calculated from ethane content y_e.

Another application of this concept appears in Chap. 11. An *optimum* isobutane composition may exist which results in minimum-cost operation. If the optimum lies within the specifications, it should be used as the set point. But under certain conditions of feed composition, heating costs, and product values, the calculated optimum could exceed specifications. Then the specifications must be imposed through selectors as in Fig. 9.13.

Altering Feedforward Controls Whenever a heat-input limit is reached, separation is no longer controllable. This naturally affects the relationship between product qualities and the material balance. The variation in the recovery factor caused by changes in V/F was cited in Fig. 8.14. Correction to the feedforward calculation was given in Eqs. (8.49) and (8.50). While the discussion associated with these equations related to operator adjustments to the V/F ratio, compensation is even more important when V/F is changed by imposed limits outside the operator's influence.

Dynamic compensation must be modified as well as the steady-state calculation. The limit removes the accelerating influence of boilup, which must then be replaced by lead action on reflux. Figure 9.14 is an attempt to combine both the steady-state and the dynamic feedforward controls for the constant-separation and variable-separation modes of operation. Heat input Q^* is set proportional to feed rate through a lag function. If heat-input measurement Q responds completely to the set point, then Q/F (and V/F) will be constant and

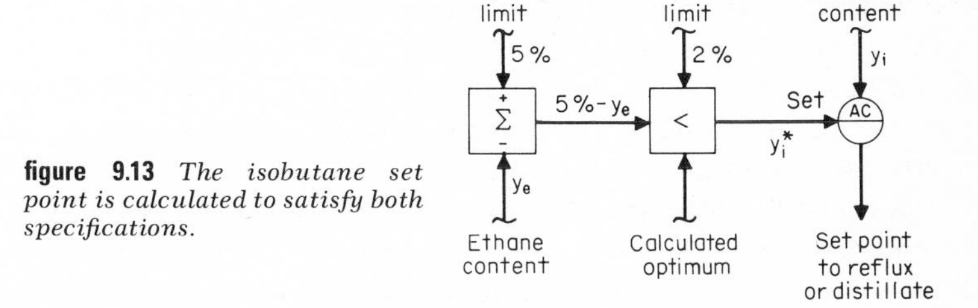

figure 9.13 *The isobutane set point is calculated to satisfy both specifications.*

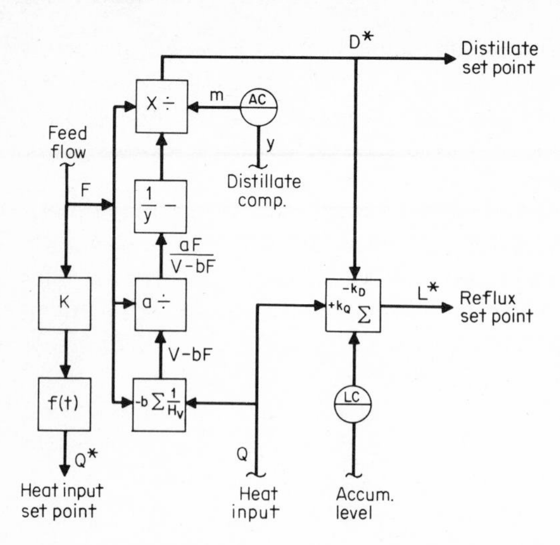

figure 9.14 *The calculations required to accommodate variable heat input complicate the control system considerably.*

distillate set point D^* will vary directly with F. But if Q fails to respond to Q^* because of any imposed limit, a correction will be applied to the multiplier-divider to make the appropriate adjustment to D^*.

Dynamic compensation is provided by manipulating reflux set point L^* by forward loops from both D^* and Q per Eq. (8.43). Because heat input cannot be relied upon, the primary adjustment for lead action is coefficient k_D on the distillate input. The lag in the forward loop to heat input would have to be long enough that distillate composition would not be significantly affected when Q is free to change. The adjustment to L^* caused by variations in Q will help to bring this about. An alternative structure is presented later in Fig. 9.16.

Loss of Quality Control Encountering a limit on heat input or reflux may or may not mean loss of quality control. Typically these two variables are the most likely to be limited, whereas the product flows are not. (An exception would be the deethanizer in Fig. 7.21 where distillate flow could be limited by availability of chilled water.) When controlling the quality of a single product, that product is usually—but not always—the manipulated variable. Then a limit on heat input (or reflux) will not prevent quality from being controlled, although alterations in the feedforward system should be made as just described.

If bottom quality is being controlled by manipulating distillate flow with boilup under accumulator level control, as in Fig. 8.5, flooding limits should be placed on *reflux*. If limits were placed on boilup, accumulator level control would have to be transferred to reflux; but of greater importance, bottom quality would lose the necessary dynamic influence of boilup and fail to respond adequately to distillate-flow manipulation.

In systems where heat input is set in ratio to distillate flow as in Fig. 8.17, a limit placed on heat input will have similar results. However, here reflux is already being used to control accumulator level, and so it cannot be limited to protect against flooding or overpressure. Then limits must be placed on boilup, culminating in loss of control over bottom-product quality. The most viable alternative in this situation would be to use the control system structure of Fig. 8.5 so that limits may be applied to *reflux* instead of boilup.

When the qualities of both products are to be controlled in a two-product column, either boilup or reflux must be manipulated by one of the quality controllers. In this case, the imposition of a limit will result in loss of that quality-control loop. A low limit will cause overpurification, which creates no problem; but the imposition of a high limit will result in underpurification and possible failure to meet specifications. The operator should be warned of this danger *before* quality is affected—the deviation alarm shown in Fig. 9.3 will give him that advance warning. He may be able to reduce column feed rate in time to avoid making any off-specification product or at least to minimize the amount produced. Although this procedure is viable for a single column, a close-coupled distillation train is more difficult to manage. Special procedures for estimating column capacity and scheduling production rate are given later in the chapter to guide the operator in managing his multicolumn system.

Occasionally, the set point to a product-quality controller will be calculated to achieve optimum operation within specification limits as in Fig. 9.13. Then loss of quality control brought about by an override *may not* mean failure to meet specifications. It is entirely possible for product quality to float between the optimum set point and the specified purity for extended intervals. In this case, there is no point in attempting to control *at* the specification limit, since operating at the boilup constraint would be the most favorable condition that can be reached. The only additional features required beyond the simple override system then would be deviation alarms between the product-quality measurement and its specification limits.

Restructuring Quality-control Loops When controlling the qualities of both products in a two-product column, one controller will manipulate a product flow and the other an energy flow. The choice of which composition controller manipulates a product and which manipulates an energy stream is contingent on considerations given in the next chapter. But when limits on energy flow are imposed due to a column or heat-transfer constraint, control of the quality assigned to that manipulated variable will be lost.

For many if not most two-product columns, one product is more valuable than the other. When control over one product's quality must be relinquished, the less valuable product is usually chosen. This may or may not be the one that manipulates the energy flow, however. So the possibility exists that the control configuration that is optimum for normal operation may be unsatisfactory under constrained operation.

A good example of this situation is the separation of styrene from ethylbenzene. Bottom-product styrene is to be controlled at 99.5 percent purity, while the ethylbenzene distillate may contain as much as 10 percent styrene. To minimize interaction between these control loops, the styrene-quality controller should manipulate boilup and the ethylbenzene-quality controller should manipulate distillate flow. When a constraint is reached, however, boilup will be limited and control of styrene quality lost. But styrene is the salable product, whereas ethylbenzene is only to be recycled for further cracking to styrene. So in constrained operation, styrene-quality control should be transferred to distillate flow and ethylbenzene-quality control sacrificed. Although this may be an inefficient mode of operation, it is still preferable to the failure of the sale product to meet specifications.

The restructuring must include several features. Bottom-product quality does not respond to distillate or reflux manipulation without the assistance of boilup, either under accumulator-level or column-pressure control. So the type of control-loop structure described in Fig. 9.14 cannot be applied in this case. To retain the necessary speed of response for control of bottom composition, boilup cannot be limited. The only choice remaining is to limit reflux and place boilup under control of accumulator level (or column pressure if the accumulator is flooded or nonexistent). Then in the *normal* mode of operation, top-product quality would be controlled by distillate flow while bottom-product quality would be controlled by reflux. Because of the direct influence of reflux flow on accumulator level (or pressure), boilup would be manipulated indirectly and thereby promote the desired response in bottom-product quality.

The imposition of a constraint on reflux would then force the system to transfer bottom-quality control to distillate flow. Because distillate flow has a similar—although reduced—effect on accumulator level (or pressure), the response of bottom quality would not be lost. Then the problem to be solved is how to transfer bottom-quality control from reflux to distillate.

Along with this transfer, the feedforward model must be adjusted for constant-reflux conditions. A set of hypothetical conditions for the styrene-ethylbenzene column is given in Table 9.1 to determine how the constraint on reflux affects the material balance.

The set of values for distillate flow is plotted against feed rate in Fig. 9.15. The departure from linearity at the point where the reflux limit is imposed is apparent. Although the relationship between D and F is nonlinear above this point, it can be modeled reasonably well with a straight line. The curvature increases with feed rate, but the loss in styrene recovery (the difference between the curve and the extension of the line) becomes so great that there is no point in increasing F beyond 0.8.

TABLE 9.1 Operating Conditions for Constrained Reflux with $x = 0.005$ and $z = 0.50$

F	V	V/F	y	D	L
0.5	1.5	3.0	0.95	0.262	1.24
0.6	1.8	3.0	0.95	0.314	1.49
0.7	2.1	3.0	0.95	0.367	1.73
0.75	2.14	2.85	0.926	0.403	1.73
0.8	2.18	2.72	0.899	0.443	1.73

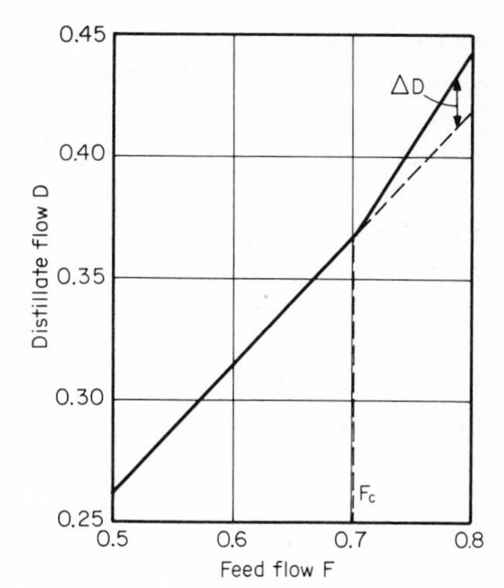

figure 9.15 *The departure from the straight line represents loss in bottom-product recovery under constrained reflux.*

The straight line itself is represented by the feedforward model:

$$D^* = m_y F \tag{9.5}$$

where m_y is the output of the distillate-composition controller. Beyond the constraint, an additional increment ΔD^* is required:

$$\Delta D^* = k(F - F_c) \tag{9.6}$$

where F_c represents the feed rate where limiting begins. This increment may be developed by the forward loop setting reflux from feed, which in the unconstrained mode generates the reflux set point:

$$L^* = m_x F \tag{9.7}$$

where m_x is the output of the bottom-composition controller.

In the constrained mode, reflux flow L is prevented from following L^*. In fact, the constrained L corresponds to the feed rate at which the limit is imposed:

$$L = m_x F_c \tag{9.8}$$

Then the difference between L^* and L can be used to generate ΔD^*:

$$L^* - L = m_x(F - F_c) \tag{9.9}$$

$$\Delta D^* = \frac{k}{m_x}(L^* - L) \tag{9.10}$$

The relationship that describes the required distillate flow for both normal and constrained operation can then be written:

$$D^* = m_y F + G(L^* - L) \tag{9.11}$$

where G is the coefficient applied to the observed departure of reflux from its projected set point.

This relationship is implemented by the system shown in Fig. 9.16. As long as there is no difference between reflux flow and its calculated set point, the two control loops function independently. But when a deviation develops, the distillate-composition controller is transferred to manual by the alarm unit, retaining its last output value, m_y. Then the distillate set point is adjusted as the sum of m_yF and the properly weighted reflux deviation per Eq. (9.11). The bottom-composition controller then adjusts distillate flow instead of reflux and is therefore able to carry on its function. Both the feedforward model and the feedback control requirements are thereby satisfied by this configuration.

Note that all the flow signals in Fig. 9.16 appear in the squared form rather than being linear as suggested by Fig. 9.15 and the equations given above. In practice, there is very little difference in form between the two cases. In the normal operating mode, D^{2*} is linear with F^2; in the constrained mode, the use of L^2 and L^{2*} actually improves the fit of the model to the material balance. Therefore, the added cost of square-root extraction for the three flow signals is not warranted.

MAXIMIZING PRODUCTIVITY

Maximum productivity demands operation against constraints. While the productivity of a single column may be maximized by the systems described earlier in this chapter, multicolumn management is more complex. For example, a single column may limit the productivity of the entire unit, depending on the configuration of that unit. But other opportunities may exist to effectively allocate feedstocks between parallel trains and even to balance the

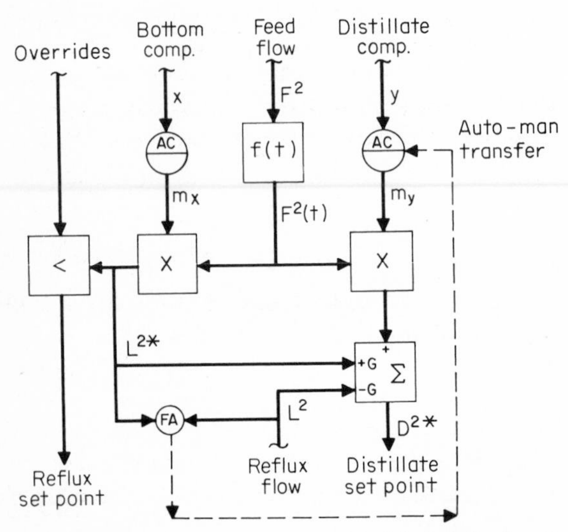

figure 9.16 *When reflux is constrained from following its calculated set point, manipulation of distillate flow is transferred to the bottom-composition controller.*

loading between columns in series by adjusting off-key components. Some of these schemes become possible when enough information is available on the locations of the constraints of all the columns in the unit.

Constraint Projection To this point, most of the constraints common to distillation have been described and systems for imposing them on the manipulated variables have been presented. For the most part, no control action was taken until the constraint was actually reached. But an equally valuable function would seem to be the prediction of which feed rate will bring about a constrained condition and which constraint will be encountered first. This concept is called "constraint projection," i.e., locating constraints by projecting current operating conditions using mathematical models of the process.

Projection of the reboiler capacity Q_r may be made, based on its current valve position m and the present heat input Q:

$$Q_r = Q + k_r(100 - m) \tag{9.12}$$

Here k_r is a coefficient selected to match the slope of Q versus m over the upper range of valve travel and m is in percent of full scale. The valve, in combination with its piping, may in fact give a nonlinear relationship between Q and m. Nonetheless, the simplicity of Eq. (9.12) is justified from the standpoint that its accuracy improves as 100 percent is approached. Furthermore, as m departs from 100 percent, the constraint becomes less meaningful.

Note that the capacity of the heat-input valve is projected as its heat-input limit Q_r expressed in Btu/h or equivalent steam flow. In most cases, this is the most recognized measure of column capacity regardless of whether the constraint exists in the reboiler, condenser, or column.

For multiple reboilers supplied by a single oil heater, a minimum-energy system like the one in Fig. 6.6 might be applied. Then the position of an individual heat-input valve would not represent an absolute limit on capacity but would be relative to that of the most open valve. The capacity of the most open valve would in turn be relative to the temperature of the oil compared to the maximum allowable temperature. For the oil heater, the limit on heat output Q_{oM} would be related to the current heat flow Q_o by the difference between the current oil temperature T_o and its limit T_{oM}:

$$Q_{oM} = Q_o + k_o(T_{oM} - T_o) \tag{9.13}$$

The contribution of the additional capacity of the heater to that of the individual valve can be included by multiplying by the ratio Q_{oM}/Q_o:

$$Q_r = \frac{Q_{oM}}{Q_o}[Q + k_r(100 - m)] \tag{9.14}$$

The condenser limit Q_c can be estimated as a valve-position limit, exactly as done with the heat-input valve, if the pressure is controlled. If the pressure is floating, the valve is normally not controlling, so condenser capacity is then estimated as a function of the difference between current pressure p and its limit p_M. Figure 9.17 shows a typical curve of condenser heat transfer versus

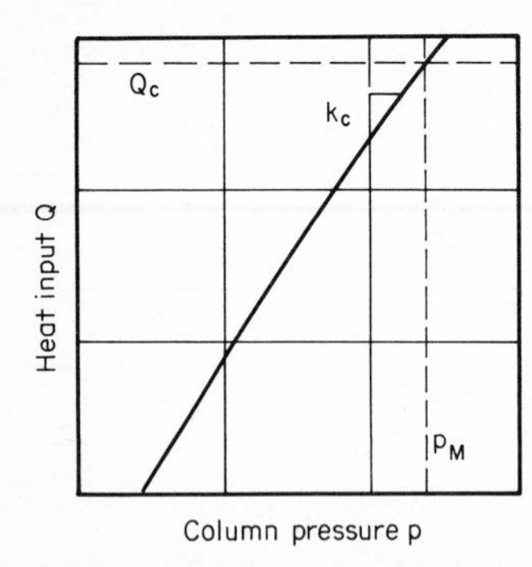

figure 9.17 *The condenser limit can be estimated from present operating conditions if the slope of the curve is known.*

pressure in a column. It is based on the abridged vapor-pressure curve of Fig. 1.6, wherein the temperature rise above the coolant temperature was plotted against vapor pressure for isobutane. A change in coolant temperature causes that curve to shift right or left so that its actual position at any time is quite variable although its slope is constant. Figure 9.17 was obtained by multiplying ΔT from Fig. 1.6 by a constant heat-transfer coefficient and area. If they are constant, then the slope of the curve in Fig. 9.17 is constant. As with the control valve, a linear approximation is sufficient:

$$Q_c = Q + k_c(p_M - p) \tag{9.15}$$

Again, errors in the slope estimate become smaller as the limit is approached.

The slope k_c could change because of fouling, but this is more likely to occur in a reboiler than a condenser. For a fan condenser, the slope will change with the number of fans in service. If the operator has not started all the fans, however, there can be no danger of a condenser limit being approached, so the question becomes academic. The same is true when rain is falling.

The differential-pressure limit may be also estimated, considering the column as a flowmeter wherein differential pressure varies with the square of boilup. Then the maximum heat input Q_d can be estimated from the ratio of the measured differential Δp to the maximum allowed, Δp_M:

$$Q_d = Q\sqrt{\frac{\Delta p_M}{\Delta p}} \tag{9.16}$$

Again, the accuracy of the estimate improves as the constraint is approached.

Differential pressure is not the only index of column flooding. When downcomers flood so that liquid is prevented from flowing down the column, its buildup on the trays does raise the differential pressure sharply. But in the case of excessive entrainment or "jet flooding," product quality may begin to deteriorate without a corresponding increase in differential pressure.

To protect against jet flooding, the estimated flood line for a column may be located as was done in Fig. 1.8. Actually, these curves represent limits of mass vapor boilup versus pressure. They must be modified by the latent heat of vaporization to generate corresponding curves of heat-input limit versus pressure.

Figure 9.18 contains a plot of the heat-input limit versus pressure for propylene, isobutane, and isopentane. Note that each mixture will have its own curve, heavier materials having a more positive slope and lighter materials exhibiting a negative slope. The limits of boilup were determined as described by Eqs. (1.22) to (1.24), with correction then applied for the variation of latent heat with pressure.

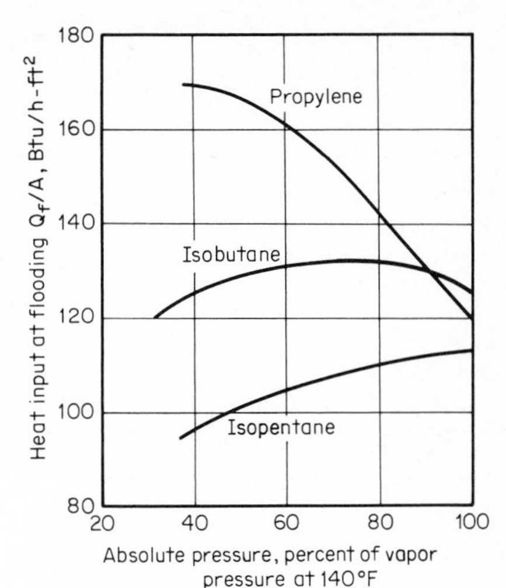

figure 9.18 *The heat-input limits for propylene, isobutane, and isopentane are compared on a relative-pressure scale.*

For simplicity, a linear flooding model can be constructed from one of these curves by estimating the flooding limit Q_f related to some established reference point Q_M and p_M:

$$Q_f = Q_M - k_f(p_M - p) \qquad (9.17)$$

The slope of the flood line is designated as k_f.

For all these components cited, maximum throughput will be achieved at the minimum attainable pressure, i.e., where the flood line and condenser constraint cross. This was illustrated by the operating window in Fig. 1.9. The window is duplicated in Fig. 9.19 using linear models for both curves.

The flood limit, i.e., where the lines cross, can be found by solving Eqs. (9.15) and (9.17) simultaneously. In Eq. (9.15), however, Q_f and p_f must be substituted for Q_c and p_M; in Eq. (9.17), p becomes p_f. For the condenser, then,

$$Q_f = Q + k_c(p_f - p) \qquad (9.18)$$

and for the column,

$$Q_f = Q_M - k_f(p_M - p_f) \qquad (9.19)$$

Combining the two by eliminating p_f yields

$$Q_f = \frac{Q_M - k_f(p_M - p + Q/k_c)}{- k_f/k_c} \qquad (9.20)$$

Using Eq. (9.20), the flood limit may be estimated for any operating condition of p and Q, given the other terms as constants.

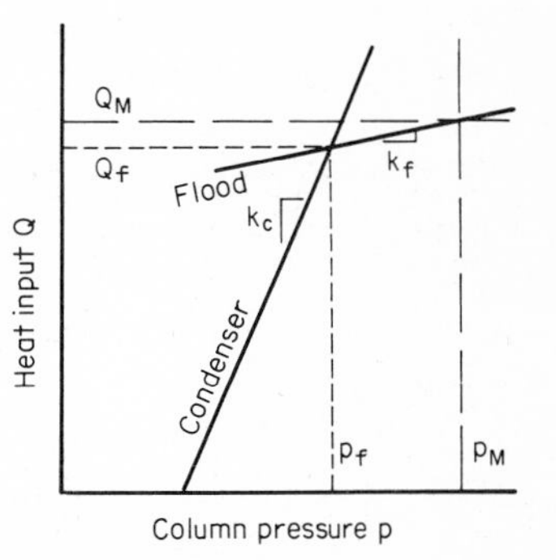

figure 9.19 *The flood limit is located at the intersection of the flood line and the condenser constraint.*

Estimating Column Capacity Once all the constraints have been located, the calculated limits must be compared to determine which will be the first likely to be imposed:

$$Q_l = \text{lowest of } (Q_r, Q_c, Q_d, Q_f) \tag{9.21}$$

Then the capacity of the column to accept more energy and hence process more feed is

$$\text{Capacity} = \frac{Q}{Q_l} \tag{9.22}$$

where Q is the current heat flow into the reboiler.

If Q/Q_l is less than 100 percent, the column has additional capacity to process feed—the margin between Q/Q_l and 100 percent. But when a constraint is reached, the override controls or self-imposed limits will prevent Q from rising above Q_l. Consequently, Eq. (9.22) is only valid up to 100 percent capacity.

What would provide a useful measure of overcapacity depends on what strategy is applied when capacity is exceeded. If control of one of the product compositions is simply forsaken, a measure of overcapacity can be obtained by comparing the current (limited) heat input (or reflux) to that needed to maintain control. In Figs. 9.2 and 9.3, the deviation between desired heat input Q^* and actual Q can be used:

$$\text{Capacity} = \frac{Q^*}{Q} \tag{9.23}$$

Or for the column with limited reflux in Fig. 9.16, the desired reflux L^* may be related to the actual value L:

$$\text{Capacity} = \frac{L^*}{L} \tag{9.24}$$

Note that neither (9.23) nor (9.24) is valid for operation *below* 100 percent capacity. To provide a relationship applicable to both operating regimes, Eq. (9.22) must be combined with (9.23) or (9.24):

$$\text{Capacity} = \frac{Q}{Q_l}\frac{Q^*}{Q} = \frac{Q^*}{Q_l} \tag{9.25}$$

During undercapacity operation, Q follows Q^*, whereas above 100 percent capacity, Q follows Q_l; therefore, (9.25) satisfies both conditions.

When the composition controller is in fact not manipulating heat input (or reflux) because of imposed limits, Q^* (or L^*) may not exactly represent the demand, however. Equation (9.2) describes how the deviation between measurement and set point for the primary controller relates to that of the secondary controller. Substituting the deviation in product quality $x - x^*$ for e in (9.2) gives

$$x - x^* = \frac{100}{P}(Q^* - Q) \tag{9.26}$$

In actual practice, a deviation $x - x^*$ can be corrected by an appropriate change $Q^* - Q$, although the proportionality is not necessarily the gain of the feedback controller, $100/P$.

Using the relationship between boilup-feed ratio and the separation factor, composition x can be related to current heat input Q:

$$\frac{Q}{F} = \beta H_V \ln \frac{y(1-x)}{x(1-y)} \tag{9.27}$$

Assuming that $1 - x$ and y are nearly 1.0, their contribution is insignificant; then for y controlled by material balance, (9.27) can be reduced to

$$\frac{Q}{F} \approx -\beta H_V \ln x(1-y) \tag{9.28}$$

To achieve a composition x^*, Q^* would be required:

$$\frac{Q^*}{F} \approx -\beta H_V \ln x^*(1-y) \tag{9.29}$$

Then the overcapacity factor Q^*/Q is the ratio of (9.29) to (9.28):

$$\frac{Q^*}{Q} = \frac{\ln x^*(1-y)}{\ln x(1-y)} \tag{9.30}$$

Alternatively, the difference between Q^* and Q may be found:

$$Q^* - Q = \beta H_V F \ln \frac{x}{x^*} \tag{9.31}$$

It can be seen that the form of Eq. (9.31) differs from that of (9.26) and they therefore cannot be equivalent expressions. For slight departures of x from x^*, their derivatives could be the same, but even this is unlikely. The derivative of (9.27) for x with respect to Q or Q/F is in fact the steady-state gain of the process. The proportional gain of the controller would be set equal to the reciprocal of the steady-state gain of the process only if there were little dynamic attenuation by a dominant lag in the process. While this is possible, it is unlikely. Consequently, the controller gain can normally be set higher than the process steady-state gain would allow, such that Q^* is likely to be greater than what would actually be required to produce x^* in a constrained situation.

For a more accurate estimate of Q^*, Eq. (9.30) could be solved directly—it requires relatively little information about the process and could be approximated for small deviations. Figure 9.20 has been prepared to show the relationship between overcapacity and the resulting composition deviation using the data in Table 9.1. Although distillate composition is uncontrolled above the reflux constraint, its deviation from the control point of $y^* = 0.95$ can be used to estimate the additional energy required to return to that point. Since the relationship seems to be reasonably linear, overcapacity estimation can be then reduced to the expression

$$\frac{Q^*}{Q} = k_x(x - x^*) \tag{9.32}$$

or

$$\frac{Q^*}{Q} = k_y(y^* - y) \tag{9.33}$$

depending on which composition is sacrificed.

If the estimate of capacity is reasonably accurate, the operator should be able to make a corresponding adjustment in feed rate to approach 100 percent capacity. If the capacity estimate were to be 87 percent, for example, the operator should be able to increase feed 13 percent (of value) without losing control. Or if 105 percent capacity is indicated, control should be restored by reducing feed rate by 5 percent (of value). All the estimates become more accurate as 100 percent capacity is approached.

Balancing Load between Columns The operating guides developed above should be very useful in helping to balance the feed streams to parallel trains of columns. While it might seem that this is the sort of task which only has to be done once, this is rarely the case. Few parallel trains are identical either in capacity or details of construction. Quite often, the second train was built after the first and is larger, even having a different source of heating and cooling. Sometimes the columns of the second train are added one at a time, in which case constraints are changed each time a new column is added. In the final analysis, the trains tend to have different limits, changing with ambient conditions, feed composition, etc., often requiring cross flow at selected points between them. Then rebalancing may be required often and the capacity estimates will be found quite useful.

Columns aligned serially within a train may also be balanced. This opportunity was presented in Chap. 3 using the multiple-column unit in Fig. 3.1 as an illustration. The deisobutanizer in that configuration was at the end of the train. The off-key components in its products, i.e., propane in the distillate and isopentane in the bottom, have a profound effect on the separation requirement if a total purity specification has to be met on each product.

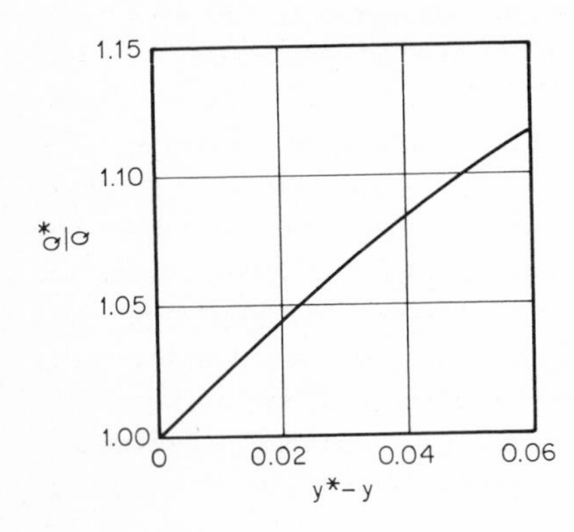

figure 9.20 *Overcapacity may be estimated from the deviation sustained in product quality.*

Thus an increase in propane content requires that less n-butane be allowed in the isobutane product if the specification of, say, 95 percent purity is to be met. Then a reduction in separation at the depropanizer would allow more propane to reach the deisobutanizer, where the separation must subsequently be increased. There may be an economic incentive to reach an optimum propane content based on its value relative to isobutane and the cost of separation in each tower. But if there is not, the amount of propane leaving the depropanizer can be adjusted to balance the loading of the two columns. By adjusting the isopentane content of the debutanizer overhead, it is also possible to balance its load against the deisobutanizer. Consequently, without adding or removing any intermediate streams all three of these columns can probably be operated at the same capacity. Following this line of reasoning, the heavy-ends columns of the same unit could also be balanced.

The load distribution for this or any multicolumn unit is heavily dependent on feed composition. So it is important for the operator to know the percent loading on each column if he is to maximize the capacity of the unit. Then he can use relationships such as those given in Eq. (9.30) and Fig. 9.20 to solve for the *composition* change which will bring about a required change in capacity. Serial balancing can then be done on a calculated basis.

Production Scheduling There is no need to wait until a constraint is reached before making an adjustment to the charge rate or to intermediate compositions. An analysis of the charge stock at any point in time can be used to predict the unadjusted loading on all the columns. Then if the estimate indicates a potential overload on any column, an intermediate composition may be found to distribute the load more equitably. As in any feedforward system, the adjustment must be programmed on a timed basis to coincide with the arrival of the new feedstock to the affected column.

No new technology is needed to reach this goal. The material balances that have been used throughout this book apply, as do the relationships between boilup and separation. Where floating pressure is applied, the effect of pressure on separation for anticipated ambient conditions should be taken into account. Then the maximum advantage will be gained from diurnal variations in condenser duty. A plant can then anticipate a token production increase every night. Columns equipped with water-cooled condensers can be adjusted to accept some of the load of columns with air-cooled condensers during the day, relinquishing it at night.

Managing a multicolumn unit in this manner may seem too complex at first thought. Although the required calculations can be made by hand for one or two columns, beyond this a computer is required. A computer can be programmed with a multicomponent model of each column in the unit, using the multicomponent material and separation equations described in Chap. 3. Having calculated all flow rates and boilup rates for each tower, the computer may adjust feed rate, cross flows, and intermediate compositions to maximize the production of the unit. Nonlinear programming is the method needed to adjust the multiple parameters against the boilup constraints successively until maximum production is achieved. A computer can converge on a solution in minutes, whereas it may take the separation unit itself hours to respond completely to a single change in feed rate. The time lags in the various columns may, in fact, prevent equilibrium from ever being reached within the diurnal cycles that affect the plant. This is a natural limit to the prospects of multicolumn optimization.

REFERENCES

1. Shinskey, F. G.: Process Control Systems with Variable Structure, *Control Eng.*, August 1974.
2. Buckley, P. S., and R. K. Cox: New Developments in Overrides for Distillation Columns, *ISA Trans.*, vol. 10, no. 4, 1971.
3. Manter, D., and R. Tressler: A Coal-Air Ratio Control System for a Cyclone Fired Steam Generator, *ISA Paper* 1-CI-61.

CHAPTER TEN

Interaction and Decoupling

Any process with more than a single control loop has the possibility of interaction between its loops. And interaction produces some strange results. When there are two valves to be manipulated by two controllers, there are two ways of connecting those single loops, one pairing more effective than the other. If one pairing produces very little interaction, i.e., the loops operate as if they were nearly independent of each other, the opposite pairing will be almost totally ineffective. It is also possible that both valves have similar influence over both controlled variables, in which case either pairing will produce essentially the same results. But because of their similar response, interaction will then be maximum, and operation of the unit will be destabilized when both controllers are in automatic.

The situation of two interacting loops is readily described using the relative-gain concept described in the following pages. Choices are limited and directions are clear. But distillation columns have as many as four to six manipulated and controlled variables. The number of possible configurations of single control loops increases factorially with the number of variables. While there are only two possible single-loop pairs of two controlled and manipulated variables, there are 6 possible pairs of three loops, 24 possible pairs of four loops, and 120 possible configurations of five single loops. This is the principal reason why the design of control systems for distillation is so difficult. Furthermore, the best pairing for one column may not be the best for another since interaction depends heavily on relative flow rates, compositions, etc. Through the use of the relative-gain concept, the degree of interaction can be estimated as a guide to the designer in structuring his control system even before the column itself has been sized.

The discussions on interaction assume an equal number of controlled and manipulated variables. There are, of course, two other possibilities: there could be an excess of either controlled or manipulated variables. The case of an excess of controlled variables has already been presented in the previous chapter. The active manipulated variables are shared among the controlled variables by simple selection or restructuring as needed to satisfy the most pressing objectives. Where there is an extra manipulated variable, it offers the possibility of adjustment to obtain some optimum economic performance of the unit. This subject is covered in the next chapter.

THE RELATIVE-GAIN CONCEPT

Earlier in the book, attention was given to process steady-state and dynamic gains and their influence over the settings of the controllers. But all this was done on a single-loop basis, i.e., as if there were no outside influences upon the loop under investigation. The presence of other control loops may alter the response of that single loop when they are closed, i.e., when their controllers are in automatic.

If the steady-state or dynamic gain of a given controlled variable in response to a given manipulated variable changes when other loops are closed, then interaction exists in the system. This interaction has an obvious effect on controller tuning. If the controller in question is tuned with all others in manual, that tuning will be incorrect when the others are placed in automatic because of their influence over process gain. Interaction can then promote instability and degrade response in relation to its severity.

It then becomes important to evaluate the expected interaction between control loops numerically, both as a guide to configuring their structure and to predict problem areas. The evaluation of loop interaction on the basis of "relative gain" has been presented by Bristol [1] and the author [2, 3]. The technique consists in comparing the process gain for a given pair of controlled and manipulated variables with all other loops open to the gain obtained with all other loops closed. The ratio of these two gains is a dimensionless term heretofore defined as the relative gain.

Determining Relative Gains Let the process gain for a given pair of controlled (c_i) and manipulated (m_j) variables be designated $\partial c_i/\partial m_j$. This gain could be evaluated either by test or by mathematical modeling in either of two situations—with all other loops open or with all of them closed. When all other loops are open, then all other manipulated variables are constant; when all other loops are closed, then all other controlled variables are considered constant. Then the relative gain λ_{ij} for the selected pair of variables c_i and m_j is defined as $\partial c_i/\partial m_j$ with all m's constant divided by $\partial c_i/\partial m_j$ with all other c's constant:

$$\lambda_{ij} \equiv \frac{\left.\dfrac{\partial c_i}{\partial m_j}\right|_m}{\left.\dfrac{\partial c_i}{\partial m_j}\right|_c} \tag{10.1}$$

The relative gains for a set of manipulated and conrolled variables are arrayed in a square matrix as shown below:

$$\Lambda = \begin{array}{c} \\ c_1 \\ c_2 \\ \\ c_i \\ \\ \\ \end{array} \begin{array}{c} \begin{array}{ccccc} m_1 & m_2 & \cdots & m_j & \cdots \end{array} \\ \left[\begin{array}{ccccc} \lambda_{11} & \lambda_{12} & \cdots & \lambda_{1j} & \cdots \\ \lambda_{21} & \lambda_{22} & \cdots & \lambda_{2j} & \cdots \\ \cdots & \cdots & \cdots & \cdots & \cdots \\ \lambda_{i1} & \lambda_{i2} & \cdots & \lambda_{ij} & \cdots \\ \cdots & \cdots & \cdots & \cdots & \cdots \end{array}\right] \\ \begin{array}{ccccc} 1.0 & 1.0 & \cdots & 1.0 & \cdots \end{array} \end{array} \begin{array}{c} \\ 1.0 \\ 1.0 \\ \\ 1.0 \\ \\ \\ \end{array}$$

The array has the very helpful property of having the numbers in every row and column summing to 1.0. This is particularly valuable in a two-loop system in that only one relative-gain term need be estimated, Another useful property of the concept is that it is unaffected by scaling or, in fact, nonlinearities. Because the process gain is, in effect, divided by itself under equivalent conditions, these factors disappear.

The significance of the relative-gain numbers can easily be demonstrated. If a process gain is the same with or without all other loops in automatic, then those other loops have no effect on it and hence there is no interaction. In this case, $\lambda = 1.0$. As λ departs from 1.0 in either direction, increasing interaction is indicated. One extreme has $\lambda = 0$. A relative gain of zero can be obtained only when the numerator in Eq. (10.1) is zero. This indicates that c_i is unaffected by m_j regardless of the presence of other loops and therefore cannot be controlled by it. In summary, pairs of variables having relative gains near 1.0 represent controllable combinations while those having relative gains approaching zero do not.

The other extreme is $\lambda = \infty$. This is indicated when the denominator in Eq. (10.1) is zero. Then the selected pair of variables is controllable only in the absence of other closed loops—the closure of another loop reduces the process gain to zero. Furthermore, the presence of ∞ in one location in the array requires two additional gains of $-\infty$ and another $+\infty$ to make all rows and columns add to 1.0. Consequently, the interaction existing in the process in this case prevents both those loops from being closed at the same time.

The process gains inserted in Eq. (10.1) are most often steady-state gains. Although they are the easiest numbers to obtain on the process, the concept is by no means limited to the steady state. For the terms to be meaningful, however, they should all be evaluated under the same conditions. If the gain terms used are dynamic in nature, they should all be evaluated at the same period or frequency for them to be compared. In the case of a 2×2 system, there are four possible control loops and hence four entries in the relative-gain array. But each loop could have a different period of oscillation. For a complete analysis of the system, the entire array should be evaluated at each of the four periods plus the steady state.

In most cases, however, evaluation of dynamic gains is unnecessary. It is obviously important that steady-state stability be guaranteed, so relative-gain determinations should always be made for the steady state. A more detailed consideration of dynamic properties is presented later in the chapter.

The relative-gain array can also be found by using only a set of gains for all loops open or only a set of gains for all other loops closed. Consider a set of gains $(\partial c_i/\partial m_j)_m$ for all loops open arrayed in a matrix designated **M**. If matrix **M** is inverted and transposed, its elements become the inverse of the gains for all other loops closed, that is, $(\partial m_j/\partial c_i)_c$. Then the relative gains λ_{ij} can be found by *multiplying* each element in the original matrix **M** by the corresponding element in its transposed inverse:

$$\lambda_{ij} = \left.\frac{\partial c_i}{\partial m_j}\right|_m \left.\frac{\partial m_j}{\partial c_i}\right|_c \tag{10.2}$$

Conversely, if process gains are known only in terms of the controlled variables, that is, $(\partial m_j/\partial c_i)_c$, they may be arrayed in a matrix identified as **C**. Then when matrix **C** is inverted and transposed, the elements become $(\partial c_i/\partial m_j)_m$. The elements of the relative-gain array are then obtained by multiplying each element in original matrix **C** by the corresponding element in its transposed inverse:

$$\lambda_{ij} = \left.\frac{\partial m_j}{\partial c_i}\right|_c \left.\frac{\partial c_i}{\partial m_j}\right|_m \tag{10.3}$$

As an illustration of this concept, consider a process wherein the controlled variables are described in terms of the manipulated variables and their associated gains:

$$c_1 = a_{11}m_1 + a_{12}m_2$$

$$c_2 = a_{21}m_1 + a_{22}m_2$$

where a_{ij} represents $(\partial c_i/\partial m_j)_m$. The array of gains is presented as

$$\mathbf{M} = \begin{array}{c|cc} & m_1 & m_2 \\ \hline c_1 & a_{11} & a_{12} \\ c_2 & a_{21} & a_{22} \end{array} \tag{10.4}$$

The multiplication of each element in **M** by the corresponding element in its transposed inverse yields $\mathbf{\Lambda}$, whose first element is

$$\lambda_{11} = \frac{a_{11}a_{22}}{a_{11}a_{22} - a_{12}a_{21}} \tag{10.5}$$

Similarly, the manipulated variables may be described in terms of the controlled variables as

$$m_1 = b_{11}c_1 + b_{12}c_2$$

$$m_2 = b_{21}c_1 + b_{22}c_2$$

In this case, the gains b_{ij} represent $(\partial m_j/\partial c_i)_c$ and are arrayed as

$$\mathbf{C} = \begin{array}{c|cc} & c_1 & c_2 \\ \hline m_1 & b_{11} & b_{12} \\ m_2 & b_{21} & b_{22} \end{array} \tag{10.6}$$

Multiplication of each element in **C** transposed by the corresponding element in **C** inverse again yields $\mathbf{\Lambda}$, whose first element is

$$\lambda_{11} = \frac{b_{11}b_{22}}{b_{11}b_{22} - b_{12}b_{21}} \tag{10.7}$$

Note that the form of (10.7) is identical to (10.5), allowing the same procedure to be followed with either set of process information.

Positive Interaction The most common type of interaction experienced in process control is called "positive interaction," where *all* relative-gain terms in the array are positive. Positive interaction is most likely to develop between dissimilar controlled variables, e.g., flow and composition, level and temperature, etc.

The example which most readily illustrates positive interaction is the blending process of Fig. 10.1. Stream D containing y fraction of a particular component is blended with stream B containing x fraction of that component to form a blend of flow F and fraction z. The total and component balances are already familiar:

$$F = D + B \tag{10.8}$$

$$Fz = Dy + Bx \tag{10.9}$$

The steady-state gain of F with respect to D, with B constant, is the derivative of (10.8):

$$\left.\frac{\partial F}{\partial D}\right|_B = 1 \tag{10.10}$$

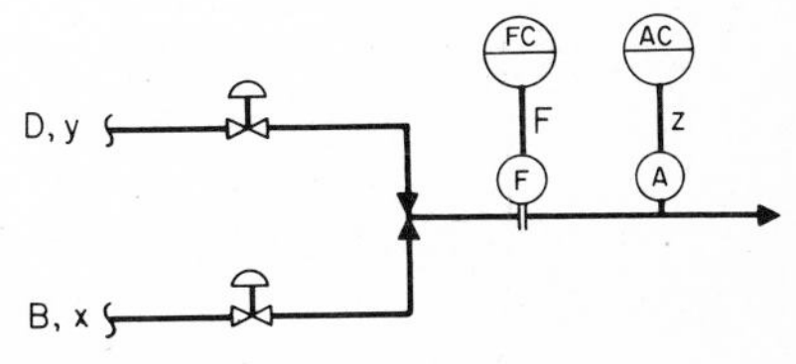

figure 10.1 *Positive interaction exists between flow and composition in this example.*

To obtain the steady-state gain of F with respect to D, with B manipulated to control z, requires that B in (10.8) be replaced by z from (10.9). Combining the two balances yields

$$F = D\,\frac{y - x}{z - x} \tag{10.11}$$

whose derivative is

$$\left.\frac{\partial F}{\partial D}\right|_z = \frac{y - x}{z - x} \tag{10.12}$$

Then the relative gain of F to D may be found:

$$\lambda_{FD} = \frac{\left.\dfrac{\partial F}{\partial D}\right|_B}{\left.\dfrac{\partial F}{\partial D}\right|_z} = \frac{z - x}{y - x} \tag{10.13}$$

When all terms in the array are entered, it appears as

$$\Lambda = \begin{array}{c|cc} & D & B \\ \hline F & \dfrac{z - x}{y - x} & \dfrac{y - z}{y - x} \\ z & \dfrac{y - z}{y - x} & \dfrac{z - x}{y - x} \end{array} \tag{10.14}$$

As z approaches y, Λ approaches

$$\begin{array}{c|cc} & D & B \\ \hline F & 1 & 0 \\ z & 0 & 1 \end{array}$$

indicating that flow is almost totally dependent on D and composition on B. But this is reasonable: if z is nearly equal to y, B must be a relatively small flow and would therefore have little influence over F. On the other hand, as z approaches x, the elements reverse and B assumes the primary influence over flow.

One lesson to be learned from this example is that the relative gains are sensitive to feed compositions and set points. Therefore, a control system that will satisfy one process may not satisfy another—each deserves individual evaluation.

In the event that z happens to be midway between y and x, all elements are equal:

$$\Lambda = \begin{array}{c|cc} & D & B \\ \hline F & 0.5 & 0.5 \\ z & 0.5 & 0.5 \end{array}$$

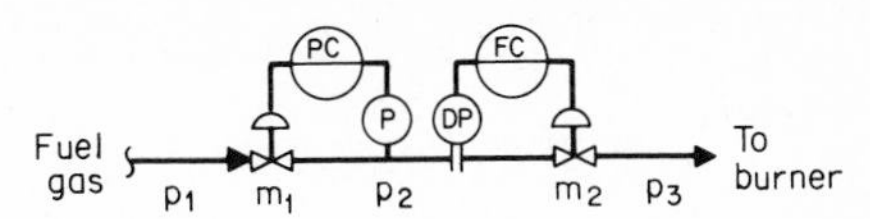

figure 10.2 *This control-loop arrangement may not be optimum, depending on the distribution of pressures in the system.*

This is the most severe interaction possible in a 2×2 system. Each valve has equal effect on each controlled variable, so it does not matter how the single loops are connected—both configurations are equally poor. A discussion on the dynamic effects of interaction is deferred until negative interaction has been presented.

An interacting system quite common to industry appears in Fig. 10.2. One valve is to control the flow of fuel to the burner while the other is to absorb fluctuations in supply and maintain pressure constant at the flowmeter. However, the loop configuration shown has an equal chance of being wrong as being right if no more information is available about the system. If the valves are equal in size, their pressure drops will be about equal and both will affect flow and pressure equally. However, it is possible that the downstream valve is larger, carrying a smaller pressure drop than the upstream valve. An analysis of this system in Ref. 2 indicates that the relative gain of flow for a particular valve is proportional to the pressure drop across that valve:

$$\lambda_{F2} = \frac{p_2 - p_3}{p_1 - p_3} \tag{10.15}$$

Then if valve 2 is larger than valve 1 and $p_2 - p_3$ is therefore smaller than $p_1 - p_2$, λ_{F2} will be less than 0.5. In this case, the configuration shown in Fig. 10.2 would be incorrect—flow would not respond particularly well to valve 2, which would in turn upset pressure. But pressure could not be controlled satisfactorily by valve 1 while inducing large flow changes. With either controller in manual, the other could operate satisfactorily; but with both in automatic, the two variables would tend to drift about erratically. Any attempt of one controller to move its variable toward the set point would cause a larger deviation in the other loop.

Negative Interaction Positive interaction can be expected when the signs of the gain terms for matrix **M** (all loops open) do not balance. In the pressure-flow example, opening valve 1 raises both flow and pressure whereas opening valve 2 raises flow and lowers pressure. There are three positive gains and one negative, and the relative gains are all positive. The same result would be obtained if there were three negative signs and one positive sign.

However, if the signs in **M** are all positive or all negative, or if there are an equal number of each, $\mathbf{\Lambda}$ will have some negative signs. If, in Eq. (10.5), all a's are positive, λ_{11} *must* be either greater than 1.0 or less than zero. Furthermore, the presence of any numbers greater than 1.0 requires a negative number in the same row and one in the same column to sum to 1.0.

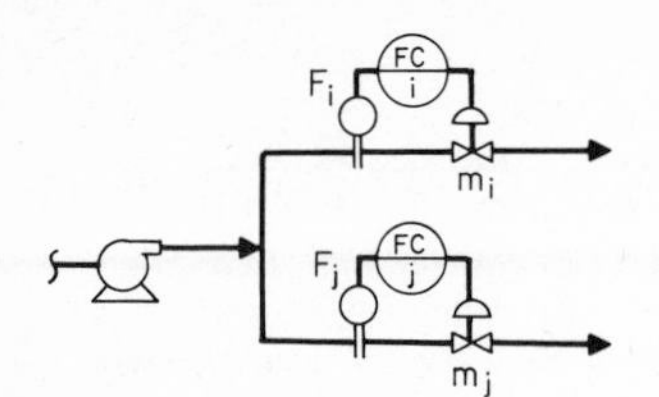

figure 10.3 *Parallel processes are capable of exhibiting negative interaction.*

For a relative gain to exceed 1.0, the denominator in Eq. (10.1) must be less than the numerator. Hence the closure of the other loops acts to *reduce* the gain of the pair being evaluated. For a relative gain to be negative, the response of a given controlled variable to the selected manipulated variable must *reverse direction* when the other loops are closed. Any attempt to close a loop characterized by a negative relative gain is inviting disaster. The action required of the subject controller (e.g., increase input–decrease output) depends on whether the other loops are in automatic or manual. Furthermore, in a two-loop system, with negative interaction, there must be two negative elements, *both* of which must be selected for control if *either* one is selected. Hence the resulting system will contain *two* conditionally stable loops, the second of which will react backward if the first is opened.

An example of negative interaction is the control of flow in parallel streams from a single source, such as a pump or compressor, as shown in Fig. 10.3. A step increase in valve m_i with controller j in manual will increase flow F_i while decreasing flow F_j. The magnitude of the decrease in F_j depends on the sensitivity of the supply pressure to variations in flow.

Each valve will then raise its own flow while reducing the other by a lesser amount:

$$F_i = a_{ii}m_i - a_{ij}m_j \tag{10.16}$$

where $a_{ii} > a_{ij}$ and, if the loops are identical,

$$F_j = a_{ii}m_j - a_{ij}m_i \tag{10.17}$$

Following the procedure developed for converting open-loop gains to relative gains as in Eq. (10.5), the relative gain of the preferred pairs is

$$\lambda_{ii} = \lambda_{jj} = \frac{a_{ii}^2}{a_{ii}^2 - a_{ij}^2} \tag{10.18}$$

For $a_{ii} > a_{ij}$, λ_{ii} must exceed 1.0 and therefore λ_{ij} must be negative. Should $a_{ii} = a_{ij}$, then $\lambda_{ii} = \infty$ and both flows cannot be controlled at the same time.

Dynamic Effects Figure 10.4 illustrates a block diagram of a linear interacting system with dynamic gains g included. Whereas the steady-state gains a may be considered constant, the dynamic gains are a function of period. Also, because of phase relationships they must be considered as vectors. Reference 2 describes the procedure for analyzing the response of an interacting system vectorially. However, the scope of the following presentation is limited to a qualitative evaluation of the dynamic effects of interaction. For purposes of discussion, the elements in the loops closed by each individual controller are called "diagonal" elements while all others are "off-diagonal."

Observe that with controller 2 in manual, the only element in the closed loop through controller 1 is the diagonal $a_{11}\mathbf{g}_{11}$. But when controller 2 is placed in automatic, controller 1 becomes enclosed in a second loop (broken lines) through $a_{21}\mathbf{g}_{21}$, the closed loop of controller 2, and $a_{12}\mathbf{g}_{12}$. Several situations are possible, depending on the response of the off-diagonal dynamic elements relative to that of the diagonal elements. If they are slow in comparison to the diagonal elements, their gains will tend to be low at the natural periods of the loops through the diagonal elements. This will tend to reduce the interaction substantially from the values indicated by steady-state gains a_{12} and a_{21}. In fact, if *either* off-diagonal gain is very low, the interaction is effectively broken. Although controller 1 could affect c_2 through $a_{21}\mathbf{g}_{21}$, for example, a low value of $a_{12}\mathbf{g}_{12}$ could minimize the reaction of the second loop.

While it is possible for the off-diagonal elements to have higher gains than the diagonal elements, this situation requires the reassignment of control loops. Then the off-diagonal elements under the new structure would have lower gains. Another possibility is that a given element may have a lower steady-state gain but a higher dynamic gain than another. This complicates the selection of loops. Reference 3 describes a process having a derivative function as one of the elements such that its response was rapid but its steady-state gain was zero. In actual practice, the system performance using this as a diagonal element was superior, although the other loop had to be closed to achieve steady-state regulation of the process.

Given that the most effective structure is applied, then the most severe interaction will be experienced when all elements are identical. This can be demonstrated with the fuel-gas system in Fig. 10.2, when the pressure drops

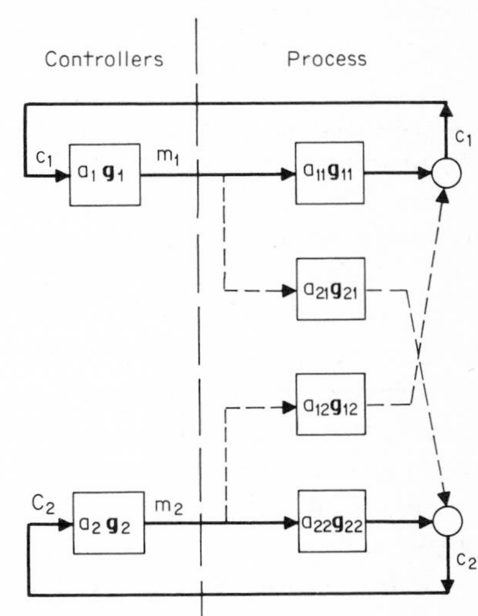

figure 10.4 *The dynamic gains* (**g**) *have considerable influence over the severity of the interaction.*

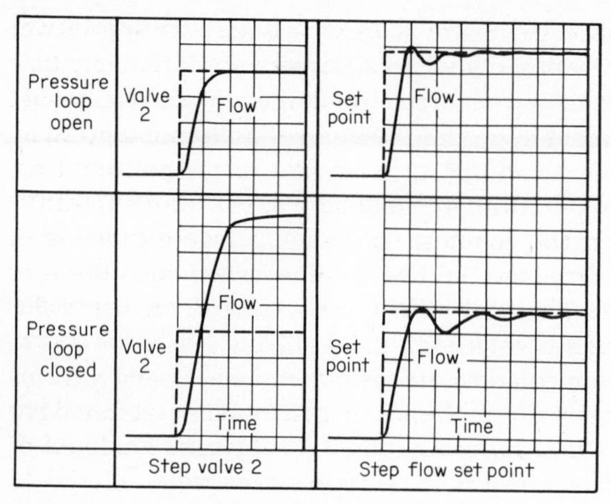

figure 10.5 *Interaction of loops with $\lambda = 0.5$ and identical dynamic response increases their natural period and required mode settings by 50 percent.*

across the two valves are equal. The dynamic response of flow by manipulation of valve 2 is described in Fig. 10.5. The left-hand set of curves compares the open-loop response of flow to valve 2 with and without pressure control. Because the response is twice as great under pressure control, a relative gain of 0.5 is indicated.

The closed-loop response of flow to a set-point change is more difficult to interpret. If the flow controller is tuned with the pressure controller in manual, the desired damped response in the upper-right position of the figure can be achieved. When the pressure loop is closed, valve 2 has twice the steady-state effect on flow, and so the response of the flow loop is altered as indicated by the lower-right record. The secondary effect is slower than the primary effect, however, in that it must traverse through *both* off-diagonal elements. So although the steady-state gain of the flow loop is doubled by the pressure controller, its dynamic gain is not. To return the flow loop to its original damping, the proportional band of the flow controller must be increased by about 50 percent of value. Furthermore, Fig. 10.5 indicates that its period of oscillation is increased about 50 percent by the pressure controller and therefore the reset time of the flow controller must also be increased by the same 50 percent. Since the two controllers are in identical situations, both must be readjusted to the same degree.

Alternatively, one of the controllers could be detuned to the point where the other could be left with its original settings. In effect, the gain of the detuned controller would be reduced sufficiently to all but eliminate the interaction.

Very often one of the loops will be much faster than the other. In the process of Fig. 10.1, the composition loop is typically much slower than the flow loop because of sampling lags and the sluggishness of the analyzer. This is particularly true when there is some volumetric capacity between the flowing streams and the analyzer. Then the mode settings required for stability of the composition loop would prevent its valve from moving rapidly. The flow controller could easily correct for any upsets from that source and would therefore appear to be free of any interaction. However, in manipulating its valve the flow controller would induce upsets in the slower composition loop.

The flow controller would respond the same with or without the composition controller in automatic. But if the composition controller were tuned with the flow controller in manual, it would have to be readjusted when the flow controller is transferred to automatic. When the composition controller moves its valve, the flow controller will readjust its valve to maintain a constant flow. Therefore, the composition controller in effect moves both valves when the flow controller is in automatic. Hence its gain must be adjusted per its relative-gain index—if $\lambda = 0.5$, its proportional band must be doubled. Because the feedback loop through the flow controller is so rapid relative to the response to the analyzer, the natural period of the composition loop is unaffected by the interaction.

Negative interaction introduces some strange dynamic behavior. The response of dynamic gains is shown in Fig. 10.6. From the open-loop response on the left, it can be seen that the off-diagonal elements act to oppose the response of the diagonal elements. When both loops are closed, this action retards the settling time of both loops considerably, as well as reducing their damping.

If the opposite pairing is selected, both loops are characterized by $\lambda = -1.0$. Then the steady-state components of the off-diagonal elements cause a reversal of response, delayed by their dynamic components. The results of such interaction appear in Fig. 10.7. The reversal caused by the interaction is similar to the inverse response described for reboilers in Chap. 6. The inversion time τ_i is similar to dead time, and hence the natural period of the loop with the reversal tends to be about $4\tau_i$. By comparing Figs. 10.6 and 10.7, it can be seen that the reversal increases the natural period of the control loop by about a factor of 4. Therefore, not only are negatively coupled loops condi-

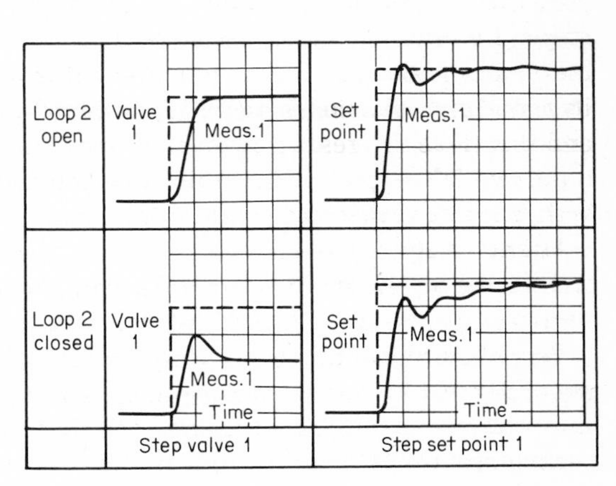

figure 10.6 *Interaction of loops with $\lambda = 2.0$ and identical dynamic response increases their settling time.*

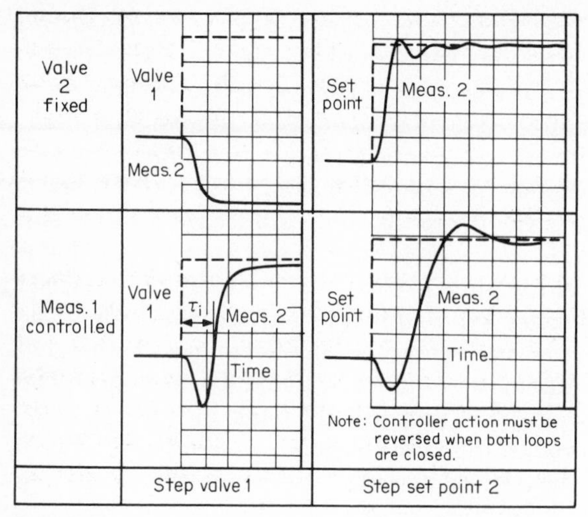

figure 10.7 *Interaction of loops with λ = −1.0 and identical dynamic response extends the period of oscillation through phase reversal.*

tionally stable (depending on whether the other controllers are in automatic) but they are much more difficult to control. The increase in natural period by the factor of 4 means that the integrated error per unit load change increases by a factor of 16. It should be obvious, then, that negatively coupled loops are to be avoided.

INTERACTIONS INVOLVING A SINGLE PRODUCT

There are three principal categories of control loops on distillation columns: flow, inventory, and product quality. Flow controllers can interact with one another in parallel service, although detuning can essentially eliminate any potential instability and no disturbance is passed on to the other loops. Flow controllers can also upset inventory and quality-control loops, and inventory loops can interact with each other and with quality-control loops. But the most serious form of interaction encountered in distillation is between individual quality-control loops. Their similar and sensitive responses can create grave stability problems, even when the other loops on the column are well regulated. Therefore, the coupling between quality-control loops is presented separately—following an examination of lesser degrees of interaction in columns.

Reducing the Matrix When controlling the quality of a single product, four variables require regulation: that quality, accumulator level, base level, and pressure. However, five valves are usually available: distillate, reflux, condenser, reboiler, and bottom product. The fifth valve would customarily set the heat load on the column—so in effect there are five controlled variables.

To estimate the relative gains for a 5 × 5 system is an ambitious undertaking. All 25 open-loop gains are required to fill matrix **M**, or corresponding gains with all other loops closed to fill matrix **C**. Then the 5 × 5 matrix must be inverted and transposed by a computer. Alternatively, the equations could be written representing controlled variables in terms of manipulated variables and again in terms of other controlled variables. Using this procedure, individual relative gains may be calculated by using Eq. (10.1). However, 16 relative gains must be found to describe the 5 × 5 matrix completely, requiring 32 differentiations.

Fortunately, the effort needed to evaluate the relative gains may be reduced somewhat by inspection. For example, if boilup rate is affected *only* by the heat-input valve, that pair of variables may be eliminated from further consideration. While it is true that the heat-input valve affects other variables besides boilup, the fact that it cannot be affected in return means that it does not interact with the other variables according to the definition of relative gain. (For a further discussion on this point, see "Partial Coupling" in Ref. 2.)

Following the same line of reasoning, column pressure is unaffected by distillate, reflux, or bottom-product flow and is therefore under sole control of the condenser valve. With all valves in manual control, a change in distillate flow *only* affects accumulator level and a change in bottom-product flow only affects base level. Entering zeros for all the unaffected controlled variables in those columns allows the reflux column in Fig. 10.8 to be completely filled in.

However, Fig. 10.8 is a false picture of the relative-gain array of a distillation column. The effectiveness of material-balance control over product quality has been demonstrated in earlier chapters, yet Fig. 10.8 shows it to be nonexistent. A control system structured per the inspection process is known to be less effective than a properly structured material-balance system.

Since product quality is the slowest responding variable in the column, it can be affected by all the other variables. As pointed out before, its response to manipulation rests heavily on the arrangement of the other, faster loops. Unfortunately, the simplicity of the relative-gain concept fails to take this factor into account.

Instead of evaluating the complete 5 × 5 array with *all* valves fixed, it is preferable to examine 2 × 2 subsets, assuming that the remaining variables are all controlled. This is a viable technique as long as each selected 2 × 2 subset is evaluated against the others for interaction and stability. For example, the relative gains of distillate and bottoms compositions to distillate flow and boilup will be evaluated as a subset and compared to the relative gains of those same compositions to bottoms flow and boilup and then to reflux and boilup.

	D	B	L	m_r	m_c
y	0	0	1	0	0
V	0	0	0	1	0
p	0	0	0	0	1
l_a	1	0	0	0	0
l_b	0	1	0	0	0

figure 10.8 *This false picture of the relative-gain array can result if inspection is followed throughout.*

Nested Control Loops To begin the evaluation of 2 × 2 subsets, the combination of pressure and accumulator level controlled by reflux and the vent valve appearing in Fig. 9.6 is examined. By the definition of relative gain, this system indicates no interaction since in the open loop, level is not affected by the vent valve nor is pressure changed by reflux. Then the question arises: Is the nested configuration of Fig. 9.6 controllable at all?

An examination of the block diagram of this system in Fig. 10.9 reveals that *both* controllers must be in automatic to form a *single* closed loop. In the level loop, for example, both process elements $\mathbf{g}_p$ and $\mathbf{g}_l$ and the pressure controller appear. If the pressure controller has proportional action alone, the level-control loop can be stable. However, its proportional band is directly related to the proportional band of the pressure controller since the two are in series. To achieve tighter control over pressure, the proportional band of the pressure controller would be reduced, but that of the level controller must then be expanded by an equivalent amount. So tighter pressure control can only be achieved at the price of poorer level control.

But if both controllers have reset action, the open-loop phase angle (with only one controller in the loop) crosses $-180°$ twice. This is described by Chestnut and Mayer [4] as a conditionally stable system. If the gain of the controllers is high enough, the loop will oscillate at its natural period, which would be in the neighborhood of 1 min for the combined dynamics of the level and pressure lags. But if the loop gain is reduced, either because of a change in process gain (as might happen with an equal-percentage vent valve) or a change in controller tuning, an oscillation may develop at a much longer period, say 20 min.

The appearance of an abnormally slow cycle like this in a level or pressure loop is an indicator of conditional stability. Unfortunately, the corrective action usually applied by the engineer—widening the proportional band—makes the loop cycle at a still longer period. In the extreme, the gain of his controller would be so low as to effectively be in manual. Then the other controller is in an open loop and will wind up. Furthermore, Ref. 4 states that the reset time of at least one of the controllers must exceed the sum of the time constants of all the other lags in the loop. In the limit, that controller would approach proportional action and the loop would become stable.

The author has observed an unnaturally long period of oscillation in the accumulator level loop for the system with the submerged condenser shown in Fig. 7.4. Because the bypass valve must change level before pressure can be affected, conditional stability is likely.

Interaction between Composition and Level At this point the interaction between distillate composition and accumulator level is worth investigating. From an open-loop analysis, there appears to be no interaction since composition is unaffected by distillate-flow changes with reflux and boilup fixed. Therefore, it would seem inappropriate to control composition with distillate flow. However, as has been said already, the loop is responsive if reflux is used to control accumulator level. The dependence of the composition loop on the performance of the level loop would appear to indicate a nested con-

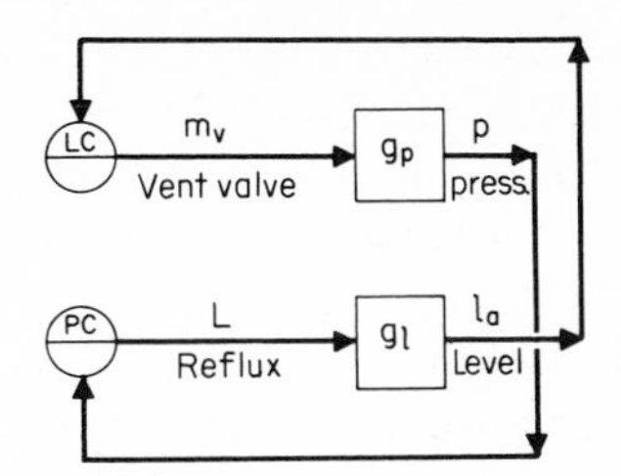

figure 10.9 *Because there is no steady-state process interaction in this system,* both *controllers must be in automatic to form a* single *closed loop.*

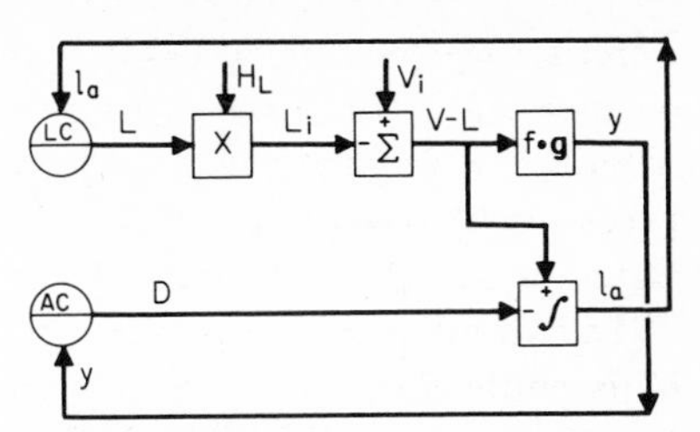

figure 10.10 *This system differs from that of Fig. 10.9 in that reflux flow affects* both *controlled variables, although distillate affects but one.*

figuration. However, an examination of the block diagram of this system in Fig. 10.10 shows how it differs from that of Fig. 10.9.

The manipulated external reflux L develops an internal-reflux flow L_i through its enthalpy H_L. This converts a given value of internal vapor rate V_i into an equivalent external vapor rate V, since $V_i - V = L_i - L$. The sensitivity of composition y to control action was given in Eq. (8.17) as $dy/d(D/F)$. However, it was assumed in that derivation that all other variables were under control and therefore that the manipulated D/F was in fact imposed on the column internals. In actual practice, composition is determined by $V_i - L_i$, which is only D in the steady state. The functional relationship $dy/d(D/F)$ is replaced by the symbol f in Fig. 10.10; a dynamic-gain component g is also indicated.

Liquid level in the accumulator is the time integral of $V - L - D$, and so an integrator appears in the block diagram. Because reflux affects both composition and level, this natural interaction makes the system behave differently from the nested configuration of Fig. 10.9. The level-control loop will function without the composition loop closed. In essence, the composition controller in manipulating D is calling for a specified value of $V - L$ to be imposed on the column by the level-control loop. The accumulator acts effectively as an integrating controller, and the level controller becomes just another element in that closed loop. Disturbances in H_L and V_i are counteracted by the level loop so that $V - L$ is held equal to the desired value of D. Because the composition loop is much slower than the level loop, the two act like the primary and secondary of a cascade system. And as in a cascade system, if the secondary controller is given too wide a proportional band or is placed in manual, the primary loop will not function; the tighter the secondary variable is controlled, the more responsive the primary loop will be.

Parallel Reboilers Parallel units in general are suspects for interaction, but parallel kettle reboilers are among the worst examples. Figure 10.11 gives the details of one such installation. The reboilers were supplied with liquid from the column base, and vapor was returned through individual lines to the column. Each reboiler had a weir to keep its tubes covered, and the liquid

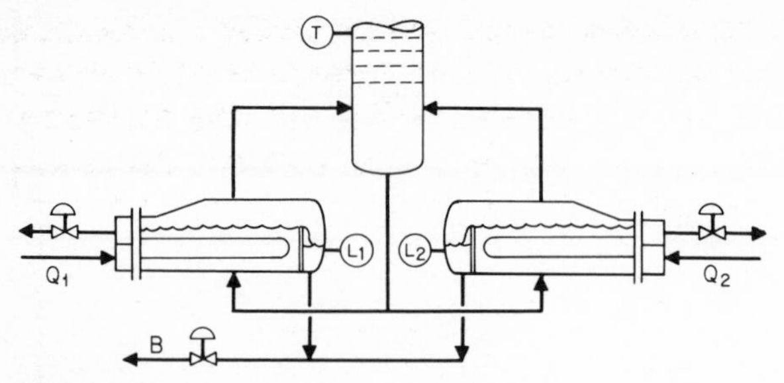

figure 10.11 *The interaction between the liquid levels in parallel kettle reboilers is particularly severe.*

level on the downstream side was to be controlled. Note that the two weir compartments were connected to a common draw-off valve. In addition to controlling the two liquid levels, bottom composition needed to be controlled as measured by a tray temperature. The three manipulated variables were hot-oil flows to each reboiler and bottom-product flow. Operators reported that they could control temperature and one level but never both levels.

The step response of the process to the manipulated variables with all loops open revealed the nature of the interaction. When bottom-product flow was increased, both levels began falling and temperature showed no response, as expected. When one heat input was increased, it made its own level drop quickly and then ramped downward; in the meantime, the other level rose the same amount as quickly and then ramped downward like the first as shown in Fig. 10.12.

Because the levels were connected together, a difference in the pressure drop of flowing vapor and liquid between the reboilers appeared as a difference in levels. So an increase in the heat input Q_1 would increase the pressure drop caused by its vapor outflow and thereby lower its liquid level while raising the other an equal amount. But the net effect of the increase in boilup with bottom flow constant is a steadily falling bottom inventory reflected in both levels.

The 3×3 relative-gain array for this system may be deduced by inspection, assuming that column pressure, distillate composition, and accumulator level are controlled satisfactorily by the condenser valve, distillate flow, and reflux. Because bottom flow has no direct effect on temperature, a zero would be entered for that element in Fig. 10.13. Then because bottom flow affects both levels identically, and both heating valves produce the same response in temperature, 0.5 could be entered for all those elements. If the relative gain of a level to its own heat-input valve could be represented by gain a, then that of the opposite level would have to be $0.5 - a$. The only question remaining is the value of a.

The initial responses of both levels to a heat-input valve are equal and opposite. Therefore, it would appear that a and $0.5 - a$ should be equal and of the opposite sign. They could be equal at $+0.25$, but to be opposite in sign they must be $\pm\infty$. The sign of a undoubtedly changes with the period of excitation as evidenced by the inverse response in Fig. 10.12. In any case, the interaction is unfavorable—two loops with relative gains of 0.5 may be closed, while the third loop must have a relative gain of either a or $0.5 - a$.

The interaction could be moderated by installing separate draw-off valves for the two reboilers. But this would require the process to be shut down. Instead, the decoupling system shown in Fig. 10.14 was installed. The level signals were averaged and used to control the draw-off valve. The temperature controller manipulated both heat-input valves, biased by the differential level controller. This last device received one level signal as a measurement input and the other as a set point. If the two were unequal, the controller could add to one heat-input valve and subtract an equal amount from the other. Thus it was capable of correcting level imbalances without affecting temperature.

This solution assumed proper selection of the remaining control loops for the column. It is possible that a better selection could be made by placing bottom-product flow under column temperature control. Then each liquid-level controller would adjust its own heating valve, although the two would still require decoupling. Figure 10.15 shows how the two liquid levels could be decoupled independent of the temperature loop. The decoupling elements are derivatives, giving a change in output only in the unsteady state. Then a step increase in m_1 would cause both heating valves to open equally, preventing the transfer of liquid from one reboiler to the other. Gradually the derivative influence would dissipate, leaving the increased heat input only on one reboiler, which should influence only that level.

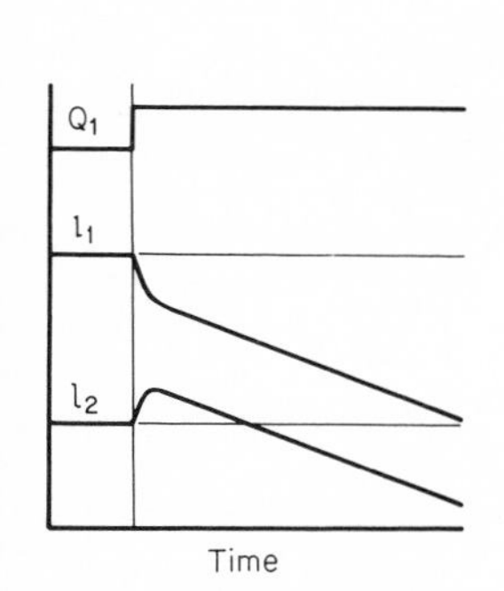

figure 10.12 *A change in one of the heat-input valves causes inverse response in the other level.*

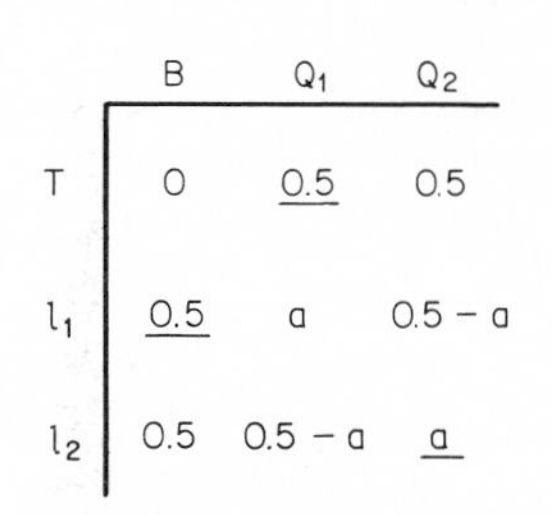

	B	Q_1	Q_2
T	0	0.5	0.5
l_1	0.5	a	0.5 − a
l_2	0.5	0.5 − a	a

figure 10.13 *Closing the third loop always destabilizes the system.*

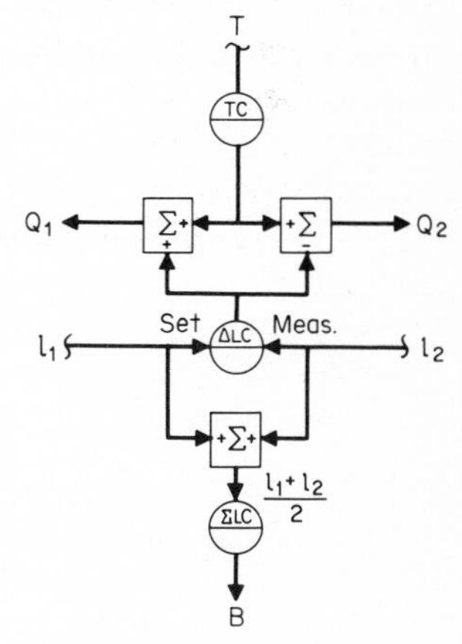

figure 10.14 *This decoupling system stabilized the parallel reboilers.*

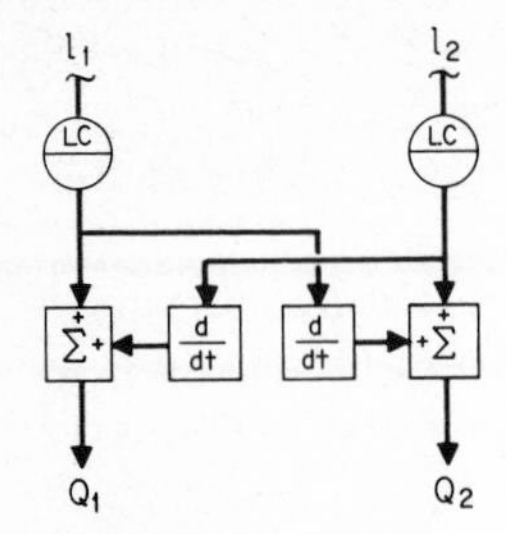

figure 10.15 *Derivative action can also be used to decouple the two liquid levels.*

INTERACTION BETWEEN COMPOSITION LOOPS

Most of the attention given to interaction in distillation columns has focused on that existing between product-quality loops. Rijnsdorp [5], Toijala and Fagervik [6], and Luyben [7] have all addressed the interaction problem and experimented with decoupling systems. The reason for this attention is the sheer difficulty encountered in attempting to control product compositions. Chapter 8 described how material-balance feedforward controls are used to protect quality from upsets in the heat balance and column feed changes. These strategies are sufficient when a single composition needs to be controlled.

But multiple composition controllers tend to upset one another as they perform their regulating functions. These interactions are most troublesome because they develop between the slowest, most sensitive loops in the plant. The penalty for incorrect loop configuring is severe—one of the control objectives has to be abandoned. So it is important to define the magnitude of the interactions quantitatively so that the most effective structure may be designed before start-up. The relative-gain technique is particularly useful here, enabling the comparison of composition subsets against each other. Then if interaction is anticipated even with the best selection of system structure, decoupling can be provided on a calculated basis.

Manipulating a Product and Separation The interaction experienced between two composition loops is, of course, dependent on the selection of manipulated variables. There are two general choices to be made—material-balance control or reflux-and-boilup manipulation. Relative gains for these choices differ fundamentally and are in fact mutually exclusive. The preferred choice—manipulating the material balance and separation—is examined now.

The first 2×2 set to be evaluated is the manipulation of D/F and V/F to control x and y in a binary column, that is, Λ_{DV}. Fortunately, the derivatives which must be evaluated to calculate λ_{yD} have already been developed. From Eq. (8.17) we have the derivative of D/F with respect to y at constant separation:

$$\left.\frac{\partial(D/F)}{\partial y}\right|_S = \frac{-D/F}{y-x} - \frac{1-D/F}{y-x}\,\frac{x(1-x)}{y(1-y)} \tag{10.19}$$

And immediately preceding (8.17) is the partial derivative of D/F with respect to y at constant x:

$$\left.\frac{\partial(D/F)}{\partial y}\right|_x = \frac{-D/F}{y-x} \tag{10.20}$$

Relative gain λ_{yD} of the subset Λ_{DV} is then

$$\lambda_{yD}(\Lambda_{DV}) = \frac{\left.\dfrac{\partial y}{\partial (D/F)}\right|_S}{\left.\dfrac{\partial y}{\partial (D/F)}\right|_x} = \frac{\left.\dfrac{\partial (D/F)}{\partial y}\right|_x}{\left.\dfrac{\partial (D/F)}{\partial y}\right|_S}$$

$$= \frac{1}{1 + \dfrac{(y - z)x(1 - x)}{(z - x)y(1 - y)}} \tag{10.21}$$

A qualitative assessment is worthwhile at this point. For all real values of x, y, and z, λ_{yD} falls between 0 and 1. As x approaches zero and y approaches unity, x and $1 - y$ dominate. Then for a purer distillate, $x > (1 - y)$ and $\lambda_{yD} < 0.5$; for a purer bottom product, $x < (1 - y)$ and $\lambda_{yD} > 0.5$.

example 10.1

From the data in Table 8.1, calculate $\lambda_{yD}(\Lambda_{DV})$ by using Eq. (10.21) evaluated at base conditions and also by combining observed open-loop gains in Eq. (10.5).

$$\lambda_{yD}(\Lambda_{DV}) = \frac{1}{1 + \dfrac{(0.95 - 0.50)0.05(1 - 0.05)}{(0.50 - 0.05)0.95(1 - 0.95)}} = 0.50$$

Values of D/F are obtained as $1 - B/F$. Then open-loop gains are estimated as $\Delta y/\Delta(D/F)$ and so forth:

$$\mathbf{M}_{DV} = \begin{array}{c|cc} & D & V \\ \hline y & -0.992 & +0.0426 \\ x & -0.806 & -0.0426 \end{array}$$

$$\lambda_{yD}(\Lambda_{DV}) = \frac{(-0.992)(-0.0426)}{(-0.992)(-0.0426) - (-0.806)(0.0426)} = 0.551$$

Another subset to be investigated is the manipulation of bottom-product flow and separation (either boilup or reflux). However, B/F is $1 - D/F$, so $\partial(B/F)$ is simply $-\partial(D/F)$. As a result, λ_{yB} of the subset Λ_{BL} is simply

$$\lambda_{yB}(\Lambda_{BL}) = \lambda_{yD}(\Lambda_{DV}) \tag{10.22}$$

This relationship promises a way out of an unfavorable relative-gain array. If, for example, the compositions in a given column were such that $\lambda_{yD}(\Lambda_{DV}) \to 0$, then distillate flow could not satisfactorily control its own composition. However, because of dynamic considerations, distillate flow would not be a good choice to control x, although $\lambda_{xD}(\Lambda_{DV}) \to 1$. But Eq. (10.22) indicates that the alternative subset Λ_{BL} would give a favorable value of λ_{xB} because $\lambda_{yB} \to 0$. Then x could be controlled with B, and y with L.

Guidelines for selecting the smaller product flow for material-balance control still apply, however. So if $D << B$, then in fact x *should* be controlled by manipulating D, and y by manipulating reflux. Dynamic responsiveness would be achieved by controlling accumulator level with heat input.

In the foregoing discussion, separation was assumed to be solely dependent on boilup or reflux, depending on the chosen subset. It is not necessary that separation be linear with V or even V/F. In fact, the inclusion of F in the D/F function is also superfluous.

The subsets examined thus far differ only in their selection of manipulated variables. With each subset Λ identified by its subscripted manipulated variables, we can summarize:

$$\Lambda_{DV} = \begin{array}{c|cc} & D & V \\ \hline y & \lambda_{yD} & 1-\lambda_{yD} \\ x & 1-\lambda_{yD} & \lambda_{yD} \end{array} \tag{10.23}$$

$$\Lambda_{DL} = \Lambda_{DV} \tag{10.24}$$

$$\Lambda_{BL} = \Lambda_{DV} \tag{10.25}$$

A subset Λ_{BV} is an unlikely prospect since it would leave base level without an appropriate manipulated variable. From the preceding considerations, it can be seen that the only relative-gain term which must be calculated in a two-product system under material-balance control is $\lambda_{yD}(\Lambda_{DV})$.

Manipulating Reflux and Boilup This subset, identified as Λ_{LV}, does not include a product flow and will therefore differ in character from those that do. When the material balance is controlled, boilup and reflux have no influence over it and can then only adjust separation. But without material-balance control, boilup and reflux affect both the separation and the material balance. Hence a more severe form of interaction can be expected.

An evaluation of λ_{yL} of the subset Λ_{VL} begins with the overhead material balance:

$$L = V - D \tag{10.26}$$

Partial derivatives of (10.26) may then be made:

$$\left.\frac{\partial(L/F)}{\partial y}\right|_V = -\left.\frac{\partial(D/F)}{\partial y}\right|_S \tag{10.27}$$

(The substitution of S for V in this expression assumes that S is totally dependent on V/F and that F is constant throughout the analysis.) The right-hand term in (10.27) is represented by Eq. (10.19). The remaining partial derivative is

$$\left.\frac{\partial(L/F)}{\partial y}\right|_x = \left.\frac{\partial(V/F)}{\partial y}\right|_x - \left.\frac{\partial(D/F)}{\partial y}\right|_x \tag{10.28}$$

The last term on the right is represented by Eq. (10.20), but the first term requires further definition:

$$\left.\frac{\partial(V/F)}{\partial y}\right|_x = \left.\frac{\partial S}{\partial y}\right|_x \frac{d(V/F)}{dS} \tag{10.29}$$

The first component may be found by differentiating the binary separation equation:

$$S = \frac{y(1-x)}{x(1-y)}$$

$$\left.\frac{\partial S}{\partial y}\right|_x = \frac{S}{y(1-y)} \tag{10.30}$$

The second component is the derivative of the logarithmic separation model of Eq. (2.32) simplified to the form

$$\frac{V}{F} = \beta \ln S \tag{10.31}$$

where β is identified as the "column characterization factor." (For any given column, β can be found simply by dividing an observed V/F by the natural logarithm of the calculated separation, whether the system is binary or multicomponent.) The derivative of (10.31) is

$$\frac{d(V/F)}{dS} = \frac{\beta}{S} \tag{10.32}$$

Then (10.28) becomes

$$\left.\frac{\partial(L/F)}{\partial y}\right|_x = \frac{\beta}{y(1-y)} + \frac{D/F}{y-x} \tag{10.33}$$

The relative gain $\lambda_{yL}(\Lambda_{LV})$ is calculated as

$$\lambda_{yL}(\Lambda_{LV}) = \frac{\left.\dfrac{\partial(L/F)}{\partial y}\right|_x}{\left.\dfrac{\partial(L/F)}{\partial y}\right|_V} = \frac{\dfrac{\beta}{y(1-y)} + \dfrac{D/F}{y-x}}{-\left.\dfrac{\partial(D/F)}{\partial y}\right|_S}$$

Substituting (10.19) into the denominator gives

$$\lambda_{yL}(\Lambda_{LV}) = \frac{1 + \dfrac{\beta(y-x)^2}{y(1-y)(z-x)}}{1 + \dfrac{(y-z)x(1-x)}{(z-x)y(1-y)}} \tag{10.34}$$

The denominator of (10.34) will be recognized as being identical to that of λ_{yD} of the subset Λ_{DV}. Then

$$\frac{\lambda_{yL}(\Lambda_{LV})}{\lambda_{yD}(\Lambda_{DV})} = 1 + \frac{\beta(y-x)^2}{y(1-y)(z-x)} \tag{10.35}$$

Because all the terms in (10.35) are positive, λ_{yL} is always greater than λ_{yD} and is always positive. (This departs from an earlier statement by the author in Ref. 8 wherein negative values were anticipated.) The actual value of λ_{yL} in any case tends to be much greater than 1.0 because of the small size of $1 - y$

relative to the other terms in (10.35). The column characterization factor β also exerts a substantial influence—more difficult separations requiring a higher value of V/F will give a higher relative gain. For very pure distillate products that are difficult to separate, such as ethylene from ethane and propylene from propane, $\lambda_{yL}(\Lambda_{LV})$ will tend to be very high.

example 10.2

From the data in Table 8.1, calculate $\lambda_{yL}(\Lambda_{LV})$ by using Eq. (10.35) evaluated at base conditions and also by combining observed open-loop gains in Eq. (10.5).

$$\beta = \frac{V/F}{\ln S} = \frac{3.0}{\ln 361} = 0.509$$

$$\frac{\lambda_{yL}(\Lambda_{LV})}{\lambda_{yD}(\Lambda_{DV})} = 1 + \frac{0.509(0.95 - 0.05)^2}{(0.95)(1 - 0.95)(0.50 - 0.05)} = 20.29$$

$$\lambda_{yL}(\Lambda_{LV}) = 0.50(20.29) = 10.14$$

Open-loop gains are estimated as $\Delta y/\Delta(L/F)$ and so forth:

$$\mathbf{M}_{LV} = \begin{array}{c|cc} & L & V \\ \hline y & 1.025 & -0.917 \\ x & 0.833 & -0.833 \end{array}$$

$$\lambda_{yL}(\Lambda_{LV}) = \frac{(1.025)(-0.833)}{(1.025)(-0.833) - (0.833)(-0.917)} = 9.46$$

The magnitude of the relative-gain numbers is of great significance. In the Λ_{DV} subset, 0.5 is the least favorable relative gain obtainable, and the dynamic interaction pictured by Fig. 10.5 is therefore the worst that could be expected. Figure 10.6 shows the dynamic response of loops coupled with a relative gain of 2.0. A comparison of the set-point responses in the two lower-right-hand curves indicates that about the same integrated absolute deviation from set point is obtained in each case. Hence a relative gain of 2.0 is roughly equivalent to 0.5 with regard to its effect on control-loop performance. Then very high values of $\lambda_{yL}(\Lambda_{LV})$ such as those estimated in Example 10.2 can be expected to result in very poor performance indeed.

The Effect of Additional Components As pointed out earlier, few separations are truly binary, and in many cases the key components in the products may not even constitute the majority. In these columns the dilution by off-key components can have a substantial effect on interaction.

Fortunately most of the groundwork for this analysis has already been laid. Since the material-balance equations for all components are identical, the partial derivatives of D/F with respect to y and to x, at constant x and y, respectively, are the same as in a binary system. But the separation factor is a two-component expression rather than the single-component binary equation. The partial of x with respect to y for either key was given in Eq. (8.26). This partial may be used to evaluate the partial of D/F with respect to y at constant S, as was done in Chap. 8, for either key:

$$\left.\frac{\partial(D/F)}{\partial y_h}\right|_S = \frac{-D/F}{y_h - x_h} - \frac{(1 - D/F)(y_l + y_h)x_l x_h}{(y_h - x_h)(x_l + x_h)y_l y_h} \tag{10.36}$$

Finally, λ_{yD} for the subset Λ_{DV} is derived as in (10.21):

$$\lambda_{yD}(\Lambda_{DV}) = \frac{1}{1 + \dfrac{(y_h - z_h)(y_l + y_h)x_l x_h}{(z_h - x_h)(x_l + x_h)y_l y_h}} \tag{10.37}$$

The relative gains for the two keys must be the same even though their values of x, y, and z differ. By material balance,

$$\frac{y_h - z_h}{z_h - x_h} = \frac{B}{D} = \frac{y_l - z_l}{z_l - x_l} \tag{10.38}$$

Consequently there is only one $\lambda_{yD}(\Lambda_{DV})$ for the column.

example 10.3

Calculate $\lambda_{yD}(\Lambda_{DV})$ for the depropanizer having the following compositions:

	Feed, %	Distillate, %	Bottoms, %
Ethane	1.2	2.9	
Propane	39.4	95.5	0.4
Isobutane	14.0	1.6	22.6
Heavier	45.4		77.0

$$\lambda_{yD}(\Lambda_{DV}) = \frac{1}{1 + \dfrac{(95.5 - 39.4)(95.5 + 1.6)0.4(22.6)}{(39.4 - 0.4)(0.4 + 22.6)95.5(1.6)}} = 0.736$$

Observe that $x_l + x_h$ has a substantial effect on λ_{yD} and therefore use of the binary equation for this column will give erroneous results.

Through a derivation similar to that given in Eqs. (10.26) to (10.35), it can be shown that, for the multicomponent system,

$$\frac{\lambda_{yL}(\Lambda_{LV})}{\lambda_{yD}(\Lambda_{DV})} = 1 + \frac{\beta(y_l - x_l)^2}{z_l - x_l}\left(\frac{1}{y_l} + \frac{1}{y_h}\right) \tag{10.39}$$

Multiple-product Columns Obtaining relative gains for control of three product compositions is much more complex. Consider the example of the sidestream column shown in Fig. 10.16. Three components are indicated: the most volatile is designated a; the least, c. For simplicity, c is assumed to be negligible in the distillate and a negligible in the bottom. The equations describing the system are

$$F = D + P + B \tag{10.40}$$

$$Fz_a = Dy_a + Pw_a \tag{10.41}$$

$$Fz_b = Dy_b + Pw_b + Bx_b \tag{10.42}$$

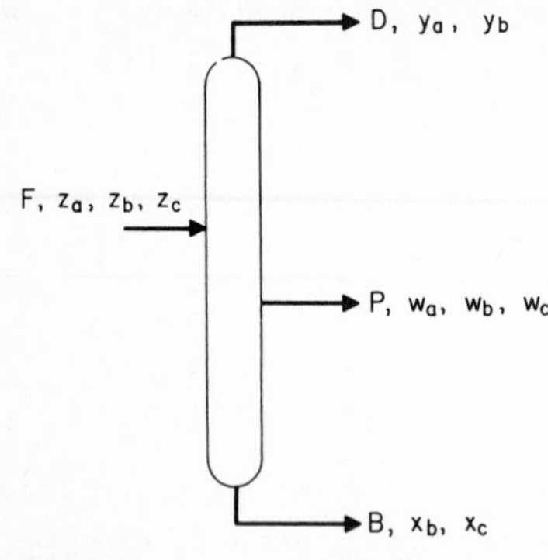

figure 10.16 *Two product flows and separation may be manipulated to control three compositions.*

$$\frac{y_a/y_b}{w_a/w_b} = S_{ab} \tag{10.43}$$

$$\frac{w_b/w_c}{x_b/x_c} = S_{bc} \tag{10.44}$$

$$\frac{V}{F} = \beta_{ab} \ln S_{ab} = \beta_{bc} \ln S_{bc} \tag{10.45}$$

Equation (10.45) recognizes that the relative volatility between components a and b differs from that between b and c and that the number of trays in the two sections of the column also differ; yet both separation factors vary in the same way with V/F.

Equations (10.40) to (10.42) may be combined to eliminate B and P, or B and D, giving

$$\frac{D}{F} = \frac{w_a(z_b - x_b) - z_a(w_b - x_b)}{w_a(y_b - x_b) - y_a(w_b - x_b)} \tag{10.46}$$

$$\frac{P}{F} = \frac{y_a(z_b - x_b) - z_a(y_b - x_b)}{y_a(w_b - x_b) - w_a(y_b - x_b)} \tag{10.47}$$

Although it is possible to obtain partial derivatives of D/F and P/F with respect to each composition with other compositions constant, the same partials evaluated at constant separation are quite complicated. Consequently, the preferred determination of relative gain would increment (10.43) to (10.47) for each selected composition independently of the others to give a set of $(\Delta m_j/\Delta c_i)_c$, identified as matrix **C**. Then $\mathbf{\Lambda}$ may be calculated from **C** by using the procedure described preceding Eq. (10.3).

example 10.4

An alkylation deisobutanizer separates a feedstock into an isobutane-rich distillate, an n-butane sidestream, and an alkylate bottom product. Let the streams and their compositions be identified by Fig. 10.16. At a V/F of 3.0, the set of compositions is

	z	y	w	x
Isobutane, a	0.80	0.95	0.01	
n-Butane, b	0.10	0.05	0.94	0.02
Heavier, c	0.10		0.05	0.98

Increment the controlled variables w_a, y_a, and x_b by +0.001, and determine matrix **C** as $\Delta y_a/\Delta(D/F)$ and so on:

C =	y_a	w_a	x_b
D	−1.12	−15.9	885
P	1.06	15.0	−9.26
V	0.117	−0.026	−0.046

Matrix C is inverted to give

$$
C^{-1} = \left[\begin{array}{ccc}
-6 \times 10^{-4} & 1.5 \times 10^{-2} & 8{,}41 \\
6.6 \times 10^{-4} & 6.6 \times 10^{-2} & -0.594 \\
1.1 \times 10^{-3} & 1.2 \times 10^{-3} & 3.3 \times 10^{-3}
\end{array}\right.
$$

Multiplication of the elements in $[C]^T$ by the corresponding elements in its inverse C^{-1} yields

$$
\Lambda = \begin{array}{c} \\ y_a \\ w_a \\ x_b \end{array}
\begin{array}{ccc}
D & P & V \\
\hline
\multicolumn{1}{|c}{-6.7 \times 10^{-4}} & 1.6 \times 10^{-2} & \underline{0.985} \\
\multicolumn{1}{|c}{-1 \times 10^{-2}} & \underline{0.995} & 1.5 \times 10^{-2} \\
\multicolumn{1}{|c}{\underline{1.011}} & -1.1 \times 10^{-5} & -10^{-6}
\end{array}
$$

Control-loop pairing for this system is quite clear from the relative gains.

If the separation in a multiproduct tower is not manipulated for composition control, then the number of controlled variables is reduced. Interaction then existing between the remaining variables is likely to differ significantly from that observed when separation is manipulated, depending on the selection of the remaining variables. Because of the complexity of the material-balance equations (10.46) and (10.47), incrementing compositions from base conditions is preferred. However, the value of the uncontrolled composition must be calculated by using separation equations (10.43) and (10.44) before D/F and P/F can be found.

example 10.5

It is decided to hold constant the V/F ratio for the column in Example 10.4, allowing y_a to go uncontrolled. Estimate the relative gains for the remaining subset. Repeat with x_b uncontrolled.

Incrementing w_a and x_b by 0.001 at constant separation gives new values of y_a and y_b which are then substituted into (10.45) and (10.46) to yield new values of D/F and P/F. The gains $\Delta w_a/\Delta(D/F)$ and so on appear below:

$$
C = \begin{array}{c} \\ D \\ P \end{array}
\begin{array}{|cc}
w_a & x_b \\
-0.025 & 5.55 \\
0.237 & -5.32
\end{array}
$$

$$
\lambda_{wD} = \frac{(-0.025)(-5.32)}{(-0.025)(-5.32) - (5.55)(0.237)} = -0.11
$$

Distillate flow should then control x_b and sidestream flow should control w_a.

Incrementing y_a and w_a by 0.001 at constant separation gives new values of $x \geqslant$ which are then substituted into (10.45) and (10.46) to yield new values of D/F and P/F. At constant separation x_b is virtually independent of w_a and y_a, so that the open-loop gains for uncontrolled x_b are the subset of the 3×3:

$$
C = \begin{array}{c} \\ D \\ P \end{array}
\begin{array}{|cc}
y_a & w_a \\
-1.12 & -15.9 \\
1.06 & 15.0
\end{array}
$$

$$
\lambda_{yD} = \frac{(-1.12)(15.0)}{(-1.12)(15.0) - (-15.9)(1.06)} = -311
$$

It should be apparent from Example 10.5 that multiproduct columns pose some serious interaction problems. Negative interaction exists between manipulated product flows, presenting the possibility of an uncontrollable situation as in the last example. Controlled and manipulated variables should be selected to yield reasonable relative gains which will allow effective pairing.

DECOUPLING SYSTEMS

When relative gains indicate the presence of a troublesome amount of interaction even with the best possible pairing, decoupling is necessary. A decoupling system is intended to produce the opposite effect of the interaction naturally existing in the process. If a given valve affects two compositions, it is possible to manipulate a second valve to offset the effect of the first on one of the compositions. Through a decoupling system, a single composition controller can manipulate two or more valves in such a way as to affect that composition alone. There are several means of achieving this objective, some of which are impractical or pose operating problems [2, 7]. But the decoupling systems described below will be found similar in structure and operation to feedforward systems.

Decoupling Positive Interaction Systems with positive interaction yield readily to decoupling. The method employed is to combine controller outputs by summation to generate the manipulated D/F and V/F as described in Fig. 10.17. The decoupling elements solve for the manipulated variables in terms of controller outputs:

$$\frac{D}{F} = d_{yD}m_y + d_{xD}m_x \tag{10.48}$$

$$\frac{V}{F} = d_{yV}m_y + d_{xV}m_x \tag{10.49}$$

where d_{ij} represents the gain of the decoupling elements. Note that the form of these equations matches that described in (10.6). Therefore, coefficients d_{ij} may equal b_{ij} in (10.6), that is, $(\partial m_j/\partial c_i)_c$.

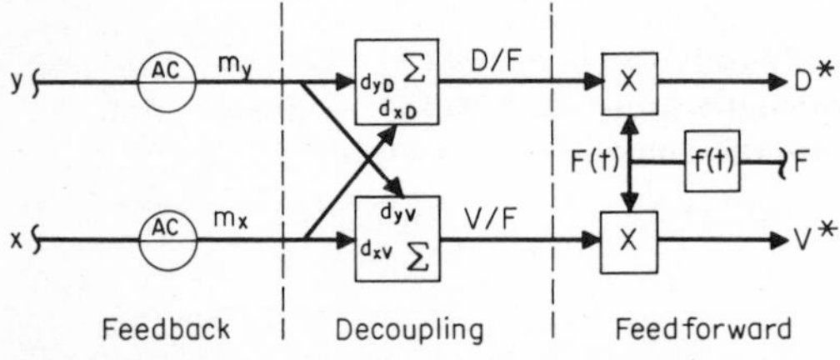

figure 10.17 *The decoupling elements combine controller outputs to generate the manipulated variables.*

Process gains in terms of the controlled variables are summarized below:

$$b_{yD} = \left.\frac{\partial(D/F)}{\partial y_i}\right|_x = \frac{-D/F}{y_i - x_i} = \frac{-(z_i - x_i)}{(y_i - x_i)^2} \tag{10.50}$$

$$b_{xD} = \left.\frac{\partial(D/F)}{\partial x_i}\right|_y = \frac{-(1 - D/F)}{y_i - x_i} = \frac{-(y_i - z_i)}{(y_i - x_i)^2} \tag{10.51}$$

$$b_{yV} = \pm \left.\frac{\partial S}{\partial y_i}\right|_x \frac{dV}{dS} = \pm S\left(\frac{1}{y_l} + \frac{1}{y_h}\right)\frac{\beta}{S} = \pm \beta\left(\frac{1}{y_l} + \frac{1}{y_h}\right) \tag{10.52}$$

$$b_{xV} = \pm \left.\frac{\partial S}{\partial x_i}\right|_y \frac{dV}{dS} = \pm S\left(\frac{1}{x_l} + \frac{1}{x_h}\right)\frac{\beta}{S} = \pm \beta\left(\frac{1}{x_l} + \frac{1}{x_h}\right) \tag{10.53}$$

Note that these equations apply to two products with any number of components. The material-balance derivatives are valid for any component as written. The signs of b_{yV} and b_{xV} will be positive for the major component in that product and negative for the key impurity. For example, $\partial x_h/\partial V$ is positive and $\partial x_l/\partial V$ is negative.

example 10.6

For the column in Example 10.3, calculate all four values of b_{ij}; V/F is 1.8. For control of the light key in both products,

$$b_{yD} = \frac{-(0.394 - 0.004)}{(0.955 - 0.004)^2} = -0.431$$

$$b_{xD} = \frac{-(0.955 - 0.394)}{(0.955 - 0.004)^2} = -0.620$$

$$\beta = \frac{V/F}{\ln S} = \frac{1.8}{\ln \dfrac{(0.955)(0.226)}{(0.004)(0.016)}} = 0.222$$

$$b_{yV} = 0.222\left(\frac{1}{0.955} + \frac{1}{0.016}\right) = 14.1$$

$$b_{xV} = -0.222\left(\frac{1}{0.226} + \frac{1}{0.004}\right) = -56.5$$

Waller (Toijola) [9] points out that two of the decoupling coefficients d_{ij} may be set equal to unity as long as they are not applied to the output of the same controller. The decoupling elements function to match one manipulated variable against the other. To adjust y without affecting x, controller y may manipulate D/F to any desired degree as long as V/F is moved by an amount $(b_{yV}/b_{yD})\Delta D/F$. Or V/F could be manipulated directly by controller y with D/F moved by an amount $(b_{yD}/b_{yV})\Delta V/F$.

Actually any arbitrary value may be assigned to one of the decoupling elements on the output of each controller—a value less than unity is desirable to restrict the range of the manipulated variable. This was done when feedback trim was applied to feedforward control. To generalize, then, decoupling coefficients should be selected such that

$$\frac{d_{yD}}{d_{yV}} = \frac{b_{yD}}{b_{yV}} \tag{10.54}$$

and

$$\frac{d_{xD}}{d_{xV}} = \frac{b_{xD}}{b_{xV}} \tag{10.55}$$

Decouplers must also be scaled as were feedforward computing elements. Only if the ranges of V/F and D/F were identical could scaling be ignored, and this is an unlikely prospect. For the top composition loop,

$$\frac{\Delta V/F}{\Delta D/F} = \frac{d_{yV}}{d_{yD}}$$

Let V/F be represented by the signal $(V/F)'$ multiplied by its range $(V/F)_M$, with the same applying to D/F. Then the scaled decoupling gains (identified by primes) for m_y are calculated by substitution:

$$\frac{\Delta(V/F)'}{\Delta(D/F)'} = \frac{d_{yV}(D/F)_M}{d_{yD}(V/F)_M} = \frac{d'_{yV}}{d'_{yD}}$$

Depending on the selection of the arbitrary decoupling coefficients, bias terms may be necessary to satisfy (10.48) and (10.49) for nominal controller output signals, for example, 50 percent.

example 10.7

For the column in Example 10.6, derive scaled equations for D/F and V/F, assigning arbitrary coefficients of 0.5 for d'_{xV} and d'_{yV}. Let $(D/F)_M = 0.5$ and $(V/F)_M = 3.0$. Scale so that m_x and m_y are both 0.5 at the given conditions.

$$d'_{yD} = d'_{yV} \frac{b_{yD}}{b_{yV}} \frac{(V/F)_M}{(D/F)_M} = 0.5 \frac{(-0.431)3.0}{(14.1)0.5} = -0.0917$$

$$d'_{xD} = d'_{xV} \frac{b_{xD}}{b_{xV}} \frac{(V/F)_M}{(D/F)_M} = 0.5 \frac{(-0.620)3.0}{(-56.5)0.5} = 0.0329$$

$$\left(\frac{D}{F}\right)' = \frac{D/F}{(D/F)_M} = \frac{0.41}{0.5} = 0.82$$

$$0.82 = 0.0329m_x - 0.0917m_y + k_D$$

$$k_D = 0.82 - 0.0329(0.5) + 0.0917(0.5) = 0.849$$

$$\left(\frac{D}{F}\right)' = 0.0329m_x - 0.0917m_y + 0.849$$

$$\left(\frac{V}{F}\right)' = \frac{V/F}{(V/F)_M} = \frac{1.8}{3.0} = 0.6$$

$$0.6 = 0.5m_x + 0.5m_y + k_V$$

$$k_V = 0.6 - 0.5(0.5) - 0.5(0.5) = 0.1$$

$$\left(\frac{V}{F}\right)' = 0.5m_x + 0.5m_y + 0.1$$

It can be seen from Example 10.7 that the choice of the arbitrary scaling factors determines all the rest of the constants. They could have been selected as +1 and −1 instead of both being +0.5. Then d'_{xD} and d'_{yD} would be twice as great and the signs of both would be positive. Bias terms k_V and k_D would then have to be recalculated.

Decoupling Negative Interaction The rules for decoupling negatively interacting systems are the same as for positively interacting systems, although the results are not necessarily the same. Process gains in terms of the controlled variables for manipulation of reflux and boilup are summarized below:

$$b_{yL} = \left.\frac{\partial(L/F)}{\partial y}\right|_x = \pm\left[\beta\left(\frac{1}{y_l} + \frac{1}{y_h}\right) + \frac{z_l - x_l}{(y_l - x_l)^2}\right] \tag{10.56}$$

$$b_{xL} = \left.\frac{\partial(L/F)}{\partial x}\right|_y = \pm\left[\beta\left(\frac{1}{x_l} + \frac{1}{x_h}\right) + \frac{y_l - z_l}{(y_l - x_l)^2}\right] \tag{10.57}$$

$$b_{yV} = \left.\frac{\partial(V/F)}{\partial y}\right|_x = \mp\beta\left(\frac{1}{y_l} + \frac{1}{y_h}\right) \tag{10.58}$$

$$b_{xV} = \left.\frac{\partial(V/F)}{\partial x}\right|_y = \mp\beta\left(\frac{1}{x_l} + \frac{1}{x_h}\right) \tag{10.59}$$

In the case of reflux and boilup manipulation, the signs of the gains are opposite; b_{yL} and b_{xL} are positive for the light key and negative for the heavy key, with the reverse for b_{yV} and b_{xV}.

Note that b_{yV} and b_{xV} are the same whether reflux or distillate is the other manipulated variable. Note also that

$$|b_{yL}| = |b_{yV}| + |b_{yD}| \tag{10.60}$$

and that

$$|b_{xL}| = |b_{xV}| + |b_{xD}| \tag{10.61}$$

Thus when $b_{yD} << b_{yV}$, b_{yL} and b_{yV} are quite similar and effective decoupling requires a high degree of precision. For the data generated in Example 10.6, $b_{yL}/b_{yV} = 0.970$ and $b_{xL}/b_{xV} = 0.989$.

This is a problem encountered by Toijala and Fagervik [6] in their digital simulation of a 10-tray binary column. Having arbitrarily selected d_{yL} and d_{xV} as 1.0, d_{yV} came out as 0.947 and d_{xL} as 0.854 for the comparatively low separation evidenced by y of 0.808 and x of 0.282. As separation is increased, the decoupling coefficients approach unity, as demonstrated by Luyben [7].

Toijala and Fagervik found that without decoupling, expanding oscillations developed with λ_{yL} of about 4. They could not be dampened until both proportional bands and both reset times were doubled. When exactly the correct amount of decoupling was applied, the loops responded independently of one another, although their periods both increased from 2 min to about 7. Surprisingly enough, reducing both the nonunity (off-diagonal) decoupling gains by 10 to 30 percent actually improved both responses by reducing their periods closer to the original 2 min. But a 50 percent reduction resulted in instability at the 2-min period.

Increasing the off-diagonal decoupling factors by 10 percent caused a large-amplitude cycle of perhaps 22 min to appear. The explanation of these responses relates to the dynamic effects of negative coupling as described in Figs. 10.6 and 10.7. With little or no decoupling, the fast cycle with slow approach to the set point appears. As decoupling is applied, loops are closed through the negative relative-gain elements, causing a phase reversal and hence a slower cycle. The higher the off-diagonal gains, the greater the effect of the negative elements. Too much decoupling is worse than too little. This limitation applies only to systems with negative interaction, and it becomes increasingly restrictive as products become purer.

Dynamic Compensation As in feedforward control, dynamic compensation can be of considerable help in improving response. And since it is possible to assign a unity gain to any two of the decoupling elements not acting on the same controller output, it is also possible to eliminate any dynamics from those elements. But then the other two elements must carry the compensation for both loops. Just as the decoupling elements were scaled to the ratio of the steady-state process gains in Eqs. (10.54) and (10.55), their dynamic compensation appears in a similar ratio.

If dynamic gains $\mathbf{g}_{yD}$ and $\mathbf{g}_{yV}$ were both represented by first-order lags, for example, their ratio would be a lead-lag function as used for dynamic compensation in feedforward systems. In a feedforward system, however, there is usually no opportunity to select which variable is to be the load and which is to be manipulated. And whether the compensator has a dominant lead depends on whether the manipulated variable acts more slowly than the load in its effect on the controlled variable.

With the decoupling system, however, a choice is available. If the overhead composition controller were to act directly on boilup and through a decoupler to distillate flow, a dominant lead would be required in the decoupler. The reasoning here is that manipulation of boilup would have a rapid effect on both y and x, which could not be corrected readily by manipulation of D. If, on the other hand, D were directly manipulated to control y, its effect on x would be delayed so that the dynamic compensation through the decoupler to boilup would take the form of a dominant lag.

Toijala and Fagervik added dynamic compensation to their off-diagonal decoupling elements. Dynamic component $\mathbf{g}_{xL}$ took the form of a fourth-order lag, while $\mathbf{g}_{yV}$ was a dead time alone. The addition of dynamic compensation restored the period of the decoupled system to that of the original loops. Nonetheless, a 30 percent reduction of both off-diagonal decoupler gains again improved the responses measurably. Dynamic compensation also broadened the stable range of decoupler gains and allowed the controllers to be more tightly adjusted. A block diagram of a decoupling system with dynamic compensation is shown in Fig. 10.18. Assignments have been made for most favorable response.

Adjusting the dynamic compensators for a decoupling system follows the same procedure as with a feedforward system. Start with both controllers in manual. Then change one output (say, m_y) manually and observe its effect on the opposite controlled variable (x). If decoupling gains are correct, x should be unaffected in the steady state. If dynamic compensation is effective, the deviation in x following the change in m_y should be minimal.

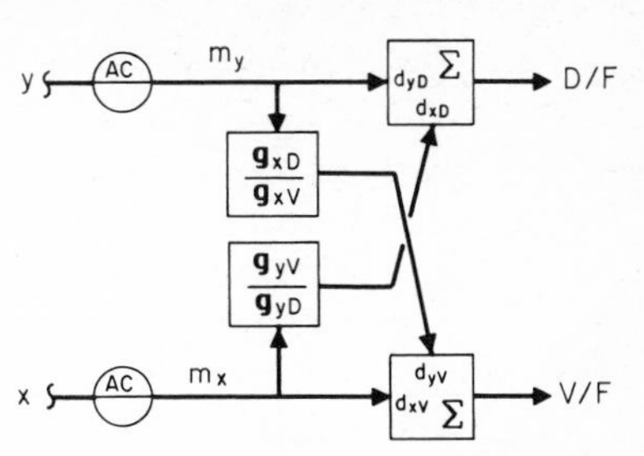

figure 10.18 *Dynamic compensation will improve the effectiveness of the decoupling system.*

Decoupling Multiple Products The atmospheric crude-oil still shown in Fig. 3.9 was cited as exhibiting rather severe interaction among the several sidestream products. These sidestreams are decoupled by using material-balance controls. Essentially, the flow rate of a light product, as demanded by its controller to meet specifications, is passed on to the next heavier product as shown in Fig. 10.19.

Vapor flow is established by the feed heater and side coolers. Liquid downflow is determined by reflux, side cooling, *and* product flow rates. Variations in the flow of a product changes the liquid flow below that point and hence the quality of heavier products, but not that of lighter products. Therefore, decoupling seems necessary in a downward direction only. If the product streams were withdrawn in the vapor phase, decoupling would have to proceed in an upward direction.

In the system shown in Fig. 10.19, the flow of lighter products is progressively summed and passed on to each successive flow controller. Each flow controller then has as its measurement its own flow plus all lighter products.

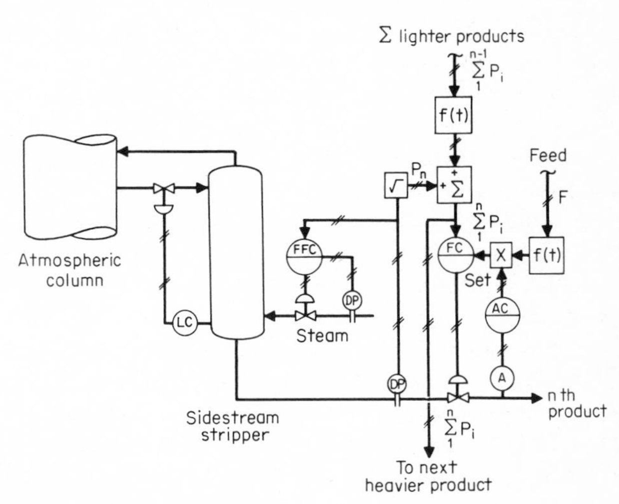

figure 10.19 *Crude-column decoupling needs to be applied in a downward direction only.*

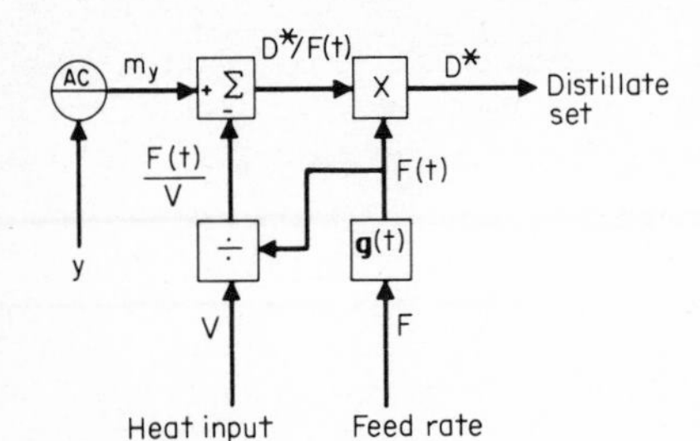

figure 10.20 *Partial decoupling using the parabolic material-balance model.*

Then a change in a light-product flow will force an equal and opposite change in the next heavier product. Dynamic compensation in the form of a lag is provided to allow time for the effect of an upset to propagate to the next heavier product. The set point for each flow controller is the dynamically compensated feed rate, trimmed by a feedback controller acting on an end-point or similar analysis. The flow of stripping steam is controlled in ratio to that of each product.

Partial Decoupling The system shown in Fig. 10.19 for decoupling multiple sidestreams has a distinct advantage in being easy to adjust, in contrast to the bidirectional systems of Figs. 10.17 and 10.18. Actually, a system that is partially decoupled, i.e. in one direction only, becomes completely stabilized. In effect, *one* of the off-diagonal elements in Fig. 10.4 can be canceled by partial decoupling, which is enough to break the second loops affecting *both* controlled variables. As a consequence, partial decoupling can accomplish most of the objectives of full decoupling with half the effort.

Figure 10.20 shows a partial decoupling system for a two-product tower wherein distillate flow is adjusted to control its composition. Variations in heat input caused by the bottom-composition controller, an override, or operator intervention, are converted into equivalent changes to distillate flow so that distillate quality may not be upset by them. The calculation needs only one adjustment: d_{VD}

$$D^* = F(t)\,[m_y - d_{VD}\,F(t)/V] \tag{10.62}$$

This equation has the advantage of being nonlinear and therefore more closely representative of true operating conditions than (10.48) or (10.49). Furthermore, it matches in form the parabolic feedforward model (8.47) developed for columns with constant boilup.

One additional advantage of partial decoupling is its dependence on measured process conditions such as boilup V and dynamically compensated feed rate $F(t)$, instead of another controller output. This fact is of enormous significance when attempting to place the system in automatic. The classical decoupling system of Fig. 10.18 requires solving for both unknowns m_y and m_x before D/F can be determined. If the overhead composition controller is placed in automatic with an arbitrary value of m_x, then D/F will be upset when m_x is later adjusted to bring x under automatic control. Further, any override of boilup, either automatically or manually imposed, will upset both compositions and affect both loop gains. However, the partial decoupling system is *always* functional and does not depend on another controller for its input. Its only disadvantage is that changes in D/F will affect x, but that upset will not be returned to the top of the column.

Partial-decoupling coefficient d_{VD} can be calculated from partial derivatives:

$$d_{VD} = \left.\frac{\partial(D/F)}{\partial(F/V)}\right|_y = \frac{\left.\frac{\partial(D/F)}{\partial x}\right|_y}{\left.\frac{\partial(V/F)}{\partial x}\right|_y}\left(\frac{V}{F}\right)^2 = \frac{b_{xD}}{b_{xV}}\left(\frac{V}{F}\right)^2 \tag{10.63}$$

Evaluation of the gains on the right are given in Eqs. (10.51) and (10.53). As with other decoupling coefficients, d_{VD} must be scaled relative to the ranges of D/F and V/F:

$$d'_{VD} = \frac{b_{xD}}{b_{xV}}\frac{(V/F)_M}{(D/F)_M}\left(\frac{V}{F}\right)^2 \tag{10.64}$$

For the column described in Example 10.7, d'_{VD} is found to be 0.213.

REFERENCES

1. Bristol, E. H.: On a New Measure of Interaction for Multivariable Process Control, *IEEE Trans. Autom. Control,* January 1966.
2. Shinskey, F. G.: "Process-Control Systems," pp. 188–202, McGraw-Hill Book Company, New York, 1967.
3. Shinskey, F. G.: What to Know about Interaction between Control Loops, *Can. Controls Instrum.,* March 1972.
4. Chestnut, H., and R. W. Mayer: "Servomechanisms and Regulating System Design," vol. 1, pp. 152–153, John Wiley & Sons, Inc., New York, 1951.
5. Rijnsdorp, J. E.: Interaction in Two-variable Control Systems for Distillation Columns, *Automatica,* vol. 1, Pergamon Press, London, 1965.
6. Toijala, K., and K. Fagervik: A Digital Simulation Study of Two-point Feedback Control of Distillation Columns, *Kem. Teollisuus,* January 1972.
7. Luyben, W. L.: Distillation Decoupling, *AIChE J.,* March 1970.
8. Shinskey, F. G.: "Process-Control Systems," p. 306, McGraw-Hill Book Company, New York, 1967.
9. Waller, K. V. T.: Decoupling in Distillation, *AIChE J.,* May 1974.

PART FIVE

Optimization

CHAPTER ELEVEN

Optimizing Control Systems

To optimize a process generally means to operate it in the best possible way consistent with certain objectives and within given constraints. But what constitutes "best" must first be defined by examining those objectives and, from them, deriving an optimum operating policy. It should come as no surprise, however, to find that for most processes optimum conditions lie beyond some constraints. Then the problem reduces to one of constraint control rather than optimization. The distinction here is that the objective function can be driven *through* a maximum or minimum only if that maximum or minimum lies within all constraints.

This chapter on optimization was relegated to the end of the book to place the subject in its proper perspective in order of both time and importance. While optimization is a desirable goal, it cannot be achieved—and, in fact, should not be attempted—until all lower-level control functions are executed. To optimize a process requires that it be moved from some existing state to a more profitable one. Yet the purpose of the product-quality controls was to maintain constant compositions. Optimization then requires that these compositions be adjusted. For this endeavor to succeed, the compositions must in fact be controlled at their optimum set points and must respond to direction rapidly enough to follow changing plant conditions. If their control is erratic and unresponsive, no optimization program, regardless of its accuracy, can hope to be successful.

PRINCIPLES OF OPTIMIZATION

To optimize a process, the engineer must first select an objective function and a variable to manipulate in a way to maximize or minimize that function. Usually it is not sufficient to solve an optimization problem just once. Any real process is affected strongly by its environment, such that its optimum operating point tends to shift with imposed conditions. Most objective functions are economic in nature; as a result, the optimum point will change with selling prices, costs, and market conditions. Then any optimizing system must be dynamic, capable of readjusting selected variables as plant conditions and economic factors dictate.

Perhaps the first consideration is the scope of the problem—how much of the plant to include. The scope, in fact, determines the nature of the objective function and the complexity of the relationships involved; it also suggests the means necessary for implementing the system.

Levels of Optimization There are several levels at which optimization may be applied, depending on the scope of plant equipment enclosed within the objective function. To begin, one might optimize the performance of a single column without regard to its impact on the rest of the plant. Since a single column is probably the smallest element in the plant capable of independent operation, this then becomes the lowest level of optimization. It is traditionally designated "local optimization."

Columns in a parallel or series configuration can, in addition, be optimized as a group. This practice avoids the possibility of excessive penalties being absorbed by one column to satisfy the local optimization of another. When the scope of the system is expanded to include all the columns in one distillation unit, we have "unit optimization." Implicit in unit optimization is the effective allocation of limited feedstocks, energy, refrigeration, etc., among the members of that unit.

Complete plant optimization involves coordinating the control of distillation units, utilities, reaction systems, furnaces, compressors, etc., to maximize the profit from the entire operation. Although this is a laudable objective, it may be more of a panacea than a realizable goal. Optimization is the end of the road—all other lower-level controls must function responsively before it can ever be approached. And the larger the segment of a plant enclosed within the scope of the optimization problem, the less the likelihood of reaching a stable optimum state. Expanding the scope of a system to enclose more units of the plant extends the time required to reach an optimum and the exposure to disturbances. Consequently, some level can be reached at which optimization is no longer effective. This limit will naturally differ with the complexity of the plant and its integrity.

The Objective Function In most cases, the objective function for optimization is monetary in nature. It could be profit or loss, operating cost, or productivity. Occasionally it can be reduced to a singular function specific to one process, such as the yield from a chemical reactor or the efficiency of a boiler or compressor. Although these specific functions are satisfactory for local optimization, their use cannot be extended to combination with other elements.

As an example, consider a stripping column exhausting into a compressor. At a given gas flow rate, there is a certain suction pressure which will maximize the efficiency of the compressor. This pressure may not represent the most efficient operating point for the column, however. Yet if the efficiencies of column and compressor were only reported on a percentage basis, there would be no way to relate the two to optimize their combined performance. If their performance were rated on a Btu basis, they could be compared only if the cost per Btu were equal for both column and compressor. Finally the problem evolves into a cost optimization, since operating cost is the common denominator between these two dissimilar pieces of equipment.

Some will argue that a profit function is most meaningful, since a plant is in operation to make a profit. However, it seems more direct to keep tabs on costs and losses, which, if minimized, will thereby constitute maximum profit within the scope of their influence. Cost functions may then be selected to limit the scope of an optimization problem. For example, one may choose to minimize product losses, energy costs, or a combination of the two, independent of production rate. By contrast, a profit function depends on production rate.

The most useful objective function for most applications is a combination of product losses and utility costs per unit feed. For example, let W represent the flow of a waste or lower-valued stream containing w fraction of the valuable component. Then Ww describes the amount of unrecovered product leaving the process. Given that the value of the component of interest *in the product* is v_P, compared to a value of v_W *in the waste* stream, we have a cost statement per unit feed:

$$\frac{\$_c}{F} = (v_P - v_W)\frac{Ww}{F} + (c_i + c_o)\frac{Q}{F} \tag{11.1}$$

The cost factors for heat input and removal are c_i and c_o, expressed in \$/Btu, where Q is the heat flow required to effect the separation.

By contrast, a profit function includes the sum of the values of all the products,whatever their worth, less the cost of the feed:

$$\$_P = \Sigma P v_P - \Sigma F v_F - \Sigma Q(c_i + c_o) \tag{11.2}$$

It is possible that under extreme conditions maximum profit may not be consistent with maximum feed rate. Consider, for example, the separation of a feed into one high-valued product and a very low or negative-valued product. As feed rate is increased, a point can be reached where any additional feed will go unrecovered and profit will begin to decrease. This point constitutes the optimum feed rate.

If specifications must be met on both products from a column, then the maximum amount of feed will be processed when these are just met while under a boilup constraint. This constitutes maximum feed rate under constrained conditions and may maximize the profit function as described by Eq. (11.2). From a control standpoint, however, this is not an optimization problem since profit cannot *pass through* a maximum but can only reach it at a constraint.

Another limitation to the profit function is its assumption of an unlimited market for products and an unlimited supply of feedstock. If, in a given case, these conditions exist, then a maximum or optimum feed rate is desirable. However, the selection of a feed rate that will meet supply and demand limits is a real consideration and, in fact, adds another dimension to the optimization problem.

Finally, in any real plant, feedstock must pass serially through a number of processing elements before a final product emerges. The local optimization of any of these elements must be independent of feed rate, which by necessity interacts with *all* the elements in the plant. Hence the cost function of Eq. (11.1) will be found to apply to the local optimization problem while the profit function may be more appropriate for unit or plant optimization.

Feedforward vs Feedback From the presentation on feedforward control in Chap. 8, its vastly superior dynamic responsiveness should be evident. Whereas feedback must search for the correct value of the manipulated variable by trial and error, feedforward arrives at the solution by direct calculation. The cost of this responsiveness is a detailed knowledge of the process, converted into computing controls. By contrast, a feedback controller can arrive at the same solution without such intelligence, but it requires a certain amount of time in an undisturbed condition.

Optimization as well as control may be conducted by either feedforward or feedback strategies. Feedforward optimization relies on a mathematical model of the process generating an off-line solution to the problem through which relationships between measured variables can be established which constitute optimum operation. For example, a cost function may depend on a certain combination of plant variables such as feed rate F, product flow D, product composition y and value v, and heat input Q along with its cost c:

$$\$_c = f_c(F, D, y, v, Q, c) \tag{11.3}$$

A certain relationship among these variables may be found which will minimize $\$_c$. One of the variables is then selected to be manipulated to optimize $\$_c$ in response to variations in the others. Then the established optimum relationship may be solved in terms of that manipulated variable, for example, Q:

$$Q_o = f_o(F, D, y, v, c) \tag{11.4}$$

Here Q_o is the heat input which minimizes $\$_c$ in response to the other variables and f_o defines that relationship which represents the minimum cost. This concept is illustrated by several examples later in the chapter.

If the cost relationship f_c is simple enough, it should be possible to differentiate $\$_c$ with respect to the selected manipulated variable, set the differential to zero, and solve for the manipulated variable:

$$\frac{d\$_c}{dQ} = \frac{df_c}{dQ}(F,D,y,v,Q,c) = 0 \tag{11.5}$$

$$Q_o = \frac{df_c}{dQ}(F,D,y,v,Q,c) + Q \tag{11.6}$$

Then the optimizing relationship f_o is

$$f_o(F,D,y,v,c) = \frac{df_c}{dQ}(F,D,y,v,Q,c) + Q \tag{11.7}$$

For a single column, this procedure will be found workable assuming that the cost function f_c is linear, quadratic, hyperbolic, or logarithmic, yielding readily to differentiation. Then the optimizing relationship will be soluble by direct calculation.

For two-column systems, a quadratic optimizing relationship appears, impeding direct solution. An approximation may then be necessary to match the true function reasonably well over a limited range. For more than two columns, or more than a single manipulated variable, the relationships become too complex either for differentiation of the cost function or direct calculation of the optimizing function. Then an iterative program is needed to increment the manipulated variables in such a way as to arrive at optimum conditions. This can be accomplished only with nonlinear programming.

Nonlinear and linear programming operate by feedback mechanisms, as does any iterative or trial-and-error procedure. If conducted directly on the process, time must be allowed for the process to settle out after each trial manipulation. Since many steps are typically required to reach an optimum level for each variable manipulated, regardless of the search method used, the progress toward optimization by feedback is slow indeed. In fact, since upsets may easily develop at any time or frequency during the search, the optimum conditions for a real plant are not likely to be reached at all by this means. Although feedback may be used for local optimization, it is not recommended. See Ref. 1 for a discussion of this subject.

Instead, the iterative procedure may be carried out on a steady-state computer *model* of the process, where individual results may be obtained in milliseconds rather than minutes or hours. Then an optimum solution may be found in seconds or minutes and be imposed on the plant with some confidence that it will at least satisfy conditions presently existing.

There are two fundamental limitations to this approach. First, it depends on the accuracy of the plant model. Although the model of the plant may in fact be optimized, there is no assurance that the plant itself is also optimized. Actually this is true of any feedforward system. As subsequent observations indicate a disparity between the plant and the model, adjustments can be made to improve the representation.

The elimination of dynamics from the model reduces its complexity and facilitates a rapid solution. But their absence constitutes a second departure from the true characteristics of the plant itself. Although a manipulation at a given point in time may represent optimum steady-state conditions at that time, its subsequent effect on a changing process could be significantly sub-optimum. One constantly changing variable is the environment surrounding the process. Passage of the sun across the sky affects condensers and may therefore cause optimum operating conditions to cycle diurnally. If the plant's serial distribution is extensive, unit optimization under these conditions may be impossible. For example, an increase in feed rate allowable at night could arrive at a downstream column the following noon, at which time it could not be accommodated. While complete plant optimization is desirable, these factors may prevent it from ever being achieved.

Selection of Variables for Manipulation A prerequisite to optimization is the existence of more than the minimum number of manipulated variables needed for regulation of the specified controlled variables. If no constraints are being applied, this requirement can be met if the number of manipulated variables exceeds the number of controlled variables. Naturally a manipulated variable is lost each time a constraint is encountered. Hence constraints place severe limits on the opportunities available for optimization.

The most common application in a fractionating plant is the local optimization of a column making a single guaranteed product. In this case, the quality of that product should be controlled by manipulating the material balance. This leaves the heat input (or reflux) free to be manipulated for optimization within the limits of flooding, etc.

When a constraint is encountered, heat input (or reflux) is the variable normally limited. Then encountering a constraint only means suboptimum operation as opposed to loss of control over product quality. But in attempting to follow the command of the optimizing program, heat input will be as close to optimum as the constraint will allow.

Allocation of utilities among several columns is another optimization problem presented later in the chapter. At this juncture, it is only necessary to point out that heating and cooling can be allocated to minimize monetary losses of all the products in the plant. Again, the selected manipulated variable is heating or cooling for each individual column.

The exact form of the manipulated variable is open to selection. In a given instance, it could be steam flow, heat flow, refrigerant flow, boilup-feed ratio, separation, or even the composition of the unspecified stream. The choice depends on the significance of the term in the cost function. In the case of a refrigeration-allocation problem, heat flow is probably the most significant variable. For local optimization, either heat flow or the composition of the unspecified product could be used, with different results. As will be seen, the use of composition eliminates the sensitivity of the system to changes in the composition of the specified product. On the other hand, the use of heat input can bring about a faster approach to optimum conditions and avoids the need for a second composition-control loop. Again, these various choices will be illustrated by examples.

SINGLE-COLUMN OPTIMIZATION

Confining the optimization effort to a single column facilitates the solution and its implementation. While there is no guarantee that what is best for a single column is best for the plant, single-column optimization is generally preferable to no optimization at all. Furthermore, the opportunities for single-column optimization far exceed those of larger scope. Consequently, these opportunities are examined in detail below since they represent the bulk of the applications which are worth pursuing in distillation units.

Optimization is only possible within constraints, so it is important to identify those conditions where constraints will allow it. The most restrictive specifications are those on product qualities. If *neither* product from a column needs to meet a particular specification, then there is complete freedom to optimize both. The more common situation is where one of the products must meet a specification which allows optimization of the *other* product composition. Although this case is more restrictive, it is also easier to define and so is covered first. A related problem, where both products are specified, is treated afterward.

Columns with Guaranteed Products The most common objective for operating a distillation column is to make a single valuable product meeting some guaranteed specification while at the same time minimizing the loss of that valuable product. In the past, these columns have been operated at the heat-input constraint, which in fact minimizes product losses for any given feed rate. However, rising energy costs have focused attention on total operating costs and have led to an exploration of possibilities for optimization.

In every one of the separations fitting the description above, a heat input exists which will minimize operating costs for any given feed rate. If the cost of energy is low in comparison to the value of the lost product, that optimum heat input may lie beyond the capacity of the column, reboiler, or condenser to accommodate it. But before that can be determined, the exact location of the optimum heat input needs to be found.

Consider the cost function of Eq. (11.1) for a column separating a binary mixture into a more valuable distillate and a less valuable bottom product. The cost function is stated as

$$\frac{\$_c}{F} = (v_D - v_B)\frac{Bx}{F} + cH_V\frac{V}{F} \tag{11.8}$$

where v_D and v_B are the values of the two products, c is the combined cost of heating and cooling, V is the boilup rate, and H_V is the heat of vaporization of the mixture, all in consistent units.

The cost function contains a pair of related variables: composition x of the lower-valued product B and the boilup-feed ratio V/F which largely determines it. One of these variables must be substituted for the other by using the separation equation:

$$\frac{V}{F} = \beta \ln S = \beta \ln \frac{y(1-x)}{x(1-y)} \tag{11.9}$$

where β is the column characterization factor. When (11.9) is substituted for V/F in (11.8), we have a cost function in two terms of x:

$$\frac{\$_c}{F} = (v_D - v_B)\frac{Bx}{F} + cH_V\beta\left(\ln\frac{y}{1-y} + \ln\frac{1-x}{x}\right) \tag{11.10}$$

This expression may be differentiated with respect to x in an effort to locate its optimum level:

$$\frac{d(\$_c/F)}{dx} = (v_D - v_B)\frac{B}{F} - \frac{cH_V\beta}{x(1-x)} \tag{11.11}$$

When the derivative is set to zero, the optimum value of x, that is, x_o, can be found:

$$x_o(1 - x_o) = \frac{cH_V\beta}{(v_D - v_B)(B/F)} \tag{11.12}$$

Although (11.12) is a quadratic in x_o, its approximate solution is readily found by letting $1 - x_o = 1$. The relative error entailed in doing so is evaluated as the difference between the approximate and exact solutions of (11.12), divided by the approximate solution:

$$\frac{x_o - x_o(1 - x_o)}{x_o} = x_o \tag{11.13}$$

In most cases, the error sustained by this approximation will not be significant. In any event, it is a systematic error that can be minimized by a correction factor C fitted to an exact solution at one point:

$$x_o \approx C\,\frac{cH_V\beta}{(v_D - v_B)(B/F)} \tag{11.14}$$

This is well within the limits of accuracy of the separation relationship for most columns.

The information required to effect the optimization using (11.14) includes the cost of energy and values of the products, along with the column characterization factor β and the latent heat of vaporization. These last two combined terms tend to vary with column pressure p in a linear manner:

$$H_V\beta = a(b + p) \tag{11.15}$$

This relationship was demonstrated for a specific column in Eq. (8.51) and Fig. 8.15, where Q/F was plotted against pressure for a constant separation. Equation (11.16) shows that Q/F is directly convertible to $H_V\beta$:

$$\frac{Q}{F} = H_V\frac{V}{F} = H_V\beta \ln S \tag{11.16}$$

Equation (11.14) also requires a measurement of B/F. This could be obtained directly from the two flowmeters or inferred from the output of the feedback controller trimming the material balance to control y. In the case that D/F is used to control y,

$$\frac{B}{F} = 1 - \frac{D}{F} \tag{11.17}$$

Using the feedback trim signal has the advantage of inherent dynamic compensation. If the two flow signals were used instead, the feed signal would require the same dynamic compensation used by the feedforward system.

However, there is some interaction within the process between x and B, If x is reduced, for example, to satisfy a new set of conditions, an increase in D/F would result because of enhanced distillate recovery and B/F would thereby be reduced. The effect of this response would be to allow a subsequent increase in x, promoting an oscillation. The feedback is negative but is delayed by the response of the other composition loop.

The feedback can be eliminated altogether by substituting this set of compositions for B/F:

$$\frac{B}{F} = \frac{y - z}{y - x} \tag{11.18}$$

This changes the cost function to read:

$$\frac{\$_c}{F} = (v_D - v_B)\left(\frac{y - z}{y - x}\right) x + cH_V\beta\left(\ln \frac{y}{1 - y} + \ln \frac{1 - x}{x}\right)$$

which is differentiated to give

$$\frac{d(\$_c/F)}{dx} = (v_D - v_B)y \frac{y - z}{(y - x)^2} - \frac{cH_V\beta}{x(1 - x)}$$

When set equal to zero and solved for x,

$$\frac{x_o(1 - x_o)}{(y - x_o)^2} = \frac{cH_V\beta}{(v_D - v_B)y(y - z)} \tag{11.19}$$

Using the same reasoning as before, the approximate value of x_o can be found by letting $1 - x_o$ and $y - x_o$ equal 1.0. This approximation will tend to be less accurate than (11.14) because $y - x_o$ will be further from unity than $1 - x_o$. Successive approximations may be necessary as indicated in the example below. This can culminate in an exact solution at one point, yielding a correction factor C that allows a close approximation at nearby points:

$$x_o \approx C \frac{cH_V\beta}{(v_D - v_B)y(y - z)} \tag{11.20}$$

example 11.1

A splitter separates a mixed-butane feed into 96 percent isobutane distillate and 5 percent isobutane bottom product at a V/F of 4.0. The distillate must meet a 95 percent specification and is more valuable by \$0.50/bbl. The combined cost of heating and cooling is \$1/10^6 Btu. The heat of vaporization of n-butane at 100°F is

28,930 Btu/bbl. Calculate x_o for feed compositions of 30, 40, and 50 percent isobutane. Determine the correction factor needed for the first approximation at 40 percent and the error sustained in using it at 30 and 50 percent:

$$\beta = \frac{V/F}{\ln S} = \frac{3.0}{\ln \dfrac{96(95)}{5(4)}} = 0.490$$

From Eq. (11.19), at $z = 0.3$,

$$\frac{x_o(1 - x_o)}{(0.96 - x_o)^2} = \frac{(1.00/10^6)(28{,}930)(0.490)}{(0.50)(0.96)(0.96 - 0.30)} = 0.447$$

Second approximation:

$$x_o = 0.0447 \frac{(0.96 - 0.0447)^2}{1 - 0.0447} = 0.0392$$

Third approximation:

$$x_o = 0.0447 \frac{(0.96 - 0.0392)^2}{1 - 0.0392} = 0.0394$$

Repeating the iterative solution for the other feed compositions gives

$z = 0.4 \qquad x_o = 0.0461$

$z = 0.5 \qquad x_o = 0.0555$

At $z = 0.4$,

$$\frac{x_o(1 - x_o)}{(0.96 - x_o)^2} = 0.0527$$

$$C = \frac{0.0461}{0.0527} = 0.875$$

At $z = 0.3$, the corrected first approximation is

$0.875(0.0447) = 0.0391 \qquad$ (versus 0.0394)

At $z = 0.5$, the corrected first approximation is

$0.875(0.0641) = 0.0561 \qquad$ (versus 0.0555)

The corrected first approximation seems perfectly adequate despite the sizable departure of the correction from unity. A plot of x_o versus z for this example is shown in Fig. 11.1.

In many cases, a bottom-product analyzer is not available, so that x_o cannot be controlled directly. Then it is possible to calculate the equivalent optimum value of V_o/F using the following expression:

$$\frac{V_o}{F} = \beta \ln \frac{y(1 - x_o)}{x_o(1 - y)} \tag{11.21}$$

example 11.2

Using the operating conditions given in Example 11.1, estimate the optimum ratio of V/F corresponding to the optimum bottom composition.

For $z = 0.3$,

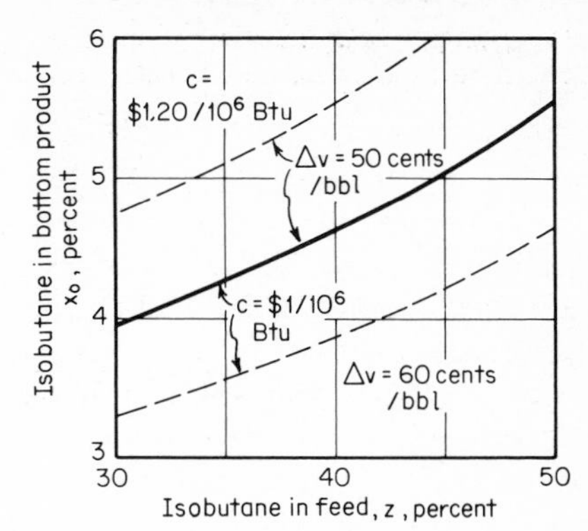

figure 11.1 *For a given set of costs and values, the optimum bottom composition can be related to feed composition.*

$$\frac{V_o}{F} = 0.490 \ln \frac{0.96(0.9606)}{0.0394(0.04)} = 3.122$$

Following the same procedure for other feed compositions gives

$$z = 0.4 \qquad \frac{V_o}{F} = 3.042$$

$$z = 0.5 \qquad \frac{V_o}{F} = 2.946$$

Observe that the relative change in V_o/F for the three cases above is much less than the relative change in x_o. As feed composition varies from 30 to 50 percent isobutane, optimum boilup decreases by only 5.7 percent while isobutane in the bottom product increases 41 percent of its original value. From this analysis, manipulation of x can locate the optimum nearly an order of magnitude more accurately than V. Additionally, V_o is much more sensitive to variations in y than is x_o.

A worthwhile investigation would be to determine the penalty for suboptimum operation and, from it, the accuracy really required of the optimizing system.

example 11.3

Calculate the cost function for the base case of the column in Example 11.1, and compare it to that of the optimum conditions for $z = 0.3$.

For $x = 0.05$ and $V/F = 3$, using Eq. (11.8):

$$\frac{\$_c}{F} = 0.50 \frac{0.96 - 0.30}{0.96 - 0.05} 0.05 + \frac{1}{10^6} (28{,}930)3.0$$

$$= 0.0181 + 0.08679 = \$0.1049/\text{bbl}$$

For $x_o = 0.0394$ and $V_o/F = 3.122$:

$$\frac{\$_c}{F} = 0.50 \frac{0.96 - 0.30}{0.96 - 0.0394} 0.0394 + \frac{1}{10^6} (28{,}930)3.122$$

$$= 0.0141 + 0.0903 = \$0.1044/\text{bbl}$$

It can be seen from Example 11.3 how shallow the cost curve is as a function of x. This indicates that relatively little return can be expected from on-line optimization and shows the unimportance of extreme accuracy in making the calculations. Holding V/F constant at 3.0 gives results not far from optimum. By contrast, reducing the isobutane purity to 95 percent by improved quality control could reduce x to 0.04 with the original V/F of 3, reducing the cost function to $0.1010/bbl. In this example, the cost reduction brought about by improved regulation is eight times as great as that possible with optimization.

Changing product values and utility costs have a pronounced impact on optimum operating conditions. From (11.19) it can be seen that a given percentage increase in the cost of heating and cooling will raise x_o by essentially the same percentage, while an increase in the product-value difference has exactly the opposite effect. Curves indicating the effects of these variations are shown in Fig. 11.1.

The case for optimizing single columns with *two* guaranteed products is virtually identical to that for a single guaranteed product. The principal difference is that optimization of the composition of the lower-valued product is only possible within its established specifications. When both products meet their specifications exactly, energy usage will be as low as possible. Increasing the energy flow can improve the purity of either or both products. But improving the purity of the lower-valued product will always increase recovery of the higher-valued product. In cases where improved recovery is worth more than the cost of additional energy, optimization is possible.

The specification on the lower-valued product is nothing more than a low limit on its purity. As long as x_o calculates to be within that limit, optimization may be carried out. When it exceeds the limit, the limit must be enforced on the composition controller. At the other extreme, sufficient boilup may not be available to reduce x to x_o—then operation at the boilup constraint is the best possible condition under the existing circumstances. A system allowing optimization within limits is shown in Fig. 9.13 and discussed in that context.

Columns without Guaranteed Products When neither product from a column is for sale but instead is used elsewhere within the plant, it is possible to optimize the compositions of both. Because neither composition need be controlled at a fixed set point, the material balance is free to be manipulated for optimization. The heat input may then remain fixed or may also be manipulated, in which case the optimization becomes a two-parameter problem.

For the simpler case where V/F is fixed, the cost statement is a sum of the losses of the two products in the opposite streams. For a problem like this to be meaningful, *both* products must incur penalties for being lost. In other words, there must be a penalty attached to the heavier component appearing in the distillate as well as to the lighter component appearing in the bottom product. If only a single loss function were used as in the previous examples, there would be no incentive to make the separation—all the feed could leave as the valuable product.

To demonstrate, consider the example of the butane splitter, which has already been introduced. But in addition to the \$0.50/bbl penalty for isobutane in the bottom product, consider a \$0.30/bbl penalty for n-butane in the distillate. The cost function excluding utilities becomes a loss function only:

$$\frac{\$_L}{F} = \Delta v_B \frac{B}{F} x + \Delta v_D \frac{D}{F}(1-y) \tag{11.22}$$

where Δv_B and Δv_D represent the bottom and distillate penalties, respectively.

To differentiate the loss function with respect to a single composition requires that the other composition be replaced by the separation-factor relationship:

$$\frac{\$_L}{F} = \Delta v_B \frac{B}{F} x + \Delta v_D \frac{D}{F} \frac{1-x}{1+x(S-1)} \tag{11.23}$$

When the loss function is differentiated with respect to x and set to zero, the following solution is obtained:

$$\frac{S}{[1+x_o(S-1)]^2} = \frac{\Delta v_B\, B}{\Delta v_D\, D}$$

which, by eliminating S, becomes

$$\frac{y_o(1-y_o)}{x_o(1-x_o)} = \frac{\Delta v_B\, B}{\Delta v_D\, D} = \frac{\Delta v_B(y_o - z)}{\Delta v_D(z - x_o)} \tag{11.24}$$

The solution gives the optimum *ratio* of the impurities, but not their individual optimum values. The latter are functions of the energy used to make the separation. Successive approximations may be required to reach a final solution. In this case, a correction factor C may be found to apply to the first approximation, allowing a direct solution of reasonable accuracy:

$$\frac{1-y_o}{x_o} \approx C \frac{\Delta v_B}{\Delta v_D} \frac{1-z}{z} \tag{11.25}$$

example 11.4

The butane splitter in Example 11.1 is assessed penalties of \$0.50/bbl for isobutane in the bottom product and \$0.30/bbl for n-butane in the distillate. Estimate the optimum ratio of these impurities for feed concentrations of 30, 40, and 50 percent isobutane. Evaluate the correction factor at 30 percent and check its accuracy at 40 and 50 percent. Then compare the loss function for the optimum compositions at 30 percent against that of the base case.

From (11.25), the first approximation for $z = 0.3$ is

$$\frac{1-y_o}{x_o} = \frac{0.50(1-0.30)}{0.30(0.30)} = 3.89$$

Using the separation factor of 456 at $V/F = 3$,

$$1 - y_o = \frac{1-x_o}{1+455x_o} = 3.89x_o$$

Next, x_o is found by trial and error to be

$$x_o = 0.0225$$

and

$$1 - y_o = 0.0870$$

Substituting y_o and x_o into (11.24) and solving for $(1 - y_o)/x_o$ gives the second approximation:

$$\frac{1 - y_o}{x_o} = \frac{0.50(0.913 - 0.3)(1 - 0.0225)}{0.30(0.3 - 0.0225)(0.913)} = 3.94$$

For $z = 0.4$, the first approximation of $(1 - y_o)/x_o$ is 2.62 and the second is 2.49. For $z = 0.5$, the first approximation gives 1.67 and the second gives 1.62.

The loss function for the base case is

$$\frac{\$_L}{F} = 0.5\,\frac{0.96 - 0.3}{0.96 - 0.05}\,0.05 + 0.3\,\frac{0.3 - 0.05}{0.96 - 0.05}\,0.04$$

$$= 0.0181 + 0.0033 = \$0.0214/\text{bbl}$$

and for the optimum compositions it is

$$\frac{\$_L}{F} = 0.5\,\frac{0.913 - 0.3}{0.913 - 0.0225}\,0.0225 + 0.3\,\frac{0.3 - 0.0225}{0.913 - 0.0225}\,0.087$$

$$= 0.00774 + 0.00813 = \$0.0159/\text{bbl}$$

For this example, at least, the first approximation seems to be sufficient without a correction factor.

Although the optimization was successful in reducing the loss function from \$0.0214/bbl to \$0.0159/bbl, the cost of heating and cooling at \$0.0868/bbl for a V/F of 3 remains a dominant factor. The second part of the problem is then to determine the optimum boilup, or the true optimum values of x and y, by including heating and cooling in the cost function:

$$\frac{\$_c}{F} = \Delta v_B \frac{B}{F} x + \Delta v_D \frac{D}{F}(1 - y) + cH_V \frac{V}{F} \tag{11.26}$$

Unfortunately, differentiation of the substituted cost function is unwieldy because of its nonlinear nature. A trial-and-error procedure is required, incrementing V/F and determining x_o and $1 - y_o$ following Example 11.4. The results of such an evaluation for the cost figures given in the examples are summarized in Table 11.1.

Note that the optimum is quite flat as a function of V/F, so there is little

TABLE 11.1 Two-parameter Optimization of the Butane Splitter for $z = 0.3$, $\Delta v_B = \$0.50/\text{bbl}$, $\Delta v_D = \$0.30/\text{bbl}$, and $c = \$1/10^6$ Btu

V/F	$1 - y_o$	x_o	$\Delta v_B Bx/F$	$\Delta v_D D(1 - y)/F$	$cH_V V/F$	$\$_c/F$
3.0	0.087	0.0225	0.0077	0.0081	0.0868	0.1027
2.5	0.139	0.036	0.0122	0.0133	0.0723	0.0979
2.3	0.168	0.043	0.0145	0.0164	0.0665	0.0975
2.0	0.219	0.056	0.0186	0.0221	0.0579	0.0985

penalty for a failure to find its exact value. The 9 percent change from 2.3 to 2.5 incurs only a 0.4 percent increase in the cost function. A similar analysis carried out for $z = 0.4$ again gives an optimum V/F in the vicinity of 2.3. The conclusion to be drawn is that V/F does not have to be adjusted continuously as feed composition changes but only needs repositioning with variations in heating and cooling costs relative to product-impurity penalties. Presumably, variations in these parameters are so infrequent that off-line calculations of V_o/F are adequate.

To control the column having no guaranteed products, V/F may be set at the optimum value as determined above. Then the ratio $(1 - y)/x$ may be controlled at the calculated optimum by manipulating the material balance. Actually, the optimum composition ratio can be maintained quite acceptably with feedforward control alone. For the optimum compositions estimated in Example 11.4, D/Fz varies only between 1.039 and 1.025 as z changes from 0.3 to 0.5. Reducing V/F from 3 to 2.3 at $z = 0.3$ requires D/Fz to increase to 1.086. Nonetheless, at any given V/F simple material-balance control will maintain the optimum product-composition ratio almost without feedback trim. This is an important consideration since measurements of *both* product compositions are necessary for the feedback loop to be closed. Due to differences in dynamic response of the two compositions to manipulation of D (or B), feedback may be sluggish at best. But feedforward should make feedback unnecessary if feed composition is included in the feedforward calculation.

Finding the Optimum Feed Rate When the feed to a column is free to be manipulated, its rate may be selected to optimize a profit function. This adds another manipulated variable to those already used. While having the opportunity to select a feed rate for an individual column is rare enough, the ability to reach an optimum profit by adjusting it is still less likely. It requires that no more than one product meet guaranteed specifications. If both products must meet guaranteed specifications, maximum profit will be achieved at maximum boilup when both products exactly meet their specifications.

With a single guaranteed product, an optimum V/F was found to minimize the combined cost of utilities and losses of product per unit of feed. Actually, that V/F could have been obtained by manipulating V, as described, or F. Assuming that the products from a column are always worth more than the feed, maximum profit can be realized only at maximum boilup.

The optimum feed rate is heavily dependent on the values of the products relative to the feed. If *both* products are worth more than the feed, the optimum feed rate will be higher than if one product is equal to or less than the value of the feed. To determine the optimum feed rate, a profit function must be written:

$$\$_P = v_D D + v_B B - v_F F - \$_o - \$_H \tag{11.27}$$

where v_D, v_B, v_F = values of designated products and feeds

$\$_o$ = fixed cost of overhead, labor, and capital write-off

$\$_H$ = cost of heating and cooling at maximum boilup

TABLE 11.2 Optimum Feed Rate for the Butane Splitter for $v_D - v_B =$ \$0.50/bbl, $v_B - v_F =$ \$0.20/bbl, and $y = 0.96$

z	F/V_M	x	D/F	$(\$_P - \$_o - \$_H)/V_M$	B/V_M
0.3	0.35	0.066	0.262	0.116	0.258
	0.40	0.127	0.207	0.121	0.317
	0.45	0.205	0.126	0.118	0.393
0.4	0.38	0.100	0.349	0.142	0.247
	0.43	0.172	0.289	0.148	0.306
	0.48	0.255	0.206	0.145	0.381
0.5	0.43	0.172	0.416	0.175	0.251
	0.48	0.255	0.348	0.179	0.313
	0.53	0.338	0.261	0.175	0.392

Feed rate can be factored from the first three terms:

$$\$_P = F\left[v_D \frac{D}{F} + v_B\left(1 - \frac{D}{F}\right) - v_F\right] - \$_o - \$_H$$

Then terms may be combined and D/F replaced by compositions to give

$$\$_P = F\left[(v_D - v_B)\frac{z - x}{y - x} + v_B - v_F\right] - \$_o - \$_H \tag{11.28}$$

For a controlled y, x will vary with V/F and hence with F. Although it is possible to differentiate $\$_P$ with respect to F, the resulting derivative contains too many functions of x to be easily solved. It is far simpler to increment F and calculate the profit resulting from its effect on x. The results of such a study are summarized in Table 11.2, normalized to maximum allowable boilup V_M. The optimum conditions are underlined.

The last column in Table 11.2 showing B/V_M is included because its value under optimum conditions is virtually constant. This seems to be characteristic for the waste product. Then if D is manipulated to control y at a specified composition, feed rate may be optimized simply by holding B at an average constant value. Figure 11.2 shows an optimizing control loop to perform this function. Although this configuration may not apply to every situation, it is given as an example of how a simple control system can solve a complex problem. Each individual case should be evaluated as was done here, for all expected load conditions, to arrive at a workable control system. As in other optimizing applications, the objective function is not terribly sensitive to changes in the manipulated variable, so precise control is not mandatory.

MULTIPLE-COLUMN OPTIMIZATION

Opportunities for multiple-column optimization are infrequent and diversified. Although it was possible to categorize three classic problems in single-column optimization, the addition of a second element to the process brings with it another dimension. Actually, what is presented here is but a sampling of the possibilities.

There are certainly more two-element optimization problems than those consisting of two columns. Combinations of a column and a compressor, a column and a reactor, etc., are familiar to both the chemical and the petroleum industries. However, reactor characteristics are so specific to the particular process involved that to speculate on such a problem in this text does not seem appropriate. So the discussion that follows is limited to two-column optimization, with a presentation on the optimum policy for allocation of limited resources among multiple users.

Two-column Optimization The classic single-column program developed the optimum composition for the unspecified key component. By contrast, the two-column problem involves the unspecified off-key component, which by definition cannot be controlled in the column from which the product is withdrawn. An example of this application as shown in Fig. 11.3 was described by the author in an earlier paper [2].

A propane product must meet a purity specification of 95 percent with less than 2 percent isobutane. Without considering the third component—ethane—this could be a single-column optimization if the value of isobutane were sufficiently greater than that of propane. However, the actual value trade-off in this column is not between isobutane and propane but between isobutane and ethane. Typically, the value of ethane is so much lower than that of the other components that its concentration should be maximized. The operating strategy for the depropanizer is then to control propane content as close to its purity specification as possible while optimizing the ethane and isobutane balance.

To reduce the isobutane content, more energy must be applied to the depropanizer. But at the same time, the ethane content is allowed to increase, which requires less energy at the deethanizer. The cost function then con-

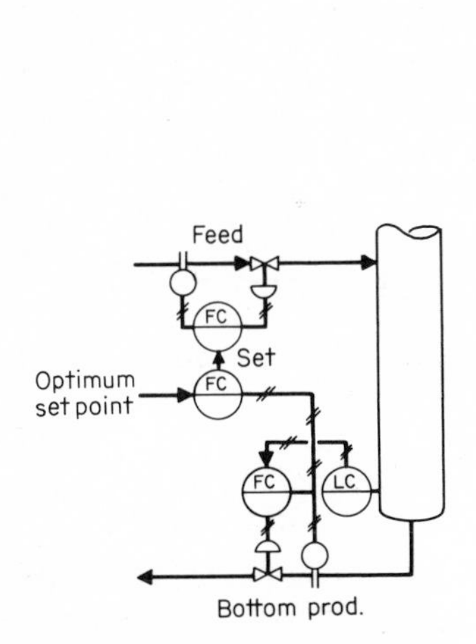

figure 11.2 *The second bottom-flow controller slowly adjusts feed rate to maximize profit.*

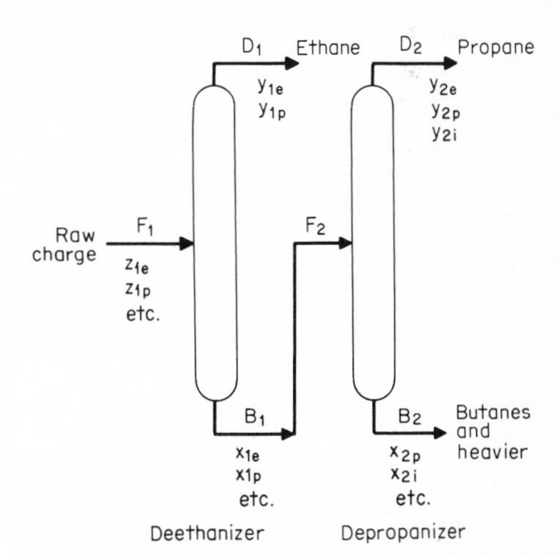

figure 11.3 *The optimum ethane content in the propane product is a function of deethanizer boilup.*

tains three terms—isobutane loss, depropanizer boilup, and deethanizer boilup:

$$\$_c = (v_i - v_e)D_2y_{2i} + cH_V(V_2 + V_1) \tag{11.29}$$

where subscripts 1 and 2 refer to the deethanizer and depropanizer and i and e refer to isobutane and ethane. Substitutions may be made for V_1 and V_2:

$$V_1 = F_1\beta_1 \ln \frac{y_{1e}/y_{1p}}{x_{1e}/x_{1p}} \tag{11.30}$$

$$V_2 = F_2\beta_2 \ln \frac{y_{2p}/y_{2i}}{x_{2p}/x_{2i}} \tag{11.31}$$

Specifications on the ethane product require that y_{1e}/y_{1p} be constant, and similarly the isobutane product requires that x_{2p}/x_{2i} be constant. Equations (11.30) and (11.31) can then be simplified to

$$V_1 = F_1\left(k_1 - \beta_1 \ln \frac{x_{1e}}{x_{1p}}\right) \tag{11.32}$$

$$V_2 = F_2\left(k_2 + \beta_2 \ln \frac{y_{2p}}{y_{2i}}\right) \tag{11.33}$$

Further reduction of (11.32) is possible. First, y_{2i} can be replaced by y_{2p} and y_{2e}:

$$y_{2i} = 1 - y_{2p} - y_{2e} \tag{11.34}$$

Next, the ratio x_{1e}/x_{1p} may be substituted for y_{2e}/y_{2p}, assuming that very little propane leaves the depropanizer base. Then

$$\frac{y_{2p}}{y_{2i}} = \frac{y_{2p}}{1 - y_{2p} - y_{2p}\, x_{1e}/x_{1p}} = \frac{1}{1/y_{2p} - 1 - x_{1e}/x_{1p}} \tag{11.35}$$

Finally, since $\ln 1 = 0$, (11.32) may be restated as

$$V_2 = F_2\left[k_2 - \beta_2 \ln \left(\frac{1}{y_{2p}} - 1 - \frac{x_{1e}}{x_{1p}}\right)\right] \tag{11.36}$$

When the same substitution is used to eliminate y_{2i} in (11.29), the remaining cost function contains only x_{1e}/x_{1p} as a controlled variable:

$$\$_c = (v_i - v_e)D_2\left[1 - y_{2p}\left(1 + \frac{x_{1e}}{x_{1p}}\right)\right] + cH_VF_2 \times \left[k_2 - \beta_2 \ln \left(\frac{1}{y_{2p}} - 1 - \frac{x_{1e}}{x_{1p}}\right)\right] + cH_VF_1\left(k_1 - \beta_1 \ln \frac{x_{1e}}{x_{1p}}\right)\Bigg] \tag{11.37}$$

Differentiation of $\$_c$ with respect to x_{1e}/x_{1p} eliminates some of the terms from consideration:

$$\frac{d\$_c}{d(x_{1e}/x_{1p})} = -(v_i - v_e)D_2y_{2p} + \frac{cH_VF_2\beta_2}{1/y_{2p} - 1 - x_{1e}/x_{1p}} - \frac{cH_VF_1\beta_1}{x_{1e}/x_{1p}} \tag{11.38}$$

When set equal to zero and solved for x_{1e}/x_{1p}, a quadratic equation is formed:

$$(v_i - v_e)D_2 y_{2p}\left(\frac{x_{1e}}{x_{1p}}\right)^2 - [(v_i - v_e)D_2(1 - y_{2p}) - cH_V(F_2\beta_2 + F_1\beta_1)]$$

$$\times \frac{x_{1e}}{x_{1p}} - cH_V F_1\beta_1\left(\frac{1}{y_{2p}} - 1\right) = 0 \quad (11.39)$$

The quadratic expression is a natural consequence of the second column's presence. The cost may be referred to the raw charge by dividing all terms by F_1. This expression remains limited in accuracy, however, because D_2 and F_2 are both functions of x_{1e}/x_{1p}.

In the analysis described in Ref. 2, the savings in deethanizer energy were small compared with the cost of depropanizer energy for an incremental increase in x_{1e}/x_{1p}. One factor was the easier separation in the deethanizer such that $\beta_1 < \beta_2$. And as ethane content was increased and isobutane content reduced, the difference in energy usage between the two columns became more pronounced. An incremental increase in an already high (4 percent) ethane concentration had much less effect on separation than the same increase in an already low (0.5 percent) isobutane concentration. As a result, the problem was solved as that of a single column generating an optimum isobutane concentration. The latter was then converted to a set point for x_{1e}/x_{1p} by substitution as in (11.35).

Allocation of Limited Resources In some plants, a single refrigeration unit serves more than one column. Refrigeration may not always be in short supply, but there are certain seasons of the year and hours of the day when it limits plant production. During these times some sort of allocation policy should be applied to optimize the use of this resource.

Consider the separations unit of the ethylene plant, represented schematically in Fig. 3.5. Assume that a common refrigeration unit supplies all the columns. Naturally, ethylene losses from each column will vary with the boilup–feed ratio of each. If refrigeration were apportioned arbitrarily, inequities might cause some columns to sustain high losses while others lost little. Because of the nonlinear relationship between losses and boilup, taking from the "poor" and giving to the "rich" is likely to cause a net increase in total plant losses. The curve of ethylene losses versus V/F for the demethanizer, as shown in Fig. 11.4, bears this out.

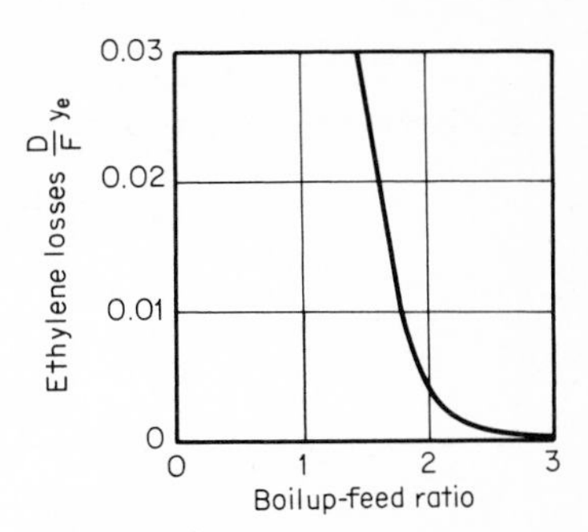

figure 11.4 *Ethylene losses with the distillate vary inversely with the boilup-feed ratio.*

Every column has a loss curve of this same shape. Because the slope becomes less steep as V/F increases, any increment applied to a column with a low V/F will have more effect on total plant losses than when applied to a column with a high V/F. Therefore, an optimum policy exists which determines the allocation of additional utilities resulting in the best net reduction in losses. Naturally, the converse is true—a reduction in resource availability should be applied to minimize the effect on total plant losses.

Reference 3 develops an optimum allocation policy for the demethanizer and ethylene fractionator using simple hyperbolic models of column-loss functions. If the ethylene losses from each column are assumed to be simply inversely proportional to boilup, then it is demonstrated that an optimum percentage of total available energy exists for each column in the unit. Unfortunately, this concept is not so readily proved with the more exact logarithmic models used here—differentiation of loss functions with respect to boilup yields some exponential terms.

However, on-line readjustment of allocation does not seem to be necessary. Because the component being lost (ethylene) is the same for all columns in the unit, variations in its value would not affect allocation. Production rate would not need to be considered since all columns would be affected equally by it. Variations in feed composition may call for an adjustment to allocation because the flow of one waste stream (e.g., fuel gas) might increase and thereby increase the losses from that column. A change in composition of the controlled product should also be considered since it will influence ethylene losses at a given V/F ratio.

If these variations are small or infrequent, the optimum allocation can be estimated off line by an accurate model and implemented as shown in Fig. 11.5. As with the other examples described earlier, the total-loss function can be expected to be rather flat in the optimum region, so that extreme accuracy in predicting the optimum allocation is unwarranted.

The compressor discharge-pressure controller determines the availability of refrigerant vapor to the reboilers. The flow to each column is set in ratio to

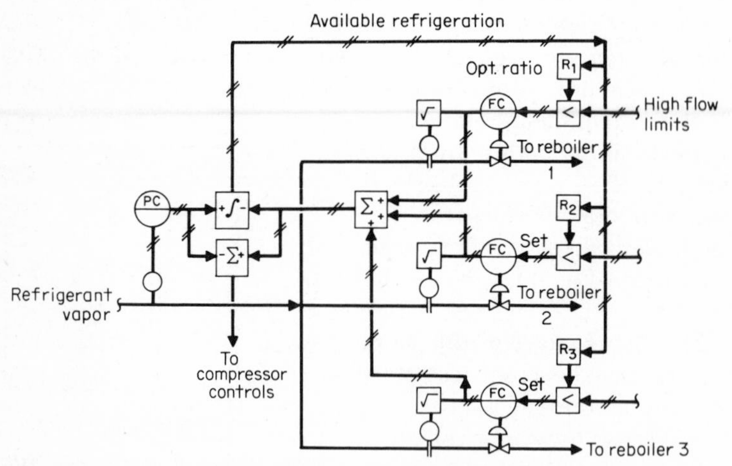

figure 11.5 *A pressure controller can set the total energy available to the columns.*

the total as determined by an off-line optimization program. However, high-flow overrides must be available in the event a flooding situation should develop or to satisfy a local optimization program. Feedback of total flow to an integrator set by the pressure controller prevents an override from upsetting the pressure loop. Should all three flow set points be overridden, the available energy cannot be used by the columns and the integrator will be unable to eliminate the deviation between its two inputs. However, a subtractor acting on this same deviation will then reduce compressor speed to control pressure. This system combines multiple-output controls with variable structuring as described in Chap. 9.

REFERENCES

1. Shinskey, F. G.: "Process-Control Systems," pp. 176–179, McGraw-Hill Book Company, New York, 1967.
2. Fauth, C. J., and F. G. Shinskey: Advanced Control for Distillation Columns, *Chem. Eng. Prog.*, June 1975.
3. Shinskey, F. G.: The Values of Process Control, *Oil Gas J.*, Feb. 18, 1974.

APPENDIX A

Tray-to-Tray Program for Binary Separations

INFORMATION REQUIRED:

- $Y[1]$ Mole fraction of more volatile component in vapor leaving top tray [1]
- D Distillate-feed ratio (estimated)
- V Vapor-feed ratio
- NF Number of trays above feed point
- NB Number of trays in column
- Z Mole fraction of more volatile component in feed
- A Relative volatility
- KD Gain of iterative feedback loop

INFORMATION REPORTED:

- $Y[N]$ Vapor composition at any tray N
- $X[N]$ Liquid composition at any tray N
- DC Calculated distillate-feed ratio

ASSUMPTIONS:

1. Vapor flow is uniform throughout the column.
2. Relative volatility is constant.

PROCEDURE: The program starts from the top of the column with a given vapor composition $Y[1]$ and calculates compositions on progressively lower trays using $L/V = 1 - D/V$. At the feed tray, feed is blended with liquid from the tray above. Below the feed tray, L is augmented by this feed and composition calculations resume using the new *L/V* until the bottom tray [*NB*] is reached. At this point, a new distillate–feed ratio *DC* is calculated using Z, *Y*[1], and *X*[*NB*]. The initial estimate of distillate–feed ratio *D* is incremented proportional to its difference from *DC*, and the procedure is repeated. Gain *KD* determines the size of the increment and hence the rate of convergence. When successive values of *DC* agree within 0.1 percent, iteration may be terminated. If convergence is too slow, increase *KD*; if divergence or oscillation results, decrease *KD*.

The program and three sample calculations follow. Note that the second solution does not converge. Gain *KD* is then reduced from 1 to 0.5, allowing the third solution to converge.

```
ON-LINE SYSTEMS V6-2053

1328     7-23-74
USER NUMBER?..3124,10
[illegible]

V3
SYSTEM?..
.TELCOMP
TELCOMP III VERSION 7.Ø4C

>LOAD FGS1
>TYPE ALL PARTS

 1.1 DEMAND Y[1],D,V,NF,NB
 1.12 DEMAND Z,A,KD
 1.13 DC=D
 1.2 X[Ø]=Y[1]
 1.25 D=D+KD*(DC-D)
 1.3 K=(1-D/V)
 1.4 DO PART 2 FOR N=1:1:NF-1
 1.5 N=NF
 1.51 DO STEP 2.1
 1.52 Y[NF+1]=Y[NF]+(1+(1-D)/V)*X[NF]-K*X[NF-1]-Z/V
 1.6 K=1+(1-D)/V
 1.7 DO PART 2 FOR N=NF+1:1:NB-1
 1.8 N=NB
 1.81 DO STEP 2.1
 1.82 DC=(Z-X[NB])/(Y[1]-X[NB])
 1.9 TYPE DC

 2.1 X[N]=Y[N]/(Y[N]+A*(1-Y[N]))
 2.2 Y[N+1]=Y[N]-K*(X[N-1]-X[N])

 3.1 DO PART 1
 3.2 DO PART 1.2
 3.25 DONE IF DC-D=Ø
 3.3 DO PART 3.2
```

```
>DO PART 3
         Y[1]=.97
            D=.5
            V=5
           NF=1Ø
           NB=2Ø
            Z=.5
            A=1.5
           KD=1
           DC=          .467754624
           DC=          .483769Ø45
           DC=          .4761191Ø1
           DC=          .47984338
           DC=          .478Ø46767
           DC=          .478917313
           DC=          .478496394
           DC=          .4787ØØ124
           DC=          .4786Ø1566
           DC@=          .478649257
           DC=          .478626
 INTERRUPTED AT STEP 1.6
>TYPE X[2Ø]
        X[2Ø]=          .Ø685255123
>DO PART 3
         Y[1]=.928
            D=.5
            V=3
           NF=1Ø
           NB=2Ø
            Z=.5
            A=1.5
           KD=1
           DC=          .473Ø87618
           DC=          .5Ø5188Ø26
           DC=          .465539269
           DC=          .512558658
           DC=          .453818415
           DC=          .522942426
           DC=          .434917993
           DC=
 INTERRUPTED AT STEP 3.25
>DO PART 3
         Y[1]=.928
            D=.5
            V=3
           NF=1Ø
           NB=2Ø
            Z=.5
            A=1.5
           KD=.5
           DC=          .473Ø87618
           DC=          .49Ø433Ø48
           DC=          .4881Ø3128
           DC=          .4883364Ø4
           DC=          .488311834
           DC=          .4883144Ø9
           DC
 INTERRUPTED AT STEP 1.9
>TYPE X[2Ø\]
         X[2]=          .8585Ø65Ø3
>TYPE X[2Ø]
        X[2Ø]=          .Ø915492516
```

APPENDIX B

Table of Symbols

Uppercase

A	Area; acid flow; amplitude
B	Bottom-product flow
C	Heat capacity; correction factor
$\mathbf{C}$	Closed-loop gain matrix
D	Distillate flow
E	Extractant flow; efficiency
F	Feed flow
G	Gas flow; gain
H	Enthalpy; Henry'slaw constant
K	Equilibrium constant
L	Reflux flow
M	Molecular weight
$\mathbf{M}$	Open-loop gain matrix
N	Number of moles
P	Product flow; proportional band
Q	Heat flow
R	Recovery; gas constant; reset time
S	Separation
T	Temperature
U	Heat-transfer coefficient
V	Vapor flow
W	Mass; mass flow
X	Abscissa; excess air
Y	Ordinate
Z	Van Laar's constant

Lowercase

a	Coefficient; gain
b	Coefficient; gain
c	Controlled variable; cost
d	Differential; decoupling coefficient
e	Deviation; 2.178
f	Fractional flow; function
g	Gravitational acceleration
$\mathbf{g}$	Dynamic gain vector
h	Head
i	Component
j	Component
k	Constant
l	Liquid level; length
m	Manipulated variable
n	Number of trays
p	Pressure
p°	Vapor pressure
q	Feed enthalpy; heat quantity
r	Set point; rangeability
s	Entropy
t	Time
u	Velocity
v	Value; fractional vapor flow; volume
w	Composition; work
x	Liquid composition
y	Vapor composition
z	Feed composition

Greek

α	Relative volatility
β	Characterization factor
γ	Activity coefficient; specific-heat ratio
Δ	Difference
∂	Partial differential
Λ	Relative-gain matrix
λ	Relative gain
π	3.1416
ρ	Density; specific gravity
$\sum$	Sum
σ	Surface tension
τ	Time constant

APPENDIX C

Conversion of Compositions

Converting mole fraction to liquid-volume fraction:

$$v_i = \frac{x_i M_i/\rho_i}{\sum x_i M_i/\rho_i} \tag{C.1}$$

where v_i = liquid volume fraction of component i
x_i = mole fraction of component i
M_i = molecular weight of component i

Converting liquid-volume fraction to weight fraction:

$$w_i = \frac{v_i \rho_i}{\sum v_i \rho_i} \tag{C.2}$$

where w_i = weight fraction of component i.

Converting liquid-volume fraction to mole fraction:

$$x_i = \frac{v_i \rho_i/M_i}{\sum v_i \rho_i/M_i} \tag{C.3}$$

Converting mole fraction to weight fraction:

$$w_i = \frac{x_i M_i}{\sum x_i M_i} \tag{C.4}$$

TABLE C.1 Molar Volumes of Light Hydrocarbons in bbl/mol at 60°F

	M, lb/mol	ρ, lb/bbl	M/ρ, bbl/mol
Ethane	30	131	0.229
Propane	44	161	0.248
Isobutane	58	197	0.295
n-Butane	58	205	0.284
Isopentane	72	218	0.330
n-Pentane	72	221	0.326
n-Hexane	86	232	0.370

APPENDIX D

Dynamic Analysis of Inverse Response

Equation (6.32) describes the response of base level l_b to changes in vapor rate dv:

$$dl_b = \frac{\tau_i}{\tau}\, dv - \frac{1}{\tau}\int dv\, dt \tag{6.32}$$

Let the vapor rate vary sinusoidally with an amplitude A and a period τ_o:

$$dv = A \sin \frac{2\pi t}{\tau_o} \tag{D.1}$$

Then

$$dl_b = \frac{\tau_i}{\tau} A \sin \frac{2\pi t}{\tau_o} - \frac{1}{\tau}\int \left(A \sin \frac{2\pi t}{\tau_o}\right) dt \tag{D.2}$$

Integrating $A \sin 2\pi t/\tau_o$ with time constant τ gives

$$\frac{1}{\tau}\int \left(A \sin \frac{t}{\tau_o}\right) = \frac{A\tau_o}{2\pi\tau} \sin \left(-90^\circ + \frac{2\pi t}{\tau_o}\right) \tag{D.3}$$

The gain of the integral term is its output amplitude divided by its input amplitude:

$$G_\int = \frac{A\tau_o/(2\pi\tau)}{A} = \frac{\tau_o}{2\pi\tau} \tag{D.4}$$

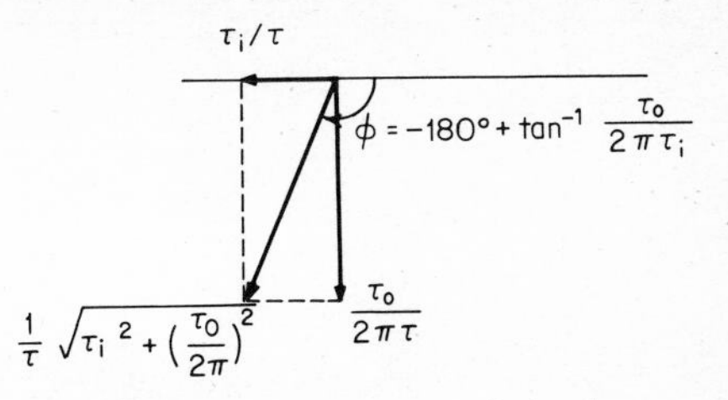

figure D.1 *Inverse response shifts the process phase lag beyond 90°.*

And its phase shift is the output phase angle less its input phase angle:

$$\phi_{\int} = -90° + \frac{2\pi t}{\tau_o} - \frac{2\pi t}{\tau_o} = -90° \tag{D.5}$$

The gain of the inverse responding term is simply τ_i/τ and its phase angle is $-180°$.

The gain and phase of the combined function is the vector sum of the inverse and integral gains, as shown in Fig. D.1:

$$G_i = \sqrt{\left(\frac{\tau_i}{\tau}\right)^2 + \left(\frac{2\pi\tau_o}{\tau}\right)^2} = \frac{1}{\tau}\sqrt{\tau_i^2 + (2\pi\tau_o)^2}$$

$$\phi_i = -180 + \tan^{-1}\frac{\tau_o/(2\pi\tau)}{\tau_i/\tau} = -180 + \tan^{-1}\frac{\tau_o}{2\pi\tau_i} \tag{D.6}$$

Input variations having a very short period encounter a gain of τ_i/τ and a phase angle of $-180°$; variations having a relatively long period produce a gain of $\tau_o/(2\pi\tau)$ and a phase angle approaching $-90°$.

Index